Annals of Mathematics Studies
Number 201

The Master Equation and the Convergence Problem in Mean Field Games

Pierre Cardaliaguet
François Delarue
Jean-Michel Lasry
Pierre-Louis Lions

PRINCETON UNIVERSITY PRESS

PRINCETON AND OXFORD

2019

Copyright © 2019 by Princeton University Press

Requests for permission to reproduce material from this work
should be sent to permissions@press.princeton.edu

Published by Princeton University Press
41 William Street, Princeton, New Jersey 08540
6 Oxford Street, Woodstock, Oxfordshire OX20 1TR

press.princeton.edu

LCCN: 2019936796
ISBN 978-0-691-19070-9
ISBN (pbk.) 978-0-691-19071-6

British Library Cataloging-in-Publication Data is available

Editorial: Vickie Kearn, Susannah Shoemaker, and Lauren Bucca
Production Editorial: Nathan Carr
Production: Erin Suydam
Publicity: Matthew Taylor and Kathryn Stevens
Copyeditor: Theresa Kornak

This book has been composed in LaTeX

Printed on acid-free paper. ∞

Printed in the United States of America

10   9   8   7   6   5   4   3   2   1

Contents

Preface

THE PURPOSE OF THIS SHORT MONOGRAPH is to address recent advances in the theory of mean field games (MFGs), which has met an amazing success since the simultaneous pioneering works by Lasry and Lions and by Caines, Huang, and Malhamé more than ten years ago. While earlier developments in the theory have been largely spread out over the last decade, issues that are addressed in this book require a new step forward in the analysis. The text evolved with the objective to provide a self-contained study of the so-called *master equation* and to answer the *convergence problem*, which has remained mainly open so far. As the writing progressed, the manuscript became longer and longer and, in the end, it turned out to be more relevant to publish the whole as a book.

There might be several reasons to explain the growing interest for MFGs. From the technical point of view, the underpinning stakes fall within several mathematical areas, including partial differential equations, probability theory, stochastic analysis, optimal control, and optimal transportation. In particular, several issues raised by the analysis of MFGs may be tackled by analytical or probabilistic tools; sometimes, they even require a subtle implementation of mixed arguments, which is precisely the case in this book. As a matter of fact, researchers from different disciplines have developed an interest in the subject, which has grown very quickly. Another explanation for the interest in the theory is the wide range of applications that it offers. While they were originally inspired by works in economics on heterogeneous agents, MFG models now appear under various forms in several domains, which include, for instance, mathematical finance, study of crowd phenomena, epidemiology, and cybersecurity.

Mean field games should be understood as games with a continuum of players, each of them interacting with the whole statistical distribution of the population. In this regard, they are expected to provide an asymptotic formulation for games with finitely many players with mean field interaction. In most of the related works, the connection between finite games and MFGs is addressed in the following manner: It is usually shown that solutions of the asymptotic problem generate an *almost equilibrium*, understood in the sense of Nash, to the corresponding finite game, the accuracy of the equilibrium getting stronger and stronger as the number of players in the finite game tends to infinity. The main purpose of this book is to focus on the converse problem, which may be formulated as follows: Do the equilibria of the finite games (if they exist) converge to a solution of the corresponding MFG as the number of players becomes very large? Surprisingly, answering this question turns out to be much more difficult than proving that any asymptotic solution generates an almost equilibrium. Though several works addressed this problem in specific cases, including the case when the equilibria of the finite player game are taken over open loop strategies, the general case when the agents play strategies in closed (Markovian) feedback form has remained open so far. The objective here is to exhibit a quite large class of MFGs for which the answer is positive and, to do so, to implement a method that is robust enough to accommodate other sets of assumption.

The intrinsic difficulty in proving the convergence of finite player equilibria may be explained as follows. When taken over strategies in closed Markovian form, Nash equilibria of a stochastic differential game with N players in a state of dimension d may be described through a system of N quasilinear parabolic partial differential equations in dimension $N \times d$, which we refer to as the *Nash system* throughout the monograph. As N becomes larger and larger, the system obviously becomes more and more intricate. In particular, it seems especially difficult to get any a priori estimate that could be helpful for passing to the limit by means of a compactness argument. The strategy developed in the book is thus to bypass any detailed study of the Nash system. Instead, we use a short cut and focus directly on the expected limiting form of the Nash system. This limiting form is precisely what we call *the master equation* in the title of the book. As a result of the symmetry inherent in the mean field structure, this limiting form is no longer a system of equations but reduces to one equation only, which makes it simpler than the Nash system. It describes the equilibrium cost to one representative player in a continuum of players. Actually, to account for the continuum of players underpinning the game, the master equation has to be set over the Euclidean space times the space of probability measures; the state variable is thus a pair that encodes both the private state of a single representative player together with the statistical distribution of the states of all the players. Most of the book is thus dedicated to the analysis of this master equation. One of the key results in the book is to show that, under appropriate conditions on the coefficients, the master equation is uniquely solvable in the classical sense for an appropriate notion of differential calculus on the space of probability measures. Among the assumptions we require, we assume the coefficients to be monotone in the direction of the measure; as demonstrated earlier by Lasry and Lions, this forces uniqueness of the solution to the corresponding MFG.

Smoothness of the solution then plays a crucial role in our program. It is indeed the precise property we use next for proving the convergence of the finite player equilibria to the solution of the limiting MFG. The key step is indeed to expand the solution of the master equation along the "equilibrium trajectories" of the finite player games, which requires enough regularity. As indicated earlier, this methodology seems to be quite sharp and should certainly work under different sets of assumptions.

Actually, the master equation was already introduced by Lions in his lectures at Collège de France. It provides an alternative formulation to MFGs, different from the earlier (and most popular) one based on a coupled forward–backward system, known as the *MFG system*, which is made of a backward Hamilton–Jacobi equation and a forward Kolmogorov equation. Part of our program in the book is then achieved by exploiting quite systematically the connection between the MFG system and the master equation: In short, the MFG system plays the role of characteristics for the master equation. In the text, we use this correspondence heavily to establish, by means of a flow method, the smoothness of the solution to the master equation.

Though the MFG system was extensively studied in earlier works, we provide in the book a detailed analysis of it in the case when players in the finite game are subject to a so-called common noise: Under the action of this common noise, both the backward and forward equations in the MFG system become stochastic, which makes it more complicated; as a result, we devote a whole chapter to addressing the solvability of the MFG system under the presence of such a common noise. Together with the study of the convergence problem, this perspective is completely new in the literature.

The book is organized in six chapters, which include a detailed introduction and are followed by an appendix. The guideline follows the aforementioned steps: Part of the book is dedicated to the analysis of the master equation, including the study of the MFG system with a common noise, and the rest concerns the convergence problem. The main results obtained in the book are collected in Chapter 2. Chapter 3 is a sort of warm-up, as it contains a preliminary analysis of the master equation in the simpler case when there is no common noise. In Chapter 4, we study the MFG system in the presence of a common noise, and the corresponding analysis of the master equation is performed in Chapter 5. The convergence problem is addressed in Chapter 6. We suggest the reader start with the Introduction, which contains in particular a formal derivation of the master equation, and then to carry on with Chapters 2 and 3. Possibly, the reader who is interested only in MFGs without common noise may skip Chapters 4 and 5 and switch directly to Chapter 6. In such a case, she/he has to set the parameter β, which stands for the intensity of the common noise throughout the book, equal to 0. The Appendix contains several results on the differential calculus on the space of probability measures together with an Itô's formula for functionals of a process taking values in the space of probability measures. We emphasize that, for simplicity, most of the analysis provided in the book is on the torus, but, as already explained, we feel that the method is robust enough to accommodate the nonperiodic setting.

To conclude, we would like to thank our colleagues from our field for all the stimulating discussions and work sessions we have shared with them. Some of them have formulated very useful comments and suggestions on the preliminary draft of the book. They include in particular Yves Achdou, Martino Bardi, Alain Bensoussan, René Carmona, Jean-François Chassagneux, Markus Fischer, Wilfrid Gangbo, Christine Grün, Daniel Lacker and Alessio Porretta. We are also very grateful to the anonymous referees who examined the various versions of the manuscript. Their suggestions helped us greatly in improving the text.

We also thank the French National Research Agency, which supported part of this work under the grant ANR MFG (ANR-16-CE40-0015-01).

Pierre Cardaliaguet, Paris
François Delarue, Nice
Jean-Michel Lasry, Paris
Pierre-Louis Lions, Paris

The Master Equation and the Convergence Problem in Mean Field Games

Chapter One

Introduction

1.1 FROM THE NASH SYSTEM TO THE MASTER EQUATION

Game theory formalizes interactions between "rational" decision makers. Its applications are numerous and range from economics and biology to computer science. In this monograph we are interested mainly in noncooperative games, that is, in games in which there is no global planner: each player pursues his or her own interests, which are partly conflicting with those of others.

In noncooperative game theory, the key concept is that of Nash equilibria, introduced by Nash in [82]. A Nash equilibrium is a choice of strategies for the players such that no player can benefit by changing strategies while the other players keep theirs unchanged. This notion has proved to be particularly relevant and tractable in games with a small number of players and action sets. However, as soon as the number of players becomes large, it seems difficult to implement in practice, because it requires that each player knows the strategies the other players will use. Besides, for some games, the set of Nash equilibria is huge and it seems difficult for the players to decide which equilibrium they are going to play: for instance, in repeated games, the Folk theorem states that the set of Nash equilibria coincides with the set of feasible and individually rational payoffs in the one-shot game, which is a large set in general (see [93]).

In view of these difficulties, one can look for configurations in which the notion of Nash equilibria simplifies. As noticed by Von Neumann and Morgenstern [96], one can expect that this is the case when the number of players becomes large and each player individually has a negligible influence on the other players: it "is a well known phenomenon in many branches of the exact and physical sciences that very great numbers are often easier to handle than those of medium size [...]. This is of course due to the excellent possibility of applying the laws of statistics and probabilities in the first case" (p. 14). Such *nonatomic games* were analyzed in particular by Aumann [10] in the framework of cooperative games. Schmeidler [91] (see also Mas-Colell [78])) extended the notion of Nash equilibria to that setting and proved the existence of pure Nash equilibria.

In the book we are interested in games with a continuum of players, in continuous time and continuous state space. Continuous time, continuous space games are often called *differential games*. They appear in optimal control problems in which the system is controlled by several agents. Such problems (for a finite number of players) were introduced at about the same time by Isaacs [59] and Pontryagin [87]. Pontryagin derived optimality conditions for these games. Isaacs, working on specific examples of two-player zero-sum differential games, computed explicitly the solution of these games and established the formal connection with the Hamilton–Jacobi equations. The rigorous justification of Isaacs ideas for general systems took some time. The main difficulty arose from from the set of strategies (or from the dependence on the cost of the players with respect to these strategies), which is much more complex than for classical games: indeed, the players have to observe the actions taken by the other players in continuous time and choose their instantaneous actions accordingly. For two-player, zero-sum differential games, the first general existence result of a Nash equilibrium was established by Fleming [39]: in this case the Nash equilibrium is unique and is called the value function (it is a function of time and space). The link between this value function and the Hamilton–Jacobi equations was made possible by the introduction of viscosity solutions by Crandall and Lions [32] (see also [33] for a general presentation of viscosity solutions). The application to zero-sum differential games are due to Evans and Souganidis [35] (for determinist problems) and Fleming and Souganidis [40] (for stochastic ones).

For non-zero-sum differential games, the situation is more complicated. One can show the existence of general Nash equilibria thanks to an adaptation of the Folk theorem: see Kononenko [64] (for differential games of first order) and Buckdahn, Cardaliaguet, and Rainer [23] (for differential games with diffusion). However, this notion of solution does not allow for dynamic programming: it lacks time consistency in general. The existence of time-consistent Nash equilibria, based on dynamic programming, requires the solvability of a strongly coupled system of Hamilton–Jacobi equations. This system, which plays a key role in this book, is here called *the Nash system*. For problems without diffusions, Bressan and Shen explain in [21, 22] that the Nash system is ill-posed in general. However, for systems with diffusions, the Nash system becomes a uniformly parabolic system of partial differential equations. Typically, for a game with N players and with uncontrolled diffusions, this backward in time system takes the form

$$\begin{cases} -\partial_t v^i(t, \boldsymbol{x}) - \mathrm{tr}(a^i(t, \boldsymbol{x}) D^2 v^N(t, \boldsymbol{x})) + \mathcal{H}^i(t, \boldsymbol{x}, Dv^1(t, \boldsymbol{x}), \ldots, Dv^N(t, \boldsymbol{x})) = 0 \\ \qquad\qquad\qquad\qquad \text{in } [0, T] \times (\mathbb{R}^d)^N, \; i \in \{1, \ldots, N\}, \\ v^i(T, \boldsymbol{x}) = G^i(\boldsymbol{x}) \qquad \text{in } (\mathbb{R}^d)^N. \end{cases} \qquad (1.1)$$

The foregoing system describes the evolution in time of the value function v^i of agent i ($i \in \{1, \ldots, N\}$). This value function depends on the positions of all the players $\boldsymbol{x} = (x_1, \ldots, x_N)$, x_i being the position of the state of player i. The second-order terms $\mathrm{tr}(a^i(t, \boldsymbol{x}) D^2 v^N(t, \boldsymbol{x}))$ formalize the noises affecting the dynamics of agent i. The Hamiltonian $\mathcal{H}^i$ encodes the cost player i has to pay to control her state and reaching some goal. This cost depends on the positions of the other players and on their strategies.

The relevance of such a system for differential games has been discussed by Star and Ho [94] and Case [30] (for first-order systems) and by Friedman [43] (1972) (for second-order systems); see also the monograph by Başar and Olsder [11] and the references therein. The well-posedness of this system has been established under some restrictions on the regularity and the growth of the Hamiltonians: See in particular the monograph by Ladyženskaja, Solonnikov, and Ural'ceva [70] and the paper by Bensoussan and Frehse [14].

As for classical games, it is natural to investigate the limit of differential games as the number of players tends to infinity. The hope is that in this limit configuration the Nash system simplifies. This notion makes sense only for time-consistent Nash equilibria, because no simplification occurs in the framework of Folk's theorem, where the player who deviates is punished by all the other players.

Games in continuous space with infinitely many players were first introduced in the economic literature (in discrete time) under the terminology of heterogeneous models. The aim was to formalize dynamic general equilibria in macroeconomics by taking into account not only aggregate variables—GDP, employment, the general price level, for example—but also the distributions of variables, say the joint distribution of income and wealth or the size distribution of firms, and to try to understand how these variables interact. We refer in particular to the pioneering works by Aiyagari [6], Huggett [58], and Krusell and Smith [65], as well as the presentation of the continuous-time counterpart of these models in [5].

In the mathematical literature, the theory of differential games with infinitely many players, known as mean field games (MFGs), started with the works of Lasry and Lions [71, 72, 74]; Huang, Caines, and Malhamé [53–57] presented similar models under the name of the certainty equivalence principle. Since then the literature has grown very quickly, not only for the theoretical aspects, but also for the numerical methods and the applications: we refer to the monographs [16, 48] or the survey paper [49].

This book focuses mainly on the derivation of the MFG models from games with a finite number of players. In classical game theory, the rigorous link between the nonatomic games and games with a large but finite number of agents is quite well-understood: one can show (1) that limits of Nash equilibria as the number of agents tends to infinity is a Nash equilibrium of the nonatomic game (Green [50]), and (2) that any optimal strategy in the nonatomic game provides an ϵ-Nash equilibrium in the game with finitely many players, provided the number of players is sufficiently large (Rashid [90]).

For MFGs, the situation is completely different. If the equivalent of question (2) is pretty well understood, problem (1) turns out to be surprisingly difficult. Indeed, passing from the MFG equilibria to the differential game with finitely many problem relies mostly on known techniques in mean field theory: this has been developed since the beginning of the theory in [54] and well studied since then (see also, for instance, [25, 62]). On the contrary, when one considers a sequence of solutions to the Nash systems with N players and one wants to let N tend to infinity, the problem becomes extremely intricate. The main reason is that, in classical game theory, this convergence comes from compactness properties of the problem; this compactness is completely absent for differential games. This issue is related to the difficulty of building time-consistent solutions for these games. A less technical way to see this is to note that there is a change of nature between the Nash system and its conjectured limit, the MFG. In the Nash system, the players observe each other, and the deviation of a single player could a priori change entirely the outcome of the game. On the contrary, in the MFG, players react only to the evolving population density and therefore the deviation of a single player has no impact at all on the system. The main purpose of this book is to explain why this limit holds despite this change of nature.

1.1.1 Statement of the Problem

To explain our result further, we first need to specify the Nash system we are considering. We assume that players control their own state and interact only through their cost function. Then the Nash system (1.1) takes the more specific form:

$$\begin{cases} -\partial_t v^{N,i}(t,\boldsymbol{x}) - \sum_{j=1}^{N} \Delta_{x_j} v^{N,i}(t,\boldsymbol{x}) - \beta \sum_{j,k=1}^{N} \mathrm{Tr} D^2_{x_j,x_k} v^{N,i}(t,\boldsymbol{x}) \\ +H(x_i, D_{x_i} v^{N,i}(t,\boldsymbol{x})) + \sum_{j \neq i} D_p H(x_j, D_{x_j} v^{N,j}(t,\boldsymbol{x})) \cdot D_{x_j} v^{N,i}(t,\boldsymbol{x}) \\ = F^{N,i}(\boldsymbol{x}) \qquad \text{in } [0,T] \times (\mathbb{R}^d)^N, \\ v^{N,i}(T,\boldsymbol{x}) = G^{N,i}(\boldsymbol{x}) \qquad \text{in } (\mathbb{R}^d)^N. \end{cases} \quad (1.2)$$

As before, the above system is stated in $[0,T] \times (\mathbb{R}^d)^N$, where a typical element is denoted by $(t,\boldsymbol{x})$ with $\boldsymbol{x} = (x_1, \ldots, x_N) \in (\mathbb{R}^d)^N$. The unknowns are the N maps $(v^{N,i})_{i \in \{1,\ldots,N\}}$ (the value functions). The data are the Hamiltonian $H : \mathbb{R}^d \times \mathbb{R}^d \to \mathbb{R}$, the maps $F^{N,i}, G^{N,i} : (\mathbb{R}^d)^N \to \mathbb{R}$, the nonnegative parameter β, and the horizon $T \geqslant 0$. In the second line, the symbol $\cdot$ denotes the inner product in $\mathbb{R}^d$.

System (1.2) describes the Nash equilibria of an N-player differential game (see Section 1.2 for a short description). In this game, the set of "optimal trajectories" solves a system of N coupled stochastic differential equations (SDEs):

$$\begin{aligned} dX_{i,t} = -D_p H\big(X_{i,t}, Dv^{N,i}(t, \boldsymbol{X}_t)\big) dt + \sqrt{2}\, dB^i_t + \sqrt{2\beta}\, dW_t, \\ t \in [0,T], \ i \in \{1, \ldots, N\}, \end{aligned} \quad (1.3)$$

where $v^{N,i}$ is the solution to (1.2) and the $((B^i_t)_{t \in [0,T]})_{i=1,\ldots,N}$ and $(W_t)_{t \in [0,T]}$ are d-dimensional independent Brownian motions. The Brownian motions $((B^i_t)_{t \in [0,T]})_{i=1,\ldots,N}$ correspond to the *individual noises*, while the Brownian motion $(W_t)_{t\in[0,T]}$ is the same for all the equations and, for this reason, is called the *common noise*. Under such a probabilistic point of view, the collection of random processes $((X_{i,t})_{t\in[0,T]})_{i=1,\dots,N}$ forms a dynamical system of interacting particles.

The aim of this book is to understand the behavior, as N tends to infinity, of the value functions $v^{N,i}$. Another, but closely related, objective of our book is to study the mean field limit of the $((X_{i,t})_{t\in[0,T]})_{i=1,\dots,N}$ as N tends to infinity.

1.1.2 Link with the Mean Field Theory

Of course, there is no chance to observe a mean field limit for (1.3) under a general choice of the coefficients in (1.2). Asking for a mean field limit certainly requires that the system has a specific symmetric structure in such a way that the players in the differential game are somewhat exchangeable (when in equilibrium). For this purpose, we suppose that, for each $i \in \{1, \dots, N\}$, the maps $(\mathbb{R}^d)^N \ni \boldsymbol{x} \mapsto F^{N,i}(\boldsymbol{x})$ and $(\mathbb{R}^d)^N \ni \boldsymbol{x} \mapsto G^{N,i}(\boldsymbol{x})$ depend only on x_i and on the empirical distribution of the variables $(x_j)_{j\neq i}$:

$$F^{N,i}(\boldsymbol{x}) = F(x_i, m_{\boldsymbol{x}}^{N,i}) \quad \text{and} \quad G^{N,i}(\boldsymbol{x}) = G(x_i, m_{\boldsymbol{x}}^{N,i}), \qquad (1.4)$$

where $m_{\boldsymbol{x}}^{N,i} = \frac{1}{N-1}\sum_{j\neq i}\delta_{x_j}$ is the empirical distribution of the $(x_j)_{j\neq i}$ and where $F, G : \mathbb{R}^d \times \mathcal{P}(\mathbb{R}^d) \to \mathbb{R}$ are given functions, $\mathcal{P}(\mathbb{R}^d)$ being the set of Borel probability measures on $\mathbb{R}^d$. Under this assumption, the solution of the Nash system indeed enjoys strong symmetry properties, which imply in particular the required exchangeability property. Namely, $v^{N,i}$ can be written in a form similar to (1.4):

$$v^{N,i}(t, \boldsymbol{x}) = v^N(t, x_i, m_{\boldsymbol{x}}^{N,i}), \quad t \in [0,T], \quad \boldsymbol{x} \in (\mathbb{R}^d)^N, \qquad (1.5)$$

for a function $v^N(t, \cdot, \cdot)$ taking as arguments a state in $\mathbb{R}^d$ and an empirical distribution of size $N-1$ over $\mathbb{R}^d$.

In any case, even under the foregoing symmetry assumptions, it is by no means clear whether the system (1.3) can exhibit a mean field limit. The reason is that the dynamics of the particles $(X_{1,t}, \dots, X_{N,t})_{t\in[0,T]}$ are coupled through the unknown solutions $v^{N,1}, \dots, v^{N,N}$ to the Nash system (1.2), whose symmetry properties (1.5) may not suffice to apply standard results from the theory of propagation of chaos. Obviously, the difficulty is that the function v^N on the right-hand side of (1.5) precisely depends on N. Part of the challenge in the text is thus to show that the interaction terms in (1.3) get closer and closer, as N tends to the infinity, to some interaction terms with a much more tractable and much more explicit shape.

To get a picture of the ideal case under which the mean-field limit can be taken, one can choose for a while $\beta = 0$ in (1.3) and then assume that the function v^N in the right-hand side of (1.5) is independent of N. Equivalently, one can replace in (1.3) the interaction function $(\mathbb{R}^d)^N \ni \boldsymbol{x} \mapsto D_pH(x_i, v^{N,i}(t, \boldsymbol{x}))$

by $(\mathbb{R}^d)^N \ni \boldsymbol{x} \mapsto b(x_i, m_{\boldsymbol{x}}^{N,i})$, for a map $b : \mathbb{R}^d \times \mathcal{P}(\mathbb{R}^d) \mapsto \mathbb{R}^d$. In such a case, the coupled system of SDEs (1.3) turns into

$$dX_{i,t} = b\Big(X_{i,t}, \frac{1}{N-1}\sum_{j\neq i}\delta_{X_{j,t}}\Big)dt + \sqrt{2}\,dB_t^i, \quad t \in [0,T],\ i \in \{1,\dots,N\},\ (1.6)$$

the second argument in b being nothing but the empirical measure of the particle system at time t. Under suitable assumptions on b (e.g., if b is bounded and Lipschitz continuous in both variables, the space of probability measures being equipped with the Wasserstein distance) and on the initial distribution of the $((X_{i,t})_{i=1,\dots,N})_{t\in[0,T]}$, both the marginal law of $(X_t^1)_{t\in[0,T]}$ (or of any other player) and the empirical distribution of the whole system converge to the solution of the McKean–Vlasov equation:

$$\partial_t m - \Delta m + \mathrm{div}\big(m\,b(\cdot, m)\big) = 0.$$

(see, among many other references, McKean [77], Sznitman [92], Méléard [79]). The standard strategy for establishing the convergence consists in a coupling argument. Precisely, if one introduces the system of N independent equations

$$dY_{i,t} = b\big(Y_{i,t}, \mathcal{L}(Y_{i,t})\big)\,dt + \sqrt{2}\,dB_t^i, \qquad t \in [0,T],\ i \in \{1,\dots,N\},$$

(where $\mathcal{L}(Y_{i,t})$ is the law of $Y_{i,t}$) with the same (chaotic) initial condition as that of the processes $((X_{i,t})_{t\in[0,T]})_{i=1,\dots,N}$, then it is known that (under appropriate integrability conditions; see Fournier and Guillin [42])

$$\sup_{t\in[0,T]} \mathbb{E}\left[|X_{1,t} - Y_{1,t}|\right] \leqslant CN^{-\frac{1}{\max(2,d)}}\left(\mathbf{1}_{\{d\neq 2\}} + \ln(1+N)\mathbf{1}_{\{d=2\}}\right).$$

In comparison with (1.6), all the equations in (1.3) are subject to the common noise $(W_t)_{t\in[0,T]}$, at least when $\beta \neq 0$. This makes a first difference between our limit problem and the above McKean–Vlasov example of interacting diffusions, but, for the time being, it is not clear how deeply this may affect the analysis. Indeed, the presence of a common noise does not constitute a real challenge in the study of McKean–Vlasov equations, the foregoing coupling argument working in that case as well, provided that the distribution of Y is replaced by its conditional distribution given the realization of the common noise. However, the key point here is precisely that our problem is not formulated as a McKean–Vlasov equation, as the drifts in (1.3) are not of the same explicit mean field structure as they are in (1.6) because of the additional dependence on N in the right-hand side of (1.5): obviously this is the second main difference between (1.3) and (1.6). This makes rather difficult any attempt to guess the precise impact of the common noise on the analysis. Certainly, as we already pointed out, the major issue in analyzing (1.3) stems from the complex nature of the underlying interactions. As the equations depend on one another through the nonlinear system (1.2), the evolution with N of the coupling between all of them is indeed much more intricate than in (1.6). And once again, on the top of that, the common noise adds another layer of difficulty. For these reasons, the convergence of both (1.2) and (1.3) has been an open question since Lasry and Lions' initial papers on MFGs [71, 72].

1.1.3 The Mean Field Game System

If one tries, at least in the simpler case $\beta = 0$, to describe—in a heuristic way—the structure of a differential game with infinitely many indistinguishable players, i.e., a "nonatomic differential game," one finds a problem in which each (infinitesimal) player optimizes his payoff, depending on the collective behavior of the others, and, meanwhile, the resulting optimal state of each of them is exactly distributed according to the state of the population. This is the "mean field game system" (MFG system):

$$
\begin{cases}
-\partial_t u - \Delta u + H(x, D_x u) = F(x, m(t)) & \text{in } [0, T] \times \mathbb{R}^d, \\
\partial_t m - \Delta m - \operatorname{div}(m D_p H(x, D_x u)) = 0 & \text{in } [0, T] \times \mathbb{R}^d, \\
u(T, x) = G(x, m(T)), \ m(0, \cdot) = m_{(0)} & \text{in } \mathbb{R}^d,
\end{cases}
\tag{1.7}
$$

where $m_{(0)}$ denotes the initial state of the population. The system consists in a coupling between a (backward) Hamilton–Jacobi equation, describing the dynamics of the value function of any of the players, and a (forward) Kolmogorov equation, describing the dynamics of the distribution of the population. In that framework, H reads as a Hamiltonian, F is understood as a running cost, and G as a terminal cost. Since its simultaneous introduction by Lasry and Lions [74] and by Huang, Caines, and Malhamé [53], this system has been thoroughly investigated: its existence, under various assumptions, can be found in [15, 25, 54–56, 62, 74, 76]. Concerning uniqueness of the solution, two regimes were identified in [74]. Uniqueness holds under Lipschitz type conditions when the time horizon T is short (or, equivalently, when H, F, and G are "small"), but, as for finite-dimensional two-point boundary value problems, it may fail when the system is set over a time interval of arbitrary length. Over long time intervals, uniqueness is guaranteed under the quite fascinating condition that F and G are monotone; i.e., if, for any measures m, m', the following holds:

$$
\int_{\mathbb{R}^d} (F(x, m) - F(x, m')) \, d(m - m')(x) \geqslant 0
$$

$$
\text{and} \quad \int_{\mathbb{R}^d} (G(x, m) - G(x, m')) \, d(m - m')(x) \geqslant 0.
\tag{1.8}
$$

The interpretation of the monotonicity condition is that the players dislike congested areas and favor configurations in which they are more scattered; see Remark 2.3.1 for an example. Generally speaking, condition (1.8) plays a key role throughout the text, as it guarantees not only uniqueness but also stability of the solutions to (1.7).

As observed, a solution to the MFG system (1.7) can indeed be interpreted as a Nash equilibrium for a differential game with infinitely many players: in that framework, it plays the role of the Schmeidler noncooperative equilibrium. A standard strategy to make the connection between (1.7) and differential games consists in inserting the optimal strategies from the Hamilton–Jacobi equation in (1.7) into finitely many player games in order to construct approximate Nash equilibria: see [54], as well as [25, 55, 56, 62]. However, although it establishes the interpretation of the system (1.7) as a differential game with infinitely many players, this says nothing about the convergence of (1.2) and (1.3).

When β is positive, the system describing Nash equilibria within a population of infinitely many players subject to the same common noise of intensity β cannot be described by a deterministic system of the same form as (1.7). Owing to the theory of propagation of chaos for systems of interacting particles (see the short remark earlier), the unknown m in the forward equation is then expected to represent the conditional law of the optimal state of any player given the realization of the common noise. In particular, it must be random. This turns the forward Kolmogorov equation into a forward stochastic Kolmogorov equation. As the Hamilton–Jacobi equation depends on m, it renders u random as well. At any rate, a key fact from the theory of stochastic processes is that the solution to an SDE must be adapted to the underlying observation, as its values at some time t cannot anticipate the future of the noise after t. At first sight, it seems to be very demanding, as u is also required to match, at time T, $G(\cdot, m(T))$, which depends on the whole realization of the noise up until T. The correct formulation to accommodate both constraints is given by the theory of backward SDEs, which suggests penalizing the backward dynamics by a martingale in order to guarantee that the solution is indeed adapted. We refer the reader to the monograph [84] for a complete account on the finite dimensional theory and to the paper [85] for an insight into the infinite dimensional case. Denoting by W "the common noise" (here, a d-dimensional Brownian motion) and by $m_{(0)}$ the initial distribution of the players at time t_0, the MFG system with common noise then takes the form (in which the unknowns are now (u_t, m_t, v_t))

$$
\begin{cases}
d_t u_t = \left[-(1+\beta)\Delta u_t + H(x, D_x u_t) - F(x, m_t) - \sqrt{2\beta}\,\mathrm{div}(v_t)\right] dt \\
\qquad\qquad + v_t \cdot dW_t, \qquad\qquad \text{in } [0, T] \times \mathbb{R}^d, \\
d_t m_t = \left[(1+\beta)\Delta m_t + \mathrm{div}\big(m_t D_p H(x, D_x u_t)\big)\right] dt \\
\qquad\qquad - \mathrm{div}(m_t \sqrt{2\beta}\, dW_t), \qquad\qquad \text{in } [0, T] \times \mathbb{R}^d, \\
u_T(x) = G(x, m_T), \ m_0 = m_{(0)}, \qquad \text{in } \mathbb{R}^d
\end{cases}
\tag{1.9}
$$

where we used the standard convention from the theory of stochastic processes that consists in indicating the time parameter as an index in random functions. As suggested immediately above, the map v_t is a random vector field that forces the solution u_t of the backward equation to be adapted to the filtration generated by $(W_t)_{t \in [0,T]}$. As far as we know, the system (1.9) has never been investigated and part of this book will be dedicated to its analysis (see, however, [27] for an informal discussion). Below, we call the system (1.9) the *MFG system with common noise*.

Note that the aggregate equations (1.7) and (1.9) (see also the master equation (1.10)) are the continuous-time analogues of equations that appear in the analysis of dynamic stochastic general equilibria in heterogeneous agent models (Aiyagari [6], Bewley [19], and Huggett [58]). In this setting, the factor β describes the intensity of "aggregate shocks," as discussed by Krusell and Smith in the seminal paper [65]. In some sense, the limit problem studied in the text is an attempt to deduce the macroeconomic models, describing the dynamics of a typical (but heterogeneous) agent in an equilibrium configuration, from the microeconomic ones (the Nash equilibria).

1.1.4　The Master Equation

Although the MFG system has been widely studied since its introduction in [74] and [53], it has become increasingly clear that this system was not sufficient to take into account the entire complexity of dynamic games with infinitely many players. A case in point is that the original system (1.7) becomes much more complex in the presence of a common noise (i.e., when $\beta > 0$); see the stochastic version (1.9). In the same spirit, we may notice that the original MFG system (1.7) does not accommodate MFGs with a major player and infinitely many small players; see [52]. And, last but not least, the main limitation is that, so far, the formulation based on the system (1.7) (or (1.9) when $\beta > 0$) has not allowed establishment of a clear connection with the Nash system (1.2).

These issues led Lasry and Lions [76] to introduce an infinite dimensional equation—the so-called "master equation"—that directly describes, at least formally, the limit of the Nash system (1.2) and encompasses the foregoing complex situations. Before writing down this equation, let us explain its main features. One of the key observations has to do with the symmetry properties, to which we already alluded, that are satisfied by the solution of the Nash system (1.2). Under the standing symmetry assumptions (1.4) on the $(F^{N,i})_{i=1,\ldots,N}$ and $(G^{N,i})_{i=1,\ldots,N}$, (1.5) says that the $(v^{N,i})_{1,\ldots,N}$ can be written into a form similar to (1.4), namely $v^{N,i}(t, \boldsymbol{x}) = v^N(t, x_i, m_{\boldsymbol{x}}^{N,i})$ (where the empirical measures $m_{\boldsymbol{x}}^{N,i}$ are defined as in (1.4)), but with the obvious but major restriction that the function v^N that appears on the right-hand side of the equality now depends on N. With such a formulation, the value function to player i reads as a function of the private state of player i and of the empirical distribution formed by the others. Then, one may guess, at least under the additional assumption that such a structure is preserved as $N \to +\infty$, that the unknown in the limit problem takes the form $U = U(t, x, m)$, where x is the position of the (typical) small player at time t and m is the distribution of the (infinitely many) other agents.

The question is then to write down the dynamics of U. Plugging $U = U(t, x_i, m_x^{N,i})$ into the Nash system (1.2), one obtains—at least formally—an equation stated in the space of measures (see Section 1.2 for a heuristic discussion). This is the so-called master equation. It takes the form

$$
\begin{cases}
-\partial_t U - (1+\beta)\Delta_x U + H(x, D_x U) - (1+\beta) \int_{\mathbb{R}^d} \mathrm{div}_y \left[D_m U \right] dm(y) \\
\quad + \int_{\mathbb{R}^d} D_m U \cdot D_p H\big(y, D_x U(\cdot, y, \cdot)\big) dm(y) \\
\quad - 2\beta \int_{\mathbb{R}^d} \mathrm{div}_x \left[D_m U \right] dm(y) - \beta \int_{\mathbb{R}^{2d}} \mathrm{Tr}\left[D_{mm}^2 U \right] dm^{\otimes 2}(y, y') \\
= F(x, m) \qquad\qquad \text{in } [0, T] \times \mathbb{R}^d \times \mathcal{P}(\mathbb{R}^d), \\
U(T, x, m) = G(x, m) \qquad \text{in } \mathbb{R}^d \times \mathcal{P}(\mathbb{R}^d),
\end{cases}
\tag{1.10}
$$

where $\partial_t U$, $D_x U$, and $\Delta_x U$ are understood as $\partial_t U(t, x, m)$, $D_x U(t, x, m)$, and $\Delta_x U(t, x, m)$; $D_x U(\cdot, y, \cdot)$ is understood as $D_x U(t, y, m)$; and $D_m U$ and $D_{mm}^2 U$ are understood as $D_m(t, x, m, y)$ and $D_{mm}^2 U(t, x, m, y, y')$.

In Eq. (1.10), $\partial_t U$, $D_x U$, and $\Delta_x U$ stand for the usual time derivative, space derivatives, and Laplacian with respect to the local variables (t, x) of the unknown U, while $D_m U$ and $D_{mm}^2 U$ are the first- and second-order derivatives with respect to the measure m. The precise definition of these derivatives is postponed to Chapter 2. For the time being, let us just note that it is related to the derivatives in the space of probability measures described, for instance, by Ambrosio, Gigli, and Savaré in [7] and by Lions in [76]. It is worth mentioning that the master equation (1.10) is not the first example of an equation studied in the space of measures—by far: for instance, Otto [83] gave an interpretation of the porous medium equation as an evolution equation in the space of measures, and Jordan, Kinderlehrer, and Otto [60] showed that the heat equation was also a gradient flow in that framework; notice also that the analysis of Hamilton–Jacobi equations in metric spaces is partly motivated by the specific case in which the underlying metric space is the space of measures (see in particular [8,36] and the references therein). The master equation is, however, the first one to combine at the same time the issue of being nonlocal, nonlinear, and of second order and, moreover, without maximum principle.

Besides the discussion in [76], the importance of the master equation (1.10) has been acknowledged by several contributions: see, for instance, the monograph [16] and the companion papers [17] and [18], in which Bensoussan, Frehse, and Yam generalize this equation to mean field type control problems and reformulate it as a partial differential equation (PDE) set on an L^2 space, and [27], where Carmona and Delarue interpret this equation as a decoupling field of forward–backward SDE in infinite dimension.

If the master equation has been discussed and manipulated thoroughly in the aforementioned references, it is mostly at a formal level: the well-posedness of the master equation has remained, to a large extent, open until now. Besides, even if the master equation has been introduced to explain the convergence of the Nash system, the rigorous justification of the convergence has not been understood.

The aim of this book is to provide an answer to both questions.

1.1.5 Well-posedness of the Master Equation

The largest part of this book is devoted to the proof of the existence and uniqueness of a classical solution to the master equation (1.10), where, by classical, we mean that all the derivatives in (1.10) exist and are continuous. To avoid issues related to boundary conditions or conditions at infinity, we work for simplicity with periodic data: the maps H, F, and G are periodic in the space variable. The state space is therefore the d-dimensional torus $\mathbb{T}^d = \mathbb{R}^d/\mathbb{Z}^d$ and $m_{(0)}$ belongs to $\mathcal{P}(\mathbb{T}^d)$, the set of Borel probability measures on $\mathbb{T}^d$. We also assume that $F, G : \mathbb{T}^d \times \mathcal{P}(\mathbb{T}^d) \to \mathbb{R}$ satisfy the monotonicity conditions (1.8) and are sufficiently "differentiable" with respect to both variables and, of course, periodic with respect to the state variable. Although the periodicity condition is rather restrictive, the extension to maps defined on the full space or to Neumann boundary conditions is probably not a major issue. At any rate, it would certainly require further technicalities.

So far, the existence of classical solutions to the master equation has been known in more restricted frameworks. Lions discussed in [76] a finite dimensional analogue of the master equation and derived conditions for this hyperbolic system to be well posed. These conditions correspond precisely to the monotonicity property (1.8), which we here assume to be satisfied by the coupling functions F and G. This parallel strongly indicates—but this should not come as a surprise—that the monotonicity of F and G should play a key role in the unique strong solvability of (1.10). Lions also explained in [76] how to get the well-posedness of the master equation without noise (no Laplacian in the equation) by extending the equation to a (fixed) space of random variables under a convexity assumption in space of the data. In [24] Buckdahn, Li, Peng, and Rainer studied equation (1.10), by means of probabilistic arguments, when there is no coupling or common noise ($F = G = 0$, $\beta = 0$) and proved the existence of a classical solution in this setting; in a somewhat similar spirit, Kolokoltsov, Li, and Yang [62] and Kolokoltsov, Troeva, and Yang [63] investigated the tangent process to a flow of probability measures solving a McKean–Vlasov equation. Gangbo and Swiech [45] analyzed the first-order master equation in short time (no Laplacian in the equation) for a particular class of Hamiltonians and of coupling functions F and G (which are required to derive from a potential in the measure argument). Chassagneux, Crisan, and Delarue [31] obtained, by a probabilistic approach similar to that used in [24], the existence and uniqueness of a solution to (1.10) without common noise (when $\beta = 0$) under the monotonicity condition (1.8) in either the nondegenerate case (as we do here) or in the degenerate setting provided that F, H, and G satisfy additional convexity conditions in the variables (x, p). The complete novelty of our result, regarding the specific question of solvability of the master equation, is the existence and uniqueness of a classical solution to the problem with common noise.

The technique of proof in [24, 31, 45] consists in finding a suitable representation of the solution: indeed a key remark in Lions [76] is that the master equation is a kind of transport equation in the space of measures and that its characteristics are, when $\beta = 0$, the MFG system (1.7). Using this idea, the main difficulty is then to prove that the candidate is smooth enough to perform the computation showing that it is a classical solution of (1.10). In [24, 31] this is obtained by linearizing systems of forward–backward SDEs, while [45] relies on a careful analysis of the characteristics of the associated first-order PDE.

Our starting point is the same: we use a representation formula for the master equation. When $\beta = 0$, the characteristics are just the solution to the MFG system (1.7). When β is positive, these characteristics become random under the action of the common noise and are then given by the solution of the MFG system with common noise (1.9).

The construction of a solution U to the master equation then relies on the method of characteristics. Namely, we *define* U by letting $U(t_0, x, m_0) := u_{t_0}(x)$, where the pair $(u_t, m_t)_{t \in [t_0, T]}$ is the solution to (1.9) when the forward equation is initialized at $m_{(0)} \in \mathcal{P}(\mathbb{T}^d)$ at time t_0, that is,

$$
\begin{cases}
d_t u_t = \left[-(1+\beta)\Delta u_t + H(x, D_x u_t) - F(x, m_t) - \sqrt{2\beta}\mathrm{div}(v_t) \right] dt \\
\qquad\qquad + v_t \cdot dW_t \qquad\qquad \text{in } [t_0, T] \times \mathbb{T}^d, \\
d_t m_t = \left[(1+\beta)\Delta m_t + \mathrm{div}\left(m_t D_p H(x, D_x u_t) \right) \right] dt \\
\qquad\qquad - \mathrm{div}(m_t \sqrt{2\beta}\, dW_t) \qquad \text{in } [t_0, T] \times \mathbb{T}^d, \\
u_T(x) = G(x, m_T),\ m_{t_0} = m_{(0)} \qquad \text{in } \mathbb{T}^d.
\end{cases}
\tag{1.11}
$$

There are two main difficult steps in the analysis. The first one is to establish the smoothness of U and the second one is to show that U indeed satisfies the master equation (1.10). To proceed, the cornerstone is to make a systematic use of the monotonicity properties of the maps F and G: basically, monotonicity prevents the emergence of singularities in finite time. Our approach seems to be very powerful, although the reader might have a different feeling because of the length of the arguments. As a matter of fact, part of the technicalities in the proof are caused by the stochastic aspect of the characteristics (1.11). As a result, we spend much effort to handle the case with a common noise (for which almost nothing has been known so far), but, in the simpler case $\beta = 0$, our strategy to handle the first-order master equation provides a much shorter proof than in the earlier works [24, 31, 45]. For this reason, we decided to display the proof in this simple context separately (Section 3).

It is worth mentioning that, although our result is the first one to address the MFG system (1.11) in the case $\beta > 0$, the existence and uniqueness of equilibria to MFGs with a common noise were already studied in the paper [29] by

Carmona, Delarue, and Lacker. Therein, the strategy is completely different, as the existence is investigated first by combining purely probabilistic arguments together with Kakutani–Fan–Glicksberg's theorem for set-valued mappings. As a main feature, existence of equilibria is proved by means of a discretization procedure of the common noise, which consists in focusing first on the case when the common noise has a finite number of outcomes. This constraint on the noise is relaxed in a second step. However, it must be stressed that the limiting solutions that are obtained in this way (for the MFG driven by the original noise) are *weak equilibria* only, which means that they may not be adapted with respect to the common source of noise. This fact is completely reminiscent of the construction of weak solutions to SDEs. Remarkably, Yamada-Watanabe's principle for weak existence and strong uniqueness to SDEs extends to mean field games with a common noise: provided that a form of strong uniqueness holds for the MFG, any weak solution is in fact strong. Generally speaking, it is shown in [29] that strong uniqueness indeed holds true for MFGs with a common noise whenever the aforementioned monotonicity condition (1.8) is satisfied. In this regard, the result of [29] is completely consistent with the one we obtain here for the solvability of (1.11), as we prove that the solutions to (1.11) are indeed adapted with respect to $(W_t)_{t \in [0,T]}$. The main difference with [29] is that we take a short cut to get the result as we directly benefit from the monotone structure (1.8) to apply a fixed-point argument with uniqueness (instead of a fixed-point argument without uniqueness like Kakutani–Fan–Glicksberg's theorem). As a result, we here get in the same time existence and uniqueness of a solution to (1.11).

1.1.6   The Convergence Result

Although most of the book is devoted to the construction of a solution to the master equation, our main (and *primary*) motivation remains to justify the mean field limit. Namely, we show that the solution of the Nash system (1.2) converges to the solution of the master equation. The main issue here is the complete lack of estimates on the solutions to this large system of Hamilton–Jacobi equations: this prevents the use of any compactness method to prove the convergence. So far, this question has been almost completely open. The convergence has been known in very few specific situations. For instance, it was proved for the ergodic MFGs (see Lasry-Lions [71], revisited by Bardi-Feleqi [13]). In this case, the Nash equilibrium system reduces to a coupled system of N equations in $\mathbb{T}^d$ (instead of N equations in $\mathbb{T}^{Nd}$ as (1.2)) and estimates of the solutions are available. Convergence is also known in the "linear-quadratic" setting, where the Nash system has explicit solutions: see Bardi [12]. Let us finally quote the nice results by Fischer [38] and Lacker [69] on the convergence of *open loop Nash equilibria* for the N-player game and the characterization of the possible limits. Therein, the authors overcome the lack of strong estimates on the solutions to the N-player game by using the notion of *relaxed controls* for which weak compactness criteria are available. The problem addressed here—concerning *closed loop Nash equilibria*—differs in a substantial way from [38, 69]: indeed, we underline the striking fact that the Nash system (1.2), which concerns equilibria in which the players observe each other, converges to an equation in which the players only need to observe the evolution of the distribution of the population. This is striking because it allows for a drastic gain of complexity: without common noise, limiting equilibria are deterministic and hence can be precomputed; in particular, the limiting strategies are distributed in the sense that players just need to update their own state to compute the equilibrium strategy; this is in contrast with the equilibrium given by (1.2), as the latter requires updating the states of all the players in the equilibrium feedback function.

Our main contribution is a general convergence result, in large time, for MFGs with common noise, as well as an estimate of the rate of convergence. The convergence holds in the following sense: for any $\boldsymbol{x} \in (\mathbb{T}^d)^N$, let $m_{\boldsymbol{x}}^N :=$ $\frac{1}{N} \sum_{i=1}^N \delta_{x_i}$; then

$$\sup_{i=1,\cdots,N} \left| v^{N,i}(t_0, \boldsymbol{x}) - U(t_0, x_i, m_{\boldsymbol{x}}^N) \right| \leqslant CN^{-1}, \qquad (1.12)$$

for a constant C independent of N, t_0, and $\boldsymbol{x}$. We also prove a mean field result for the optimal solutions (1.3): if the initial conditions of the $((X_{i,\cdot}))_{i=1,\ldots,N}$ are i.i.d. and with the same law $m_{(0)} \in \mathcal{P}(\mathbb{T}^d)$, then

$$\mathbb{E}\left[\sup_{t \in [0,T]} |X_{i,t} - Y_{i,t}| \right] \leqslant CN^{-\frac{1}{d+8}}, \qquad (1.13)$$

where the $((Y_{i,t})_{i=1,\ldots,N})_{t \in [0,T]}$ are the solutions to the McKean–Vlasov SDE

$$dY_{i,t} = -D_p H \big(Y_{i,t}, D_x U \big(t, Y_{i,t}, \mathcal{L}(Y_{i,t}|W) \big) \big) dt$$
$$+ \sqrt{2} dB_t^i + \sqrt{2\beta}\, dW_t, \quad t \in [t_0, T],$$

with the same initial condition as the $((X_{i,t})_{i=1,\ldots,N})_{t \in [0,T]}$. Here U is the solution of the master equation and $\mathcal{L}(Y_{i,t}|W)$ is the conditional law of $Y_{i,t}$ given the realization of the whole path W. Since the $((Y_{i,t})_{t \in [0,T]})_{i=1,\ldots,N}$ are conditionally independent given W, (1.13) shows that (conditional) propagation of chaos holds for the N-Nash equilibria.

The technique of proof consists in testing the solution U of the master equation (1.10) as nearly a solution to the N-Nash system (1.2). On the model of (1.4), a natural candidate for being an approximate solution to the N-Nash system is indeed

$$u^{N,i}(t, \boldsymbol{x}) = U\big(t, x_i, m_{\boldsymbol{x}}^{N,i} \big), \quad t \in [0,T], \ \boldsymbol{x} \in (\mathbb{T}^d)^N.$$

Taking advantage of the smoothness of U, we then prove that the "proxies" $(u^{N,i})_{i=1,\ldots,N}$ almost solve the N-Nash system (1.2) up to a remainder term that vanishes as N tends to ∞. As a byproduct, we deduce that the $(u^{N,i})_{i=1,\ldots,N}$ get closer and closer to the "true solutions" $(v^{N,i})_{i=1,\ldots,N}$ when N tends to ∞, which yields (1.12). As the reader may notice, the convergence property (1.12) holds in supremum norm, which is a very strong fact.

It is worth mentioning that the monotonicity properties (1.4) play no role in our proof of the convergence. However, surprisingly, the uniform parabolicity of the MFG system is a key ingredient of the proof. On the one hand, in the uniformly parabolic setting, the convergence holds under the sole assumption that the master equation has a classical solution (plus structural Lipschitz continuity conditions on the coefficients). On the other hand, we do not know if one can dispense with the parabolicity condition.

1.1.7   Conclusion and Further Prospects

The fact that the existence of a classical solution to the master equation suffices to prove the convergence of the Nash system demonstrates the deep interest of the master equation, when regarded as a mathematical concept in its own right. Considering the problem from a more abstract point of view, the master equation indeed captures the evolution of the time-dependent semigroup generated by the Markov process formed, on the space of probability measures, by the forward component of the MFG system (1.11). Such a semigroup is said to be *lifted* as the corresponding Markov process has $\mathcal{P}(\mathbb{T}^d)$ as state space. In other words, the master equation is a nonlinear PDE driven by a Markov generator acting on functions defined on $\mathcal{P}(\mathbb{T}^d)$. The general contribution of our book is thus to show that any classical solution to the master equation accommodates a given perturbation of the lifted semigroup and that the information enclosed in such a classical solution suffices to determine the distance between the semigroup and its perturbation. Obviously, as a perturbation of a semigroup on the space of probability measures, we are here thinking of a system of N interacting particles, exactly as that formed by the Nash equilibrium of an N-player game.

Identifying the master equation with a nonlinear PDE driven by the Markov generator of a lifted semigroup is a key observation. As already pointed out, the Markov generator is precisely the operator, acting on functions from $\mathcal{P}(\mathbb{T}^d)$ to $\mathbb{R}$, generated by the forward component of the MFG system (1.11). Put differently, the law of the forward component of the MFG system (1.11), which resides in $\mathcal{P}(\mathcal{P}(\mathbb{T}^d))$, satisfies a forward Kolmogorov equation, also referred to as a "master equation" in physics. This says that "our master equation" is somehow the *dual* (in the sense that it is driven by the adjoint operator) of the "master equation" that would describe, according to the terminology used in physics, the law of the Nash equilibrium for a game with infinitely many players (in which case the Nash equilibrium itself is a distribution). We stress that this interpretation is very close to the point of view developed by Mischler and Mouhot, [80] in order to investigate Kac's program (except that, differently from ours, Mischler and Mouhot's work investigates uniform propagation of chaos over an infinite time horizon; we refer to the companion paper by Mischler,

Mouhot, and Wennberg [81] for the analysis, based on the same technology, of mean field models in finite time). Therein, the authors introduce the evolution equation satisfied by the (*lifted*) semigroup, acting on functions from $\mathcal{P}(\mathbb{R}^d)$ to $\mathbb{R}$, generated by the d-dimensional Boltzmann equation. According to our terminology, such an evolution equation is a "master equation" on the space of probability measures, but it is linear and of the first order while ours is nonlinear and of the second order (meaning second order on $\mathcal{P}(\mathbb{T}^d)$).

In this perspective, we also emphasize that our strategy for proving the convergence of the N-Nash system relies on a similar idea to that used in [80] to establish the convergence of Kac's jump process. Whereas our approach consists in inserting the solution of the master equation into the N-Nash system, Mischler and Mouhot's point of view is to compare the semigroup generated by the N-particle Kac's jump process, which operates on symmetric functions from $(\mathbb{R}^d)^N$ to $\mathbb{R}$ (or equivalently on empirical distributions of size N), with the *limiting lifted* semigroup, when acting on the same class of symmetric functions from $(\mathbb{R}^d)^N$ to $\mathbb{R}$. Clearly, the philosophy is the same, except that, in our setting, the "limiting master equation" is nonlinear and of second order (which renders the analysis more difficult) and is set over a finite time horizon only (which does not ask for uniform in time estimates). It is worth mentioning that similar ideas have been explored by Kolokoltsov in the monograph [61] and developed, in the McKean–Vlasov framework, in the subsequent works [62] and [63] in collaboration with his coauthors.

Of course, these parallels raise interesting questions, but we refrain from comparing these different works in a more detailed way: this would require to address more technical questions regarding, for instance, the topology used on the space of probability measures and the regularity of the various objects in hand; clearly, this would distract us from our original objective. We thus feel better to keep the discussion at an informal level and to postpone a more careful comparison to future works on the subject.

We complete the introduction by pointing out possible generalizations of our results. For simplicity of notation, we work in the autonomous case, but the results remain unchanged if H or F is time dependent provided that the coefficients F, G, and H, and their derivatives (whenever they exist), are continuous in time and that the various quantitative assumptions we put on F, G, and H hold uniformly with respect to the time variable. We can also remove the monotonicity condition (1.8) provided that the time horizon T is assumed to be small enough. The reason is that the analysis of the smoothness of U relies on the solvability and stability properties of the forward–backward system (1.11) and of its linearized version: as for finite-dimensional two-point boundary value problems, Lipschitz type conditions on the coefficients (and on their derivatives since we are also dealing with the linearized version) are sufficient whenever T is small enough.

As already mentioned, we also choose to work in the periodic framework. We expect similar results under other type boundary conditions, like the entire space $\mathbb{R}^d$ or Neumann boundary conditions.

Notice also that our results can be generalized without much difficulty to the *stationary setting*, corresponding to infinite horizon problems. This framework is particularly meaningful for economic applications. In this setting the Nash system takes the form

$$
\left\{
\begin{aligned}
&rv^{N,i}(\boldsymbol{x}) - \sum_{j=1}^{N} \Delta_{x_j} v^{N,i}(\boldsymbol{x}) - \beta \sum_{j,k=1}^{N} \mathrm{Tr} D^2_{x_j,x_k} v^{N,i}(\boldsymbol{x}) + H(x_i, D_{x_i} v^{N,i}(\boldsymbol{x})) \\
&\quad + \sum_{j \neq i} D_p H(x_j, D_{x_j} v^{N,j}(\boldsymbol{x})) \cdot D_{x_j} v^{N,i}(\boldsymbol{x}) = F^{N,i}(\boldsymbol{x}) \qquad \text{in } (\mathbb{R}^d)^N,
\end{aligned}
\right.
$$

where $r > 0$ is interpreted as a discount factor. The corresponding master equation is

$$
\left\{
\begin{aligned}
&rU - (1+\beta)\Delta_x U + H(x, D_x U) \\
&\quad -(1+\beta)\int_{\mathbb{R}^d} \mathrm{div}_y \left[D_m U \right] dm(y) + \int_{\mathbb{R}^d} D_m U \cdot D_p H\big(y, D_x U(y, \cdot)\big) dm(y) \\
&\quad -2\beta \int_{\mathbb{R}^d} \mathrm{div}_x \left[D_m U \right] dm(y) - \beta \int_{\mathbb{R}^{2d}} \mathrm{Tr} \left[D^2_{mm} U \right] dm^{\otimes 2}(y, y') = F(x, m) \\
&\qquad \text{in } \mathbb{R}^d \times \mathcal{P}(\mathbb{R}^d),
\end{aligned}
\right.
$$

where the unknown is the map $U = U(x, m)$, and with the same convention of notation as in (1.10). One can again solve this system by using the method of (infinite dimensional) characteristics, paying attention to the fact that these characteristics remain time dependent. The MFG system with common noise takes the form (in which the unknown are now (u_t, m_t, v_t))

$$
\left\{
\begin{aligned}
&d_t u_t = \big[r u_t - (1+\beta)\Delta u_t + H(x, D_x u_t) - F(x, m_t) - 2\beta \mathrm{div}(v_t) \big] dt \\
&\qquad + v_t \cdot \sqrt{2\beta}\, dW_t \qquad\qquad \text{in } [0, +\infty) \times \mathbb{R}^d \\
&d_t m_t = \big[(1+\beta)\Delta m_t + \mathrm{div}\big(m_t D_p H(m_t, D_x u_t)\big) \big] dt - \mathrm{div}(m_t \sqrt{2\beta}\, dW_t), \\
&\qquad\qquad\qquad\qquad \text{in } [0, +\infty) \times \mathbb{R}^d \\
&m_0 = m_{(0)} \qquad \text{in } \mathbb{R}^d, \ (u_t)_t \text{ bounded a.s.}
\end{aligned}
\right.
$$

Lastly, we point out that, even though we do not address this question in the book, our work could be used later for numerical purposes. Solving numerically MFGs is indeed a delicate issue and, so far, numerical methods have been regarded mostly in the case without common noise: We refer to the works of Achdou and his coauthors; see, for instance [1–3] for discretization schemes of the MFG system (1.7). Owing to obvious issues of complexity, the case with common noise seems especially challenging. A case in point is that the system (1.11) is an infinite-dimensional fully coupled forward–backward system, which could be thought, at the discrete level, as an infinite-dimensional equation expanding along all the possible branches of the tree generated by a discrete random walk. Although our work does not provide any clue for bypassing these complexity issues, we guess that our theoretical results—both the representation of the equilibria in the form of the MFG system system (1.11) and through the master equation (1.10) and their regularity—could be useful for a numerical analysis.

1.1.8   Organization of the Text and Reading Guide

We present our main results in Chapter 2, where we also explain the notation, state the assumption, and rigorously define the notion of derivative on the space of measures. The well-posedness of the master equation is proved in Chapter 3 when $\beta = 0$. Unique solvability of the MFG system with common noise is discussed in Chapter 4. Results obtained in Chapter 4 are implemented in Chapter 5 to derive the existence of a classical solution to the master equation in the general case. The last chapter is devoted to the convergence of the Nash system. In the Appendix, we revisit the notion of derivative on the space of probability measures and discuss some useful auxiliary properties.

We strongly recommend that the reader starts with Section 1.2 and with Chapters 2 and 3. Section 1.2 provides heuristic arguments for the construction of a solution to the master equation; this might be really helpful to understand the key results of the book. The complete proof of existence for the first-order case (i.e., without common noise) is the precise aim of Chapter 3; we feel it really accessible.

The reader who is more interested in the analysis of the convergence problem than in the study of the case with common noise may directly skip to Chapter 6; to make things easier, she/he may follow the computations of Chapter 6 by letting $\beta = 0$ therein (i.e., no common noise). In fact, we suggest that, even if she/he is interested in the case with common noise, the reader also follow this plan, especially if she/he is not keen on probability theory and stochastic calculus; at a second time, she/he can go back to Chapters 4 and 5, which are more technical. In these latter two chapters, the reader who is really interested in MFGs with common noise will find new results: The analysis of the MFG system with common noise is mostly the aim of Chapter 4; if needed, the reader may return to Section 3.1, Proposition 3.1.1, for a basic existence result in the case without common noise. The second-order master equation (with common noise) is investigated in Chapter 5, but requires the well-posedness of the MFG system with common noise as stated in Theorem 4.3.1.

The reader should be aware of some basics of stochastic calculus (mostly Itô's formula) to follow the computations of Chapter 6. Chapters 4 and 5 are partly inspired from the theory of backward stochastic differential equations; although this might not be necessary, the reader may have a look at the two monographs [84, 97] for a complete overview of the subject and at the textbook [88] for an introduction.

Of course, the manuscript borrows considerably from the PDE literature and in particular from the theory of Hamilton–Jacobi equations; the reason is that a solution to an MFG is defined as a fixed point of a mapping taking as inputs the optimal trajectories of a family of optimal stochastic control problems. As for the connection between stochastic optimal control problems and Hamilton–Jacobi equations, we refer the reader to the monographs [41, 66]. Some PDE regularity estimates are used quite often in the text, especially for linear and nonlinear second-order parabolic equations; most of them are taken from well-known books on the subject, among which are [70] and [75].

Lastly, the reader will also find in the book results that may be useful for other purposes: Derivatives in the space of measures are discussed in Section 2.2 (definition and basic results) and in Section A.1 of the Appendix (link with Lions' approach); a chain rule (Itô's formula) for functions defined on the space of measures, when taken along the solution of a stochastic Kolmogorov equation, is derived in Section A.3 of the Appendix.

1.2   INFORMAL DERIVATION OF THE MASTER EQUATION

Before stating our main results, it is worthwhile explaining the meaning of the Nash system and the heuristic derivation of the master equation from the Nash system and its main properties. We hope that this (by no means rigorous) presentation might help the reader to be acquainted with our notation and the main ideas of proof. To emphasize the informal aspect of the discussion, we state all the ideas in $\mathbb{R}^d$, without bothering about the boundary issues (whereas in the rest of the text we always work with periodic boundary conditions).

1.2.1   The Differential Game

The Nash system (1.2) arises in differential game theory. Differential games are just optimal control problems with many (here N) players. In this game, player i ($i = 1, \ldots, N$) controls her/his state $(X_{i,t})_{t \in [0,T]}$ through her/his control $(\alpha_{i,t})_{t \in [0,T]}$. The state $(X_{i,t})_{t \in [0,T]}$ evolves according to the SDE:

$$dX_{i,t} = \alpha_{i,t}dt + \sqrt{2}\,dB_t^i + \sqrt{2\beta}\,dW_t, \qquad X_{t_0} = x_{i,0}. \qquad (1.14)$$

Recall that the d-dimensional Brownian motions $((B_t^i)_{t\in[0,T]})_{i=1,\ldots,N}$ and $(W_t)_{t\in[0,T]}$ are independent, $(B_t^i)_{t\in[0,T]}$ corresponding to the *individual noise* (or *idiosyncratic noise*) of player i and $(W_t)_{t\in[0,T]}$ being the *common noise*, which affects all the players. Controls $((\alpha_{i,t})_{t\in[0,T]})_{i=1,\ldots,N}$ are required to be progressively measurable with respect to the filtration generated by all the noises. Given an initial condition $\boldsymbol{x}_0 = (x_{1,0}, \ldots, x_{N,0}) \in (\mathbb{R}^d)^N$ for the whole system at time t_0, each player aims at minimizing the cost functional:

$$J_i^N\big(t_0, \boldsymbol{x}_0, (\alpha_{j,\cdot})_{j=1,\dots,N}\big)$$

$$= \mathbb{E}\left[\int_{t_0}^{T}\big(L(X_{i,s},\alpha_{i,s}) + F^{N,i}(\boldsymbol{X}_s)\big)\,ds + G^{N,i}(\boldsymbol{X}_T)\right],$$

where $\boldsymbol{X}_t = (X_{1,t},\dots,X_{N,t})$ and where $L : \mathbb{R}^d \times \mathbb{R}^d \to \mathbb{R}$, $F^{N,i} : \mathbb{R}^{Nd} \to \mathbb{R}$ and $G^{N,i} : \mathbb{R}^{Nd} \to \mathbb{R}$ are given Borel maps. For each player i, in order to assume that the other players are indistinguishable, we shall suppose, as in (1.4), that $F^{N,i}$ and $G^{N,i}$ are of the form

$$F^{N,i}(\boldsymbol{x}) = F(x_i, m_{\boldsymbol{x}}^{N,i}) \qquad \text{and} \qquad G^{N,i}(\boldsymbol{x}) = G(x_i, m_{\boldsymbol{x}}^{N,i}).$$

In the above expressions, $F, G : \mathbb{R}^d \times \mathcal{P}(\mathbb{R}^d) \to \mathbb{R}$, where $\mathcal{P}(\mathbb{R}^d)$ is the set of Borel measures on $\mathbb{R}^d$. The Hamiltonian of the problem is related to L by the formula

$$\forall (x,p) \in \mathbb{R}^d \times \mathbb{R}^d, \qquad H(x,p) = \sup_{\alpha \in \mathbb{R}^d}\{-\alpha \cdot p - L(x,\alpha)\}.$$

Let now $(v^{N,i})_{i=1,\dots,N}$ be the solution to (1.2). By Itô's formula, it is easy to check that $(v^{N,i})_{i=1,\dots,N}$ corresponds to an optimal solution of the problem in the sense of Nash, i.e., a *Nash equilibrium* of the game. Namely, the feedback strategies

$$\big(\alpha_i^*(t,\boldsymbol{x}) := -D_p H(x_i, D_{x_i} v^{N,i}(t,\boldsymbol{x}))\big)_{i=1,\dots,N} \tag{1.15}$$

provide a feedback Nash equilibrium for the game:

$$v^{N,i}\big(t_0, \boldsymbol{x}_0\big) = J_i^N\big(t_0, \boldsymbol{x}_0, (\alpha_{j,\cdot}^*)_{j=1,\dots,N}\big) \leqslant J_i^N\big(t_0, \boldsymbol{x}_0, \alpha_{i,\cdot}, (\hat{\alpha}_{j,\cdot}^*)_{j\neq i}\big)$$

for any $i \in \{1,\dots,N\}$ and any control $\alpha_{i,\cdot}$, progressively measurable with respect to the filtration generated by $((B_t^j)_{j=1,\dots,N})_{t\in[0,T]}$ and $(W_t)_{t\in[0,T]}$. In the left-hand side, $\alpha_{j,\cdot}^*$ is an abuse of notation for the process $(\alpha_j^*(t, X_{j,t}))_{t\in[0,T]}$, where $(X_{1,t},\dots,X_{N,t})_{t\in[0,T]}$ solves the system of SDEs (1.14) when $\alpha_{j,t}$ is precisely given under the implicit form $\alpha_{j,t} = \alpha_j^*(t, X_{j,t})$. Similarly, in the right-hand side, $\hat{\alpha}_j^*$, for $j \neq i$, denotes $(\alpha_j^*(t, X_{j,t}))_{t\in[0,T]}$, where $(X_{1,t},\dots,X_{N,t})_{t\in[0,T]}$ now solves the system of SDEs (1.14) for the given $\alpha_{i,\cdot}$, the other $(\alpha_{j,t})_{j\neq i}$'s being given under the implicit form $\alpha_{j,t} = \alpha_j^*(t, X_{j,t})$. In particular, system (1.3), in which all the players play the optimal feedback (1.15), describes the dynamics of the optimal trajectories.

1.2.2 Derivatives in the Space of Measures

To describe the limit of the maps $(v^{N,i})$, let us introduce—in a completely informal manner—a notion of derivative in the space of measures $\mathcal{P}(\mathbb{R}^d)$. A rigorous description of the notion of derivative used in this book is given in Section 2.2.

In the following discussion, we argue as if all the measures had a density. Let $U : \mathcal{P}(\mathbb{R}^d) \to \mathbb{R}$. *Restricting* the function U to the elements m of $\mathcal{P}(\mathbb{R}^d)$ that have a density in $L^2(\mathbb{R}^d)$ and assuming that U is defined in a neighborhood $\mathcal{O} \subset L^2(\mathbb{R}^d)$ of $\mathcal{P}(\mathbb{R}^d) \cap L^2(\mathbb{R}^d)$, we can use the Hilbert structure on $L^2(\mathbb{R}^d)$. We denote by $\delta U/\delta m$ the gradient of U in $L^2(\mathbb{R}^d)$, namely

$$\frac{\delta U}{\delta m}(p)(q) - \lim_{\varepsilon \to 0} \frac{1}{\varepsilon}\Big(U(p + \varepsilon q) - U(p)\Big), \quad p \in \mathcal{O}, \ q \in L^2(\mathbb{R}^d).$$

Of course, we can identify $[\delta U/\delta m](p)$ with an element of $L^2(\mathbb{R}^d)$, which we denote by $\mathbb{R}^d \ni y \mapsto [\delta U/\delta m](p, y) \in \mathbb{R}$. Then, the duality product $[\delta U/\delta m](p)(q)$ reads as the inner product $\langle [\delta U/\delta m](p, \cdot), q(\cdot) \rangle_{L^2(\mathbb{R}^d)}$. Similarly, we denote by $\delta^2 U/\delta m^2(p)$ the second-order derivative of U at $p \in L^2(\mathbb{R}^d)$ (which can be identified with a symmetric bilinear form on $L^2(\mathbb{R}^d)$ and hence with a symmetric function $\mathbb{R}^d \times \mathbb{R}^d \ni (y, y') \mapsto [\delta^2 U/\delta m^2](p, y, y') \in \mathbb{R}$ in $L^2(\mathbb{R}^d \times \mathbb{R}^d)$):

$$\frac{\delta U}{\delta m}(p)(q, q') = \lim_{\varepsilon \to 0} \frac{1}{\varepsilon}\Big(\frac{\delta U}{\delta m}(p + \varepsilon q)(q') - \frac{\delta U}{\delta m}(p)(q')\Big), \quad p \in \mathcal{O}, \ q, q' \in L^2(\mathbb{R}^d).$$

We then set, when possible,

$$D_m U(m, y) = D_y \frac{\delta U}{\delta m}(m, y), \quad D^2_{mm} U(m, y, y') = D^2_{y, y'} \frac{\delta^2 U}{\delta m^2}(m, y, y'). \quad (1.16)$$

To explain the meaning of $D_m U$, let us compute the action of U onto the push-forward of a measure m by the flow an ordinary differential equation driven by a smooth vector field. For a given smooth vector field $B : \mathbb{R}^d \to \mathbb{R}^d$ and an absolutely continuous probability measure $m \in \mathcal{P}(\mathbb{R}^d)$ with a smooth density, let $(m(t))_{t \geq 0} = (\mathbb{R}^d \ni x \mapsto m(t, x))_{t \geq 0}$ be the solution to

$$\begin{cases} \dfrac{\partial m}{\partial t} + \mathrm{div}(Bm) = 0, \\ m_0 = m. \end{cases}$$

Provided that $[\partial m/\partial t](t, \cdot)$ lives in $L^2(\mathbb{R}^d)$, this expression directly gives

$$\begin{aligned} \frac{d}{dh} U(m(h))_{|h=0} &= \langle \frac{\delta U}{\delta m}, -\mathrm{div}(Bm) \rangle_{L^2(\mathbb{R}^d)} \\ &= \int_{\mathbb{R}^d} D_m U(m, y) \cdot B(y) \, dm(y), \end{aligned} \quad (1.17)$$

where we used an integration by parts in the last equality.

Another way to understand these derivatives is to project the map U to the finite dimensional space $(\mathbb{R}^d)^N$ via the empirical measure: if $\boldsymbol{x} = (x_1, \ldots, x_N) \in$

$(\mathbb{R}^d)^N$, let $m_{\boldsymbol{x}}^N := (1/N)\sum_{i=1}^N \delta_{x_i}$ and set $u^N(\boldsymbol{x}) = U(m_{\boldsymbol{x}}^N)$. Then one checks the following relationships (see Proposition 6.1.1): for any $j \in \{1,\dots,N\}$,

$$D_{x_j}u^N(\boldsymbol{x}) = \frac{1}{N}D_mU(m_{\boldsymbol{x}}^N, x_j), \tag{1.18}$$

$$D^2_{x_j,x_j}u^N(\boldsymbol{x}) = \frac{1}{N}D_y\,[D_mU]\,(m_{\boldsymbol{x}}^N, x_j) + \frac{1}{N^2}D^2_{mm}U(m_{\boldsymbol{x}}^N, x_j, x_j) \tag{1.19}$$

while, if $j \neq k$,

$$D^2_{x_j,x_k}u^N(\boldsymbol{x}) = \frac{1}{N^2}D^2_{mm}U(m_{\boldsymbol{x}}^N, x_j, x_k). \tag{1.20}$$

1.2.3 Formal Asymptotic of the $(v^{N,i})$

Provided that (1.2) has a unique solution, each $v^{N,i}$, for $i = 1,\dots,N$, is symmetric with respect to permutations on $\{1,\dots,N\}\backslash\{i\}$ and, for $i \neq j$, the role played by x^i in $v^{N,i}$ is the same as the role played by x^j in $v^{N,j}$ (see Section 6.2). Therefore, it makes sense to expect, in the limit $N \to +\infty$,

$$v^{N,i}(t, \boldsymbol{x}) \simeq U(t, x_i, m_{\boldsymbol{x}}^{N,i})$$

where $U : [0,T] \times \mathbb{R}^d \times \mathcal{P}(\mathbb{R}^d) \to \mathbb{R}$. Starting from this *ansatz*, our aim is now to provide heuristic arguments explaining why U should satisfy (1.10). The sense in which the $(v^{N,i})_{i=1,\dots,N}$ actually converge to U is stated in Theorem 2.4.8 and the proof given in Chapter 6.

The informal idea is to assume that $v^{N,i}$ is already of the form $U(t, x_i, m_{\boldsymbol{x}}^{N,i})$ and to plug this expression into the equation of the Nash equilibrium (1.2): the time derivative and the derivative with respect to x_i are understood in the usual sense, while the derivatives with respect to the other variables are computed by using the relations in the previous section.

The terms $\partial_t v^{N,i}$ and $H(x_i, D_{x_i}v^{N,i})$ easily become $\partial U/\partial t$ and $H(x, D_x U)$. We omit for a while the second-order terms and concentrate on the expression (see the second line in (1.2)):

$$\sum_{j \neq i} D_p H(x_j, D_{x_j}v^{N,j}) \cdot D_{x_j}v^{N,i}\,.$$

Note that $D_{x_j}v^{N,j}$ is just like $D_x U(t, x_j, m_{\boldsymbol{x}}^{N,j})$. In view of (1.18),

$$D_{x_j}v^{N,i} \simeq \frac{1}{N-1}D_m U(t, x_i, m_{\boldsymbol{x}}^{N,i}, x_j),$$

and the sum over j is like an integration with respect to $m_{\boldsymbol{x}}^{N,i}$. So we find, ignoring the difference between $m_{\boldsymbol{x}}^{N,i}$ and $m_{\boldsymbol{x}}^{N,j}$,

$$\sum_{j\neq i} D_p H(x_j, D_{x_j} v^{N,j}) \cdot D_{x_j} v^{N,i}$$

$$\simeq \int_{\mathbb{R}^d} D_p H(y, D_x U(t, m_{\boldsymbol{x}}^{N,i}, y)) \cdot D_m U(t, x_i, m_{\boldsymbol{x}}^{N,i}, y) dm_{\boldsymbol{x}}^{N,i}(y).$$

We now study the term $\displaystyle\sum_{j=1}^{N} \Delta_{x_j} v^{N,i}$ (see the first line in (1.2)). As $\Delta_{x_i} v^{N,i} \simeq \Delta_x U$, we need to analyze the quantity $\displaystyle\sum_{j\neq i} \Delta_{x_j} v^{N,i}$. In view of (1.19), we expect

$$\sum_{j\neq i} \Delta_{x_j} v^{N,i} \simeq \frac{1}{N-1} \sum_{j\neq i} \operatorname{div}_y [D_m U] (t, x_i, m_{\boldsymbol{x}}^{N,i}, x_j)$$

$$+ \frac{1}{(N-1)^2} \sum_{j\neq i} \operatorname{Tr} [D_{mm}^2 U] (t, x_i, m_{\boldsymbol{x}}^{N,i}, x_j, x_j)$$

$$= \int_{\mathbb{R}^d} \operatorname{div}_y [D_m U] (t, x_i, m_{\boldsymbol{x}}^{N,i}, y) dm_{\boldsymbol{x}}^{N,i}(y)$$

$$+ \frac{1}{N-1} \int_{\mathbb{R}^d} \operatorname{Tr} [D_{mm}^2 U] (t, x_i, m_{\boldsymbol{x}}^{N,i}, y, y) dm_{\boldsymbol{x}}^{N,i}(y),$$

where we can drop the last term, as it is of order $1/N$.

Let us finally discuss the limit of the term $\displaystyle\sum_{k,l=1}^{N} \operatorname{Tr}(D_{x_j,x_k}^2 v^{N,i})$ (see the first line in (1.2)) that we rewrite

$$\Delta_{x_i} v^{N,i} + 2 \sum_{k\neq i} \operatorname{Tr}(D_{x_i} D_{x_k} v^{N,i}) + \sum_{k,l\neq i} \operatorname{Tr}(D_{x_k,x_l}^2 v^{N,i}). \qquad (1.21)$$

The first term gives $\Delta_x U$. Using (1.18), the second one becomes

$$2 \sum_{k\neq i} \operatorname{Tr}(D_{x_i} D_{x_k} v^{N,i}) \simeq \frac{2}{N-1} \sum_{k\neq i} \operatorname{Tr} [D_x D_m U] (t, x_i, m_{\boldsymbol{x}}^{N,i}, x_k)$$

$$= 2 \int_{\mathbb{R}^d} \operatorname{div}_x [D_m U] (t, x_i, m_{\boldsymbol{x}}^{N,i}, y) dm_{\boldsymbol{x}}^{N,i}(y).$$

As for the last term in (1.21), we have by (1.20):

$$
\sum_{k,l \neq i} \mathrm{Tr}\big(D^2_{x_k,x_l} v^{N,i}\big) \simeq \frac{1}{(N-1)^2} \sum_{k,l \neq i} \mathrm{Tr}\left[D^2_{mm}U\right](t, x_i, m_{\boldsymbol{x}}^{N,i}, x_j, x_k)
$$

$$
= \int_{\mathbb{R}^d} \int_{\mathbb{R}^d} \mathrm{Tr}\left[D^2_{mm}U\right](t, x_i, m_{\boldsymbol{x}}^{N,i}, y, y') dm_{\boldsymbol{x}}^{N,i}(y) dm_{\boldsymbol{x}}^{N,i}(y').
$$

Collecting the above relations, we expect that the Nash system

$$
\begin{cases}
-\dfrac{\partial v^{N,i}}{\partial t} - \displaystyle\sum_{j=1}^{N} \Delta_{x_j} v^{N,i} - \beta \sum_{k,l=1}^{N} \mathrm{Tr}\big(D^2_{x_k,x_l} v^{N,i}\big) + H(x_i, D_{x_i} v^{N,i}) \\[2ex]
\qquad + \displaystyle\sum_{j \neq i} D_p H(x_j, D_{x_j} v^{N,j}) \cdot D_{x_j} v^{N,i} = F(x_i, m_{\boldsymbol{x}}^{N,i}), \\[2ex]
v^{N,i}(T, \boldsymbol{x}) = G(x_i, m_{\boldsymbol{x}}^{N,i}),
\end{cases}
$$

has for limit

$$
\begin{cases}
-\dfrac{\partial U}{\partial t} - \Delta_x U - \displaystyle\int_{\mathbb{R}^d} \mathrm{div}_y\left[D_m U\right] dm(y) + H(x, D_x U) \\[2ex]
-\beta\left(\Delta_x U + 2 \displaystyle\int_{\mathbb{R}^d} \mathrm{div}_x\left[D_m U\right] dm(y) + \int_{\mathbb{R}^d} \mathrm{div}_y\left[D_m U\right] dm(y) \right. \\[2ex]
\qquad\qquad \left. + \displaystyle\int_{\mathbb{R}^{2d}} \mathrm{Tr}\left[D^2_{mm}U\right] dm^{\otimes 2}(y, y') \right) \\[2ex]
+ \displaystyle\int_{\mathbb{R}^d} D_m U \cdot D_p H\big(y, D_x U(\cdot, y, \cdot)\big) dm(y) = F(x, m) \\[2ex]
U(T, x, m) = G(x, m).
\end{cases}
$$

This is the master equation. Note that there are only two genuine approxima-
tions in the foregoing computation. One is where we dropped the term of order
$1/N$ in the computation of the sum $\sum_{j \neq i} \Delta_{x_j} v^{N,i}$. The other one was at the
very beginning, when we replaced $D_x U(t, x_j, m_{\boldsymbol{x}}^{N,j})$ by $D_x U(t, x_j, m_{\boldsymbol{x}}^{N,i})$. This is
again of order $1/N$.

1.2.4   The Master Equation and the MFG System

We complete this informal discussion by explaining the relationship between
the master equation and the MFG system. This relation plays a central role in
the text. It is indeed the cornerstone for constructing a solution to the master
equation via a method of (infinite dimensional) characteristics.

We proceed as follows. Assuming that *the value function* of the MFG system
is regular—while it is part of the challenge to prove that it is indeed smooth—we
show that it solves the master equation.

We start with the first-order case, i.e., $\beta = 0$, as it is substantially easier. For any $(t_0, m_{(0)}) \in [0, T] \times \mathcal{P}(\mathbb{R}^d)$, let us define the value function $U(t_0, \cdot, m_{(0)})$ as

$$U(t_0, x, m_{(0)}) := u(t_0, x) \qquad \forall x \in \mathbb{R}^d,$$

where (u, m) is a solution of the MFG system (1.7) with the initial condition $m(t_0) = m_{(0)}$ at time t_0. We claim that U is a solution of the master equation (1.10) with $\beta = 0$. As indicated, we check the claim assuming that U is smooth, although the main difficulty comes from the fact that this has to be proved. We note that, by its very definition, U must satisfy

$$U(t, x, m(t)) = u(t, x) \qquad \forall (t, x) \in [t_0, T] \times \mathbb{R}^d.$$

Using the equation satisfied by m (and provided that $\partial_t m$ can be regarded as an $L^2(\mathbb{R}^d)$ valued function), the time derivative of the left-hand side at t_0 is given by

$$
\begin{aligned}
\partial_t u(t_0, x) &= \partial_t U + \left\langle \frac{\delta U}{\delta m}, \partial_t m \right\rangle_{L^2(\mathbb{R}^d)} \\
&= \partial_t U + \left\langle \frac{\delta U}{\delta m}, \Delta m + \mathrm{div}\left(m D_p H(\cdot, D_x U)\right) \right\rangle_{L^2(\mathbb{R}^d)} \qquad (1.22) \\
&= \partial_t U \\
&\quad + \int_{\mathbb{R}^d} \Big(\mathrm{div}_y\left[D_m U\right] - D_m U \cdot D_p H(y, D_x U(\cdot, y, \cdot)) \Big) dm_{(0)}(y),
\end{aligned}
$$

where the function U and its derivatives are evaluated at time t_0 and at the measure argument $m_{(0)}$; with the exception of the last term in the right-hand side, they are evaluated at point x in space; the auxiliary variable in $D_m U$ is always equal to y. Recalling the equation satisfied by u, we also have

$$
\begin{aligned}
\partial_t u(t_0, x) &= -\Delta u(t_0, x) + H\big(x, D_x u(t_0, x)\big) - F(x, m_{(0)}) \\
&= -\Delta_x U + H(x, D_x U) - F(x, m_{(0)}).
\end{aligned}
$$

This shows that

$$
\partial_t U + \int_{\mathbb{R}^d} \Big(\mathrm{div}_y\left[D_m U\right] - D_m U \cdot D_p H\big(y, D_x U(\cdot, y, \cdot)\big) \Big) dm_{(0)}(y)
$$
$$
= -\Delta_x U + H(x, D_x U) - F(x, m_{(0)}).
$$

Rearranging the terms, we deduce that U satisfies the master equation (1.10) with $\beta = 0$ at $(t_0, \cdot, m_{(0)})$.

For the second-order master equation ($\beta > 0$) the same principle applies except that, now, the MFG system becomes stochastic. Let $(t_0, m_{(0)}) \in [0, T] \times$

$\mathcal{P}(\mathbb{R}^d)$ and (u_t, m_t, v_t) be a solution of the MFG system with common noise (1.11). We set as before

$$U(t_0, x, m_{(0)}) := u_{t_0}(x) \qquad \forall x \in \mathbb{R}^d,$$

and notice that

$$U(t, x, m_t) = u_t(x) \qquad \forall (t, x) \in [t_0, T] \times \mathbb{R}^d.$$

Assuming that U is smooth enough, we have, by Itô's formula for Banach-valued processes and by the equation satisfied by m:

$$
\begin{aligned}
d_t u_t(x) = \Big\{ \partial_t U &+ \Big\langle \frac{\delta U}{\delta m}, (1 + \beta)\Delta m_t + \mathrm{div}\big(m_t D_p H(\cdot, D_x U)\big) \Big\rangle_{L^2(\mathbb{R}^d)} \\
&+ \beta \sum_{i=1}^{d} \Big\langle \frac{\delta^2 U}{\delta m^2} D_{x_i} m_t, D_{x_i} m_t \Big\rangle_{L^2(\mathbb{R}^d)} \Big\} dt \\
&- \sqrt{2\beta} \sum_{i=1}^{d} \Big\langle \frac{\delta U}{\delta m}, D_{x_i} m_t \Big\rangle_{L^2(\mathbb{R}^d)} dW_t^i,
\end{aligned}
\tag{1.23}
$$

where, as before, the function U and its derivatives are evaluated at time t and at the measure argument m_t; with the exception of $D_x U$ in the right-hand side, they are evaluated at point x in space.

In comparison with the first-order formula (1.22), equation (1.23) involves two additional terms: The stochastic term on the third line derives directly from the Brownian part in the forward part of (1.11) while the second-order term on the second line is reminiscent of the second-order term that appears in the standard Itô calculus. We provide a rigorous proof of (1.23) in Section 5.

Using (1.16), we obtain

$$
\begin{aligned}
d_t u_t(x) \\
= \Big\{ \partial_t U & \\
+ \int_{\mathbb{R}^d} &\Big((1 + \beta)\mathrm{div}_y\,[D_m U] - D_m U \cdot D_p H(\cdot, D_x U(\cdot, y, \cdot)) \Big) dm_t(y) \\
+ \beta \int_{\mathbb{R}^d \times \mathbb{R}^d} &\mathrm{Tr}\big[D_{mm}^2 U\big] dm_t^{\otimes 2}(y, y') \Big\} dt \\
+ \Big(\int_{\mathbb{R}^d} &D_m U\, dm_t(y) \Big) \cdot \sqrt{2\beta}\, dW_t.
\end{aligned}
\tag{1.24}
$$

On the other hand, by the equation satisfied by u, we have

$$
\begin{aligned}
d_t u_t(x) &= \big\{-(1+\beta)\Delta u_t + H(x, Du_t) - F(x, m_t) - \sqrt{2\beta}\,\mathrm{div}(v_t)\big\}dt \\
&\quad + v_t \cdot dW_t \\
&= \big\{-(1+\beta)\Delta_x U + H(x, D_x U) - F(x, m_t) - \sqrt{2\beta}\,\mathrm{div}(v_t)\big\}dt \\
&\quad + v_t \cdot dW_t,
\end{aligned}
\tag{1.25}
$$

where, on the right-hand side, u_t and v_t and their derivatives are evaluated at point x.

Identifying the absolutely continuous part and the martingale part, we find

$$
\begin{aligned}
\partial_t U &+ \int_{\mathbb{R}^d} \Big((1+\beta)\mathrm{div}_y\,[D_m U] - D_m U \cdot D_p H(\cdot, D_x U(\cdot, y, \cdot))\Big) dm_t(y) \\
&+ \beta \int_{\mathbb{R}^d \times \mathbb{R}^d} \mathrm{Tr}\big[D^2_{mm} U\big] dm_t^{\otimes 2}(y, y') \\
&= -(1+\beta)\Delta_x U + H(x, D_x U) - F(x, m_t) - \sqrt{2\beta}\,\mathrm{div}(v_t)
\end{aligned}
\tag{1.26}
$$

and

$$
\sqrt{2\beta} \int_{\mathbb{R}^d} D_m U\, dm_t(y) = v_t.
$$

Inserting the latter identity in the former one, we derive the master equation. Note that, compared with the first-order setting (i.e., $\beta = 0$), one faces here the additional issue that, so far, there has not been any solvability result for (1.9) and that the regularity of the map U—which is defined through (1.9)—is much more involved to investigate than in the first-order case.

Chapter Two

Presentation of the Main Results

IN THIS CHAPTER we collect our main results. We first state the notation used in this book, specify the notion of derivatives in the space of measures, and describe the assumptions on the data.

2.1   NOTATIONS

Throughout the book, $\mathbb{R}^d$ denotes the d-dimensional Euclidean space, with norm $|\cdot|$, the scalar product between two vectors $a, b \in \mathbb{R}^d$ being written $a \cdot b$. We work in the d-dimensional torus (i.e., periodic boundary conditions) that we denote $\mathbb{T}^d := \mathbb{R}^d / \mathbb{Z}^d$. When N is a (large) integer, we use bold symbols for elements of $(\mathbb{T}^d)^N$: for instance, $\boldsymbol{x} = (x_1, \ldots, x_N) \in (\mathbb{T}^d)^N$.

The set $\mathcal{P}(\mathbb{T}^d)$ of Borel probability measures on $\mathbb{T}^d$ is endowed with the Monge–Kantorovich distance

$$\mathbf{d}_1(m, m') = \sup_\phi \int_{\mathbb{T}^d} \phi(y) \, d(m - m')(y),$$

where the supremum is taken over all Lipschitz continuous maps $\phi : \mathbb{T}^d \to \mathbb{R}$ with a Lipschitz constant bounded by 1; see also Section A.1 in the Appendix. The notion of convergence associated with this distance is the weak convergence of measures.[1] If m belongs to $\mathcal{P}(\mathbb{T}^d)$ and $\phi : \mathbb{T}^d \to \mathbb{T}^d$ is a Borel map, then $\phi \sharp m$ denotes the push-forward of m by ϕ, i.e., the Borel probability measure such that $[\phi \sharp m](A) = m(\phi^{-1}(A))$ for any Borel set $A \subset \mathbb{T}^d$. When the probability measure m is absolutely continuous with respect to the Lebesgue measure, we use the same letter m to denote its density. Namely, we write $m : \mathbb{T}^d \ni x \mapsto m(x) \in \mathbb{R}_+$. Besides, we often consider flows of time-dependent measures of the form $(m(t))_{t \in [0,T]}$, with $m(t) \in \mathcal{P}(\mathbb{T}^d)$ for any $t \in [0, T]$. When, at each time $t \in [0, T]$, $m(t)$ is absolutely continuous with respect to the Lebesgue measure on $\mathbb{T}^d$, we identify $m(t)$ with its density and sometimes denote by $m : [0, T] \times \mathbb{T}^d \ni (t, x) \mapsto m(t, x) \in \mathbb{R}_+$ the collection of the densities. In all

---

[1]For a presentation of the set $\mathcal{P}(\mathbb{T}^d)$ as a metric space, we refer the reader to the monographs [7, 89, 95], for instance.

the examples considered in the text that follows, such an m has a time–space continuous version and, implicitly, we identify m with it.

If $\phi : \mathbb{T}^d \to \mathbb{R}$ is sufficiently smooth and $\ell = (\ell_1, \ldots, \ell_d) \in \mathbb{N}^d$, then $D^\ell \phi$ stands for the derivative $\frac{\partial^{\ell_1}}{\partial x_1^{\ell_1}} \cdots \frac{\partial^{\ell_d}}{\partial x_d^{\ell_d}} \phi$. The order of derivation $\ell_1 + \cdots + \ell_d$ is denoted by $|\ell|$. Given $e \in \mathbb{R}^d$, we also denote by $\partial_e \phi$ the directional derivative of ϕ in the direction e. For $n \in \mathbb{N}$ and $\alpha \in (0,1)$, $\mathcal{C}^{n+\alpha}$ is the set of maps ϕ for which $D^\ell \phi$ is defined and α-Hölder continuous for any $\ell \in \mathbb{N}^d$ with $|\ell| \leqslant n$. We set

$$\|\phi\|_{n+\alpha} := \sum_{|\ell| \leqslant n} \sup_{x \in \mathbb{T}^d} |D^\ell \phi(x)| + \sum_{|\ell|=n} \sup_{x \neq x'} \frac{|D^\ell \phi(x) - D^\ell \phi(x')|}{|x - x'|^\alpha}.$$

In the second term, x and x' are implicitly taken in $\mathbb{R}^d$ and ϕ is seen as a periodic function. Equivalently, x and x' may be taken in the torus, in which case the Euclidean distance between x and x' must be replaced by the torus distance between x and x', namely $\inf_{k \in \mathbb{Z}} |x - x' + k|$. The dual space of $\mathcal{C}^{n+\alpha}$ is denoted by $(\mathcal{C}^{n+\alpha})'$ with norm

$$\forall \rho \in (\mathcal{C}^{n+\alpha})', \qquad \|\rho\|_{-(n+\alpha)} := \sup_{\|\phi\|_{n+\alpha} \leqslant 1} \langle \rho, \phi \rangle_{(\mathcal{C}^{n+\alpha})', \mathcal{C}^{n+\alpha}}.$$

To simplify notation, we often abbreviate the expression $\langle \rho, \phi \rangle_{(\mathcal{C}^{n+\alpha})', \mathcal{C}^{n+\alpha}}$ as $\langle \rho, \phi \rangle_{n+\alpha}$.

If a smooth map ψ depends on two space variables, e.g., $\psi = \psi(x, y)$, and $m, n \in \mathbb{N}$ are the order of differentiation of ψ with respect to x and y respectively, we set

$$\|\psi\|_{(m,n)} := \sum_{|\ell| \leqslant m, |\ell'| \leqslant n} \|D^{(\ell, \ell')} \psi\|_\infty,$$

and, if moreover the derivatives are Hölder continuous,

$$\|\psi\|_{(m+\alpha, n+\alpha)}$$
$$:= \|\psi\|_{(m,n)} + \sum_{|\ell|=m, |\ell'|=n} \sup_{(x,y) \neq (x',y')} \frac{|D^{(\ell,\ell')} \phi(x,y) - D^{(\ell,\ell')} \phi(x',y')|}{|x - x'|^\alpha + |y - y'|^\alpha},$$

with the same convention as before for the distance in the second term in the right-hand side. The notation is generalized in an obvious way to mappings depending on three or more variables.

If now the (sufficiently smooth) map ϕ depends on time and space, i.e., $\phi = \phi(t, x)$, we say that $\phi \in \mathcal{C}^{l/2, l}$ (where $l = n + \alpha$, $n \in \mathbb{N}$, $\alpha \in (0, 1)$) if $D_x^\ell D_t^j \phi$ exists for any $\ell \in \mathbb{N}^d$ and $j \in \mathbb{N}$ with $|\ell| + 2j \leqslant n$ and is α-Hölder in x and $\alpha/2$-Hölder in t. Writing D^ℓ for D_x^ℓ, we set

$$\|\phi\|_{n/2+\alpha/2, n+\alpha} := \sum_{|\ell|+2j \leqslant n} \|D^\ell D_t^j \phi\|_\infty + \sum_{|\ell|+2j=n} \langle\langle D^\ell D_t^j \phi \rangle\rangle_{x,\alpha} + \langle\langle D^\ell D_t^j \phi \rangle\rangle_{t,\alpha/2}$$

with

$$\langle\langle D^\ell D_t^j \phi \rangle\rangle_{x,\alpha} := \sup_{t, x \neq x'} \frac{|D^\ell D_t^j \phi(t,x) - D^\ell D_t^j \phi(t,x')|}{|x - x'|^\alpha},$$

$$\langle\langle D^\ell D_t^j \phi \rangle\rangle_{t,\alpha} := \sup_{t \neq t', x} \frac{|D^\ell D_t^j \phi(t,x) - D^\ell D_t^j \phi(t',x)|}{|t - t'|^\alpha}.$$

We also say that ϕ belongs to $\mathcal{C}^{1,2}$ if $\partial_t \phi$ and $D^2 \phi$ exist and are continuous in time and space variables.

If X, Y are random variables on a probability space $(\Omega, \mathcal{A}, \mathbb{P})$, $\mathcal{L}(X)$ is the law of X and $\mathcal{L}(Y|X)$ is the conditional law of Y given X. Recall that, whenever X and Y take values in Polish spaces (say $\mathcal{S}_X$ and $\mathcal{S}_Y$ respectively), we can always find a regular version of the conditional law $\mathcal{L}(Y|X)$, that is, a mapping $q : \mathcal{S}_X \times \mathcal{B}(\mathcal{S}_Y) \to [0,1]$ such that:

- for each $x \in \mathcal{S}_X$, $q(x, \cdot)$ is a probability measure on $\mathcal{S}_Y$ equipped with its Borel σ-field $\mathcal{B}(\mathcal{S}_Y)$;
- for any $A \in \mathcal{B}(\mathcal{S}_Y)$, the mapping $\mathcal{S}_X \ni x \mapsto q(x, A)$ is Borel measurable;
- $q(X, \cdot)$ is a version of the conditional law of X given Y, in the sense that

$$\mathbb{E}[f(X,Y)] = \int_{\mathcal{S}_X} \left(\int_{\mathcal{S}_Y} f(x,y) q(x, dy) \right) d(\mathcal{L}(X))(x)$$

$$= \mathbb{E}\left[\int_{\mathcal{S}_Y} f(X,y) q(X, dy) \right],$$

for any bounded Borel measurable mapping $f : \mathcal{S}_X \times \mathcal{S}_Y \to \mathbb{R}$.

2.2   DERIVATIVES

One of the striking features of the master equation is that it involves derivatives of the unknown with respect to the measure. In this book, we use two notions of derivatives. The first one, denoted by $\frac{\delta U}{\delta m}$, is, roughly speaking, the L^2 derivative when one looks at the restriction of $\mathcal{P}(\mathbb{T}^d)$ to densities in $L^2(\mathbb{T}^d)$. It is widely used in linearization procedures. The second one, denoted by $D_m U$, is more intrinsic and is related to the so-called Wasserstein metric on $\mathcal{P}(\mathbb{T}^d)$. It can be introduced as in Ambrosio, Gigli, and Savaré [7] by defining a kind of manifold structure on $\mathcal{P}(\mathbb{T}^d)$ or, as in Lions [76], by embedding $\mathcal{P}(\mathbb{T}^d)$ into an $L^2(\Omega, \mathbb{T}^d)$ space of random variables. Here we introduce this latter notion in a slightly different way, as the derivative in space of $\frac{\delta U}{\delta m}$. In the Appendix we briefly compare the different notions.

2.2.1 First-Order Derivatives

Definition 2.2.1. *We say that* $U : \mathcal{P}(\mathbb{T}^d) \to \mathbb{R}$ *is* $\mathcal{C}^1$ *if there exists a continuous map* $\frac{\delta U}{\delta m} : \mathcal{P}(\mathbb{T}^d) \times \mathbb{T}^d \to \mathbb{R}$ *such that, for any* $m, m' \in \mathcal{P}(\mathbb{T}^d)$,

$$\lim_{s \to 0^+} \frac{U((1-s)m + sm') - U(m)}{s} = \int_{\mathbb{T}^d} \frac{\delta U}{\delta m}(m, y) d(m' - m)(y).$$

Note that $\frac{\delta U}{\delta m}$ is defined up to an additive constant. We adopt the normalization convention

$$\int_{\mathbb{T}^d} \frac{\delta U}{\delta m}(m, y) dm(y) = 0. \tag{2.1}$$

For any $m \in \mathcal{P}(\mathbb{T}^d)$ and any signed measure μ on $\mathbb{T}^d$, we will use interchangeably the notations $\frac{\delta U}{\delta m}(m)(\mu)$ and $\int_{\mathbb{T}^d} \frac{\delta U}{\delta m}(m, y) d\mu(y)$.

Note also that

$$\forall m, m' \in \mathcal{P}(\mathbb{T}^d), \quad U(m') - U(m)$$
$$= \int_0^1 \int_{\mathbb{T}^d} \frac{\delta U}{\delta m}((1-s)m + sm', y) \, d(m' - m)(y) ds. \tag{2.2}$$

Let us explain the relationship between the derivative in the aforementioned sense and the Lipschitz continuity of U in $\mathcal{P}(\mathbb{T}^d)$. If $\frac{\delta U}{\delta m} = \frac{\delta U}{\delta m}(m, y)$ is Lipschitz continuous with respect to the second variable with a Lipschitz constant bounded independently of m, then U is Lipschitz continuous: indeed, by (2.2),

$$|U(m') - U(m)| \leqslant \int_0^1 \left\| D_y \frac{\delta U}{\delta m}((1-s)m + sm', \cdot) \right\|_\infty ds \, \mathbf{d}_1(m, m')$$
$$\leqslant \sup_{m''} \left\| D_y \frac{\delta U}{\delta m}(m'', \cdot) \right\|_\infty \mathbf{d}_1(m, m').$$

This leads us to define the "intrinsic derivative" of U.

Definition 2.2.2. *If* $\frac{\delta U}{\delta m}$ *is of class* $\mathcal{C}^1$ *with respect to the second variable, the intrinsic derivative* $D_m U : \mathcal{P}(\mathbb{T}^d) \times \mathbb{T}^d \to \mathbb{R}^d$ *is defined by*

$$D_m U(m, y) := D_y \frac{\delta U}{\delta m}(m, y).$$

The expression $D_m U$ can be understood as a derivative of U along vector fields:

Proposition 2.2.3. *Assume that* U *is* $\mathcal{C}^1$, *with* $\frac{\delta U}{\delta m}$ *being* $\mathcal{C}^1$ *with respect to* y, *and that* $D_m U$ *is continuous in both variables. Let* $\phi : \mathbb{T}^d \to \mathbb{R}^d$ *be a Borel measurable and bounded vector field. Then,*

$$\lim_{h \to 0} \frac{U((id + h\phi)\sharp m) - U(m)}{h} = \int_{\mathbb{T}^d} D_m U(m, y) \cdot \phi(y) \, dm(y).$$

Proof. Let us set $m_{h,s} := s(id + h\phi)\sharp m + (1 - s)m$. Then,

$$U((id + h\phi)\sharp m) - U(m)$$

$$= \int_0^1 \int_{\mathbb{T}^d} \frac{\delta U}{\delta m}(m_{h,s}, y)d((id + h\phi)\sharp m - m)(y)\, ds$$

$$= \int_0^1 \int_{\mathbb{T}^d} (\frac{\delta U}{\delta m}(m_{h,s}, y + h\phi(y)) - \frac{\delta U}{\delta m}(m_{h,s}, y))dm(y)\, ds$$

$$= h \int_0^1 \int_{\mathbb{T}^d} \int_0^1 D_m U(m_{h,s}, y + th\phi(y)) \cdot \phi(y)\, dt dm(y)\, ds.$$

Dividing by h and letting $h \to 0$ gives the result thanks to the continuity of $D_m U$. $\square$

Note also that, if $U : \mathcal{P}(\mathbb{T}^d) \to \mathbb{R}$ and $\frac{\delta U}{\delta m}$ is $\mathcal{C}^2$ in y, then $D_y D_m U(m, y)$ is a symmetric matrix, as

$$D_y D_m U(m, y) = D_y \left(D_y \frac{\delta U}{\delta m} \right)(m, y) = \text{Hess}_y \frac{\delta U}{\delta m}(m, y).$$

2.2.2   Second-Order Derivatives

If, for a fixed $y \in \mathbb{T}^d$, the map $m \mapsto \frac{\delta U}{\delta m}(m, y)$ is $\mathcal{C}^1$, then we say that U is $\mathcal{C}^2$ and denote by $\frac{\delta^2 U}{\delta m^2}$ its derivative. (Take note that y is fixed. At this stage, nothing is said about the smoothness in the direction y.) By Definition 2.2.1 we have that $\frac{\delta^2 U}{\delta m^2} : \mathcal{P}(\mathbb{T}^d) \times \mathbb{T}^d \times \mathbb{T}^d \to \mathbb{R}$ with

$$\frac{\delta U}{\delta m}(m', y) - \frac{\delta U}{\delta m}(m, y) = \int_0^1 \int_{\mathbb{T}^d} \frac{\delta^2 U}{\delta m^2}((1 - s)m + sm', y, y')\, d(m' - m)(y').$$

If U is $\mathcal{C}^2$ and if $\frac{\delta^2 U}{\delta m^2} = \frac{\delta^2 U}{\delta m^2}(m, y, y')$ is $\mathcal{C}^2$ in the variables (y, y'), then we set

$$D^2_{mm}U(m, y, y') := D^2_{y,y'} \frac{\delta^2 U}{\delta m^2}(m, y, y').$$

We note that $D^2_{mm}U$ maps $\mathcal{P}(\mathbb{T}^d) \times \mathbb{T}^d \times \mathbb{T}^d$ into $\mathbb{R}^{d \times d}$. The next statement asserts that $D^2_{mm}U$ enjoys the classical symmetries of second-order derivatives.

Lemma 2.2.4. *Assume that $\frac{\delta^2 U}{\delta m^2}$ is jointly continuous in all the variables. Then*

$$\frac{\delta^2 U}{\delta m^2}(m, y, y')$$

$$= \frac{\delta^2 U}{\delta m^2}(m, y', y) + \frac{\delta U}{\delta m}(m, y) - \frac{\delta U}{\delta m}(m, y'), \quad m \in \mathcal{P}(\mathbb{T}^d), \; y, y' \in \mathbb{T}^d.$$

In the same way, if $\frac{\delta U}{\delta m}$ is C^1 in the variable y and $\frac{\delta^2 U}{\delta m^2}$ is also C^1 in the variable y, $D_y \frac{\delta^2 U}{\delta m^2}$ being jointly continuous in all the variables, then, for any fixed $y \in \mathbb{T}^d$, the map $m \mapsto D_m U(m, y)$ is C^1 and

$$D_y \frac{\delta^2 U}{\delta m^2}(m, y, y') = \frac{\delta}{\delta m}(D_m U(m, y))(y'), \quad m \in \mathcal{P}(\mathbb{T}^d), \ y, y' \in \mathbb{T}^d,$$

while, if $\frac{\delta^2 U}{\delta m^2}$ is also C^2 in the variables (y, y'), then, for any fixed $y \in \mathbb{T}^d$, the map $\frac{\delta}{\delta m}(D_m U(\cdot, y))$ is C^1 in the variable y' and

$$D_m\big(D_m U(\cdot, y)\big)(m, y') = D_{mm}^2 U(m, y, y').$$

Proof. *First step.* We start with the proof of the first claim. By continuity, we just need to show the result when m has a smooth positive density. Let $\mu, \nu \in L^\infty(\mathbb{T}^d)$, such that $\int_{\mathbb{T}^d} \mu = \int_{\mathbb{T}^d} \nu = 0$, with a small enough norm so that $m + s\mu + t\nu$ is a probability density for any $(s, t) \in [0, 1]^2$.

Since U is C^2, the mapping $\mathcal{U} : [0, 1]^2 \in (s, t) \mapsto U(m + s\mu + n\nu)$ is twice differentiable and, by the standard Schwarz theorem, $D_t D_s \mathcal{U}(s, t) = D_s D_t \mathcal{U}(s, t)$, for any $(s, t) \in [0, 1]^2$. Notice that

$$D_t D_s \mathcal{U}(s, t) = \int_{[\mathbb{T}^d]^2} \frac{\delta^2 U}{\delta m^2}(m + s\mu + t\nu, y, y')\mu(y)\nu(y') \, dy \, dy'$$

$$D_s D_t \mathcal{U}(s, t) = \int_{[\mathbb{T}^d]^2} \frac{\delta^2 U}{\delta m^2}(m + s\mu + t\nu, y', y)\mu(y)\nu(y') \, dy \, dy'.$$

Choosing $s = t = 0$, we find

$$\int_{[\mathbb{T}^d]^2} \frac{\delta^2 U}{\delta m^2}(m, y, y')\mu(y)\nu(y') \, dy \, dy' = \int_{[\mathbb{T}^d]^2} \frac{\delta^2 U}{\delta m^2}(m, y', y)\mu(y)\nu(y') \, dy \, dy',$$

for any $\mu, \nu \in L^\infty(\mathbb{T}^d)$, such that $\int_{\mathbb{T}^d} \mu = \int_{\mathbb{T}^d} \nu = 0$. Hence, for any m smooth with a positive density, there exist maps ϕ_1, ϕ_2 such that

$$\frac{\delta^2 U}{\delta m^2}(m, y, y') = \frac{\delta^2 U}{\delta m^2}(m, y', y) + \phi_1(y) + \phi_2(y'), \quad y, y' \in \mathbb{T}^d. \qquad (2.3)$$

To identify ϕ_1 and ϕ_2, we recall the normalization condition

$$\int_{\mathbb{T}^d} \frac{\delta U}{\delta m}(m, y)m(y) \, dy = 0.$$

Taking the derivative with respect to m of this equality, we find

$$\int_{\mathbb{T}^d} \frac{\delta^2 U}{\delta m^2}(m, y, y')m(y)dy + \frac{\delta U}{\delta m}(m, y') = 0, \quad y' \in \mathbb{T}^d.$$

So integrating (2.3) with respect to $m(y)$ and using the normalization condition:

$$-\frac{\delta U}{\delta m}(m, y') = \int_{\mathbb{T}^d} \frac{\delta^2 U}{\delta m^2}(m, y, y')m(y)\, dy = 0 + \int_{\mathbb{T}^d} \phi_1(y)m(y)\, dy + \phi_2(y').$$

Integrating in the same way (2.3) with respect to $m(y')$:

$$0 = -\frac{\delta U}{\delta m}(m, y) + \phi_1(y) + \int_{\mathbb{T}^d} \phi_2(y')m(y')\, dy'.$$

Finally integrating (2.3) with respect to $m(y)m(y')$:

$$0 = \int_{\mathbb{T}^d} \phi_1(y)m(y)\, dy + \int_{\mathbb{T}^d} \phi_2(y')m(y')\, dy'$$

Combining the three last equalities leads to

$$\phi_1(y) + \phi_2(y') = \frac{\delta U}{\delta m}(m, y) - \frac{\delta U}{\delta m}(m, y')$$

which, in view of (2.3), gives the desired result.

Second step. The proof is the same for the second assertion, except that now we have to consider the mapping $\mathcal{U}' : [0, 1] \times \mathbb{T}^d \ni (t, y) \mapsto \frac{\delta U}{\delta m}(m + t\mu, y)$, for a general probability measure $m \in \mathcal{P}(\mathbb{T}^d)$ and a general finite signed measure μ on $\mathbb{T}^d$, such that $\mu(\mathbb{T}^d) = 0$ and $m + \mu$ is a probability measure. (In particular, $m + t\mu = (1 - t)m + t(m + \mu)$ is also a probability measure for any $t \in [0, 1]$.) By assumption, $\mathcal{U}'$ is $\mathcal{C}^1$ in each variable t and y with

$$D_t \mathcal{U}'(t, y) = \int_{\mathbb{T}^d} \frac{\delta^2 U}{\delta m^2}(m + t\mu, y, y')d\mu(y'), \quad D_y \mathcal{U}'(t, y) = D_m U(m + t\mu, y).$$

In particular, $D_t \mathcal{U}'$ is $\mathcal{C}^1$ in y and

$$D_y D_t \mathcal{U}'(t, y) = \int_{\mathbb{T}^d} D_y \frac{\delta^2 U}{\delta m^2}(m + t\mu, y, y')\mu(y')\, dy'.$$

By assumption, $D_y D_t \mathcal{U}'$ is jointly continuous and, by the standard Schwarz theorem, the mapping $D_y \mathcal{U}'$ is differentiable in t, with

$$D_t (D_y \mathcal{U}')(t, y) = D_t (D_m U(m + t\mu, y)) = \int_{\mathbb{T}^d} D_y \frac{\delta^2 U}{\delta m^2}(m + t\mu, y, y')\mu(y')\, dy'.$$

Integrating in t, this shows that

$$D_m U(m + \mu, y) - D_m U(m, y) = \int_0^1 \int_{\mathbb{T}^d} D_y \frac{\delta^2 U}{\delta m^2}(m + t\mu, y, y')\mu(y')\, dy'\, dt.$$

Choosing $\mu = m' - m$, for another probability measure $m' \in \mathcal{P}(\mathbb{T}^d)$ and noticing that

$$\int_{\mathbb{T}^d} D_y \frac{\delta^2 U}{\delta m^2}(m, y, y')dm(y') = 0,$$

we complete the proof of the second claim.

For the last assertion, one just need to take the derivative in y in the second one.   □

2.2.3   Comments on the Notions of Derivatives

As several concepts of derivatives have been used in the mean field game theory, we now discuss the link between these notions. For simplicity, we argue as if our state space was $\mathbb{R}^d$ and not $\mathbb{T}^d$, as most results have been stated in this context. (We refer to the Appendix for an exposition on $\mathbb{T}^d$.)

A first idea consists in looking at the restriction of the map U to the subset of measures with a density that is in $L^2(\mathbb{R}^d)$, and taking the derivative of U in the $L^2(\mathbb{R}^d)$ sense. This is partially the point of view adopted by Lions in [76] and followed by Bensoussan, Frehse, and Yam [16]. In the context of smooth densities, this is closely related to our first and second derivatives $\frac{\delta U}{\delta m}$ and $\frac{\delta^2 U}{\delta m^2}$.

Many works on mean field games (as in Buckdahn, Li, Peng, and Rainer [24]; Carmona and Delarue [27]; Chassagneux, Crisan, and Delarue [31]; Gangbo and Swiech [45]) make use of an idea introduced by Lions in [76]. It consists in working in a sufficiently rich probability space $(\Omega, \mathcal{A}, \mathbb{P})$ and in looking at maps $U : \mathcal{P}(\mathbb{R}^d) \to \mathbb{R}$ through their lifting to $L^2(\Omega, \mathcal{A}, \mathbb{P}, \mathbb{R}^d)$ defined by

$$\widetilde{U}(X) = U(\mathcal{L}(X)) \qquad \forall X \in L^2(\Omega, \mathbb{R}^d),$$

where $\mathcal{L}(X)$ is the law of X. It is clear that the derivative of $\widetilde{U}$—if it exists—enjoys special properties because $\widetilde{U}(X)$ depends only on the law of X and not on the full random variable. As explained in [76], if $\widetilde{U}$ is continuously differentiable, then its gradient at some point $X_0 \in L^2(\Omega, \mathcal{A}, \mathbb{P}, \mathbb{R}^d)$ can be written as

$$\nabla \widetilde{U}(X_0) = \partial_\mu U(\mathcal{L}(X_0))(X_0),$$

where $\partial_\mu U : \mathcal{P}(\mathbb{R}^d) \times \mathbb{R}^d \ni (m, x) \mapsto \partial_\mu U(m)(x) \in \mathbb{R}^d$. We explain in the Appendix that the maps $\partial_\mu U$ and $D_m U$ introduced in Definition 2.2.2 coincide, as soon as one of the two derivatives exists. Let us also underline that this concept of derivative is closely related to the notion introduced by Ambrosio, Gigli, and Savaré [7] in a more general setting.

2.3   ASSUMPTIONS

Throughout the book, we assume that $H : \mathbb{T}^d \times \mathbb{R}^d \to \mathbb{R}$ is smooth, globally Lipschitz continuous, and satisfies the coercivity condition:

$$0 < D_{pp}^2 H(x, p) \leqslant C I_d \qquad \text{for } (x, p) \in \mathbb{T}^d \times \mathbb{R}^d. \tag{2.4}$$

We also always assume that the maps $F, G : \mathbb{T}^d \times \mathcal{P}(\mathbb{T}^d) \to \mathbb{R}$ are globally Lipschitz continuous and monotone: for any $m, m' \in \mathcal{P}(\mathbb{T}^d)$,

$$\begin{aligned} \int_{\mathbb{T}^d} (F(x, m) - F(x, m'))d(m - m')(x) &\geqslant 0, \\ \int_{\mathbb{T}^d} (G(x, m) - G(x, m'))d(m - m')(x) &\geqslant 0. \end{aligned} \tag{2.5}$$

Note that assumption (2.5) implies that $\frac{\delta F}{\delta m}$ and $\frac{\delta G}{\delta m}$ satisfy the following monotonicity property (explained for F):

$$\int_{\mathbb{T}^d} \int_{\mathbb{T}^d} \frac{\delta F}{\delta m}(x, m, y)\mu(x)\mu(y)dxdy \geqslant 0$$

for any centered measure μ. Throughout the book, the conditions (2.4) and (2.5) are in force.

Next we describe assumptions that might differ according to the results. Let us fix $n \in \mathbb{N}$ and $\alpha \in (0, 1)$. We set (with the notation introduced in Section 2.1)

$$\text{Lip}_n(\frac{\delta F}{\delta m}) := \sup_{m_1 \neq m_2} (\mathbf{d}_1(m_1, m_2))^{-1} \left\| \frac{\delta F}{\delta m}(\cdot, m_1, \cdot) - \frac{\delta F}{\delta m}(\cdot, m_2, \cdot) \right\|_{(n+\alpha, n+\alpha)}$$

and use the analogue notation for G. We call $(\mathbf{HF1}(n))$ the following regularity conditions on F:

$$(\mathbf{HF1}(n)) \qquad \sup_{m \in \mathcal{P}(\mathbb{T}^d)} \left(\|F(\cdot, m)\|_{n+\alpha} + \left\| \frac{\delta F(\cdot, m, \cdot)}{\delta m} \right\|_{(n+\alpha, n+\alpha)} \right)$$
$$+ \text{Lip}_n(\frac{\delta F}{\delta m}) < \infty,$$

and $(\mathbf{HG1}(n))$ the analogue condition on G:

$$(\mathbf{HG1}(n)) \qquad \sup_{m \in \mathcal{P}(\mathbb{T}^d)} \left(\|G(\cdot, m)\|_{n+\alpha} + \left\| \frac{\delta G(\cdot, m, \cdot)}{\delta m} \right\|_{(n+\alpha, n+\alpha)} \right)$$
$$+ \text{Lip}_n(\frac{\delta G}{\delta m}) < \infty.$$

We use similar notation when dealing with second-order derivatives:

$$
\mathrm{Lip}_n(\frac{\delta^2 F}{\delta m^2})
$$

$$
:= \sup_{m_1 \neq m_2} (\mathbf{d}_1(m_1, m_2))^{-1} \left\| \frac{\delta^2 F}{\delta m^2}(\cdot, m_1, \cdot, \cdot) - \frac{\delta^2 F}{\delta m^2}(\cdot, m_2, \cdot, \cdot) \right\|_{(n+\alpha, n+\alpha, n+\alpha)},
$$

and call **(HF2(n))** (respectively **(HG2(n)))** the second-order regularity conditions on F:

$$
\textbf{(HF2(n))} \quad \sup_{m \in \mathcal{P}(\mathbb{T}^d)} \left(\|F(\cdot, m)\|_{n+\alpha} + \left\| \frac{\delta F(\cdot, m, \cdot)}{\delta m} \right\|_{(n+\alpha, n+\alpha)} \right)
$$

$$
+ \sup_{m \in \mathcal{P}(\mathbb{T}^d)} \left\| \frac{\delta^2 F(\cdot, m, \cdot, \cdot)}{\delta m^2} \right\|_{(n+\alpha, n+\alpha, n+\alpha)} + \mathrm{Lip}_n(\frac{\delta^2 F}{\delta m^2}) < \infty.
$$

and on G:

$$
\textbf{(HG2(n))} \quad \sup_{m \in \mathcal{P}(\mathbb{T}^d)} \left(\|G(\cdot, m)\|_{n+\alpha} + \left\| \frac{\delta G(\cdot, m, \cdot)}{\delta m} \right\|_{(n+\alpha, n+\alpha)} \right)
$$

$$
+ \sup_{m \in \mathcal{P}(\mathbb{T}^d)} \left\| \frac{\delta^2 G(\cdot, m, \cdot, \cdot)}{\delta m^2} \right\|_{(n+\alpha, n+\alpha, n+\alpha)} + \mathrm{Lip}_n(\frac{\delta^2 G}{\delta m^2}) < \infty.
$$

Example 2.3.1. Assume that F is of the form

$$
F(x, m) = \int_{\mathbb{R}^d} \Phi(z, (\rho \star m)(z)) \rho(x - z) \, dz,
$$

where $\star$ denotes the usual convolution product (in $\mathbb{R}^d$) and where $\Phi : \mathbb{T}^d \times [0, +\infty) \to \mathbb{R}$ is a smooth map that is nondecreasing with respect to the second variable and ρ is a smooth, even function with compact support. Then F satisfies the monotonicity condition (2.5) as well as the regularity conditions **(HF1(n))** and **(HF2(n))** for any $n \in \mathbb{N}$. Observe that F is periodic in x since Φ is periodic in the first argument and m belongs to $\mathcal{P}(\mathbb{T}^d)$.

Proof. Let us first note that, for any $m, m' \in \mathcal{P}(\mathbb{T}^d)$,

$$
\int_{\mathbb{T}^d} (F(x, m) - F(x, m')) d(m - m')(x)
$$

$$
= \int_{\mathbb{T}^d} [\Phi(y, \rho \star m(y)) - \Phi(y, \rho \star m'(y))] (\rho \star m(y) - \rho \star m'(y)) \, dy \geqslant 0,
$$

since ρ is even and Φ is nondecreasing with respect to the second variable. So F is monotone. Writing $\Phi = \Phi(x, \theta)$, the derivatives of F are given by

$$\frac{\delta F}{\delta m}(x, m, y) = \sum_{k \in \mathbb{Z}^d} \int_{\mathbb{R}^d} \frac{\partial \Phi}{\partial \theta}(z, \rho \star m(z)) \rho(x - z) \rho(z - y - k) \, dz$$

and

$$\frac{\delta^2 F}{\delta m^2}(x, m, y, y')$$
$$= \sum_{k, k' \in \mathbb{Z}^d} \int_{\mathbb{R}^d} \frac{\partial^2 \Phi}{\partial \theta^2}(z, \rho \star m(z)) \rho(z - y - k) \rho(z - y' - k') \rho(x - z) \, dz.$$

Then **(HF1(n))** and **(HF2(n))** hold because of the smoothness of ρ. $\square$

2.4 STATEMENT OF THE MAIN RESULTS

The book contains two main results: on the one hand, the well-posedness of the master equation, and, on the other hand, the convergence of the Nash system with N players as N tends to infinity. We start by considering the first-order master equation ($\beta = 0$), because, in this setting, the approach is relatively simple (Theorem 2.4.2). To handle the second-order master equation, we build solutions to the mean field game system with common noise, which play the role of "characteristics" for the master equation (Theorem 2.4.3). Our first main result is Theorem 2.4.5, which states that the master equation has a unique classical solution under our regularity and monotonicity assumptions on H, F, and G. Once we know that the master equation has a solution, we can use this solution to build approximate solutions for the Nash system with N-players. This yields to our main convergence results, either in term of functional terms (Theorem 2.4.8) or in term of optimal trajectories (Theorem 2.4.9).

2.4.1 First-Order Master Equation

We first consider the first-order master equation (or master equation without common noise):

$$
\begin{cases}
-\partial_t U(t, x, m) - \Delta_x U(t, x, m) + H(x, D_x U(t, x, m)) \\
\quad - \int_{\mathbb{T}^d} \operatorname{div}_y \left[D_m U \right] (t, x, m, y) \, dm(y) \\
\quad + \int_{\mathbb{T}^d} D_m U(t, x, m, y) \cdot D_p H(y, D_x U(t, y, m)) \, dm(y) \\
\quad = F(x, m) \qquad \text{in } [0, T] \times \mathbb{T}^d \times \mathcal{P}(\mathbb{T}^d), \\
U(T, x, m) = G(x, m) \qquad \text{in } \mathbb{T}^d \times \mathcal{P}(\mathbb{T}^d).
\end{cases}
\tag{2.6}
$$

We call it the first-order master equation because it contains only first-order derivatives with respect to the measure variable. Let us first explain the notion of solution.

Definition 2.4.1. *We say that a map $U : [0, T] \times \mathbb{T}^d \times \mathcal{P}(\mathbb{T}^d) \to \mathbb{R}$ is a classical solution to the first-order master equation if*

- U *is continuous in all its arguments (for the $\mathbf{d}_1$ distance on $\mathcal{P}(\mathbb{T}^d)$), is of class C^2 in x and C^1 in time (the derivatives of order 1 in time and space and of order 2 in space being continuous in all the arguments);*
- U *is of class C^1 with respect to m, the first-order derivative*

$$[0, T] \times \mathbb{T}^d \times \mathcal{P}(\mathbb{T}^d) \times \mathbb{T}^d \ni (t, x, m, y) \mapsto \frac{\delta U}{\delta m}(t, x, m, y),$$

  being continuous in all the arguments, $\delta U / \delta m$ being twice differentiable in y, the derivatives being continuous in all the arguments;
- U *satisfies the master equation (2.6).*

Theorem 2.4.2. *Assume that F, G, and H satisfy (2.4) and (2.5) in Section 2.3, and that $(\mathbf{HF1}(n+1))$ and $(\mathbf{HG1}(n+2))$ hold for some $n \geqslant 1$ and some $\alpha \in (0, 1)$. Then the first-order master equation (2.6) has a unique solution U.*

Moreover, U is C^1 (in all variables), $\frac{\delta U}{\delta m}$ is continuous in all variables and $U(t, \cdot, m)$ and $\frac{\delta U}{\delta m}(t, \cdot, m, \cdot)$ are bounded in $C^{n+2+\alpha}$ and $C^{n+2+\alpha} \times C^{n+1+\alpha}$ respectively, independently of (t, m). Finally, $\frac{\delta U}{\delta m}$ is Lipschitz continuous with respect to the measure variable:

$$\sup_{t \in [0,T]} \sup_{m_1 \neq m_2} (\mathbf{d}_1(m_1, m_2))^{-1} \left\| \frac{\delta U}{\delta m}(t, \cdot, m_1, \cdot) - \frac{\delta U}{\delta m}(t, \cdot, m_2, \cdot) \right\|_{(n+2+\alpha, n+\alpha)}$$
$$< \infty.$$

Chapter 3 is devoted to the proof of Theorem 2.4.2. We also discuss in this chapter the link between the solution U and the derivative of the solution of a Hamilton–Jacobi equation in the space of measures.

The proof of Theorem 2.4.2 relies on the representation of the solution in terms of the mean field game system: for any $(t_0, m_0) \in [0, T) \times \mathcal{P}(\mathbb{T}^d)$, the mean field game (MFG) system is the system of forward–backward equations:

$$\begin{cases} -\partial_t u - \Delta u + H(x, Du) = F(x, m(t)) & \text{in } (t_0, T) \times \mathbb{T}^d, \\ \partial_t m - \Delta m - \text{div}(m D_p H(x, Du)) = 0 & \text{in } (t_0, T) \times \mathbb{T}^d, \\ u(T, x) = G(x, m(T)), \; m(t_0, \cdot) = m_0 & \text{in } \mathbb{T}^d. \end{cases} \quad (2.7)$$

As recalled in the text that follows (Proposition 3.1.1), under suitable assumptions on the data, there exists a unique solution (u, m) to the above system. Our aim is to show that the map U defined by

$$U(t_0, \cdot, m_0) := u(t_0, \cdot) \quad (2.8)$$

is a solution to (2.6). The starting point is the obvious remark that, for U defined by (2.8) and for any $h \in [0, T - t_0]$,

$$u(t_0 + h, \cdot) = U(t_0 + h, \cdot, m(t_0 + h)).$$

Taking the derivative with respect to h and letting $h = 0$ shows that U satisfies (2.6) provided that it is smooth (as explained at the end of Chapter 1).

The main issue is to prove that the map U defined by (2.8) is sufficiently smooth to perform the above computation. To prove the differentiability of the map U, we use a flow method and differentiate the MFG system (2.7) with respect to the measure argument m_0. The derivative system then reads as a *linearized system* initialized with a signed measure. Fixing a solution (u, m) to (2.7) and allowing for a more singular Schwartz distribution $\mu_0 \in (\mathcal{C}^{n+1+\alpha}(\mathbb{T}^d))'$ as initial condition (instead of a signed measure), the *linearized system*, with (v, μ) as unknown, takes the form

$$\begin{cases} -\partial_t v - \Delta v + D_p H(x, Du) \cdot Dv = \dfrac{\delta F}{\delta m}(x, m(t))(\mu(t)) & \text{in } (t_0, T) \times \mathbb{T}^d, \\ \partial_t \mu - \Delta \mu - \text{div}\big(\mu D_p H(x, Du)\big) - \text{div}\big(m D_{pp}^2 H(x, Du) Dv\big) = 0 & \text{in } (t_0, T) \times \mathbb{T}^d, \\ v(T, x) = \dfrac{\delta G}{\delta m}(x, m(T))(\mu(T)), \ \mu(t_0, \cdot) = \mu_0 & \text{in } \mathbb{T}^d, \end{cases}$$

where v and u are evaluated at (t, x). We prove that v can be interpreted as the directional derivative of U in the direction μ_0:

$$v(t_0, x) = \int_{\mathbb{T}^d} \frac{\delta U}{\delta m}(t_0, x, m_0, y)\mu_0(y) dy.$$

Note that this shows at the same time the differentiability of U and the regularity of its derivative. For this reason the introduction of the directional derivative appears extremely useful in this context.

2.4.2   The Mean Field Game System with Common Noise

As explained in the previous section, the characteristics of the first-order master equation (2.6) are the solution to the mean field game system (2.7). The analogous construction for the second-order master equation (with $\beta > 0$) yields a system of stochastic partial differential equations, the *mean field game system with common noise*. Given an initial distribution $m_0 \in \mathcal{P}(\mathbb{T}^d)$ at an initial time $t_0 \in [0, T]$, this system reads[2]

$$\begin{cases} d_t u_t = \big[-(1+\beta)\Delta u_t + H(x, Du_t) - F(x, m_t) - \sqrt{2\beta}\text{div}(v_t)\big] dt \\ \qquad + v_t \cdot dW_t, \\ d_t m_t = \big[(1+\beta)\Delta m_t + \text{div}\big(m_t D_p H(\cdot, Du_t)\big)\big] dt \qquad\qquad (2.9) \\ \qquad - \sqrt{2\beta}\text{div}(m_t \, dW_t), \quad \text{in } [t_0, T] \times \mathbb{T}^d, \\ m_{t_0} = m_0, \ u_T(x) = G(x, m_T) \quad \text{in } \mathbb{T}^d. \end{cases}$$

---

[2]To emphasize the random nature of the functions u and m, the time variable is now indicated as a subscript, as often done in the theory of stochastic processes.

Here $(W_t)_{t\in[0,T]}$ is a given d-dimensional Brownian motion, generating a filtration $(\mathcal{F}_t)_{t\in[0,T]}$. The solution is the process $(u_t, m_t, v_t)_{t\in[0,T]}$, adapted to $(\mathcal{F}_t)_{t\in[t_0,T]}$, where, for each $t \in [t_0, T]$, v_t is a vector field that ensures the solution (u_t) to the backward equation to be adapted to the filtration $(\mathcal{F}_t)_{t\in[t_0,T]}$. Up to now, the well-posedness of this system has never been investigated, but it is reminiscent of the theory of forward–backward stochastic differential equations in finite dimension, see, for instance, the monograph [84].

To analyze (2.9), we take advantage of the additive structure of the common noise and perform the (formal) change of variable:

$$\tilde{u}_t(x) = u_t(x + \sqrt{2\beta}W_t), \quad \tilde{m}_t(x) = m_t(x + \sqrt{2\beta}W_t), \quad x \in \mathbb{T}^d, \quad t \in [0, T].$$

Setting $\tilde{H}_t(x, p) = H(x+\sqrt{2\beta}W_t, p)$, $\tilde{F}_t(x, m) = F(x+\sqrt{2\beta}W_t, m)$ and $\tilde{G}(x, m) = G(x + \sqrt{2\beta}W_T, m)$ and invoking the Itô–Wentzell formula (see Chapter 4 for a more precise account together with Subsection A.3.1 in the Appendix), the pair $(\tilde{u}_t, \tilde{m}_t)_{t\in[t_0,T]}$ formally satisfies the system:

$$\begin{cases} d_t\tilde{u}_t = \left[-\Delta\tilde{u}_t + \tilde{H}_t(\cdot, D\tilde{u}_t) - \tilde{F}_t(\cdot, m_t)\right]dt + d\tilde{M}_t, \\ d_t\tilde{m}_t = \left[\Delta\tilde{m}_t + \operatorname{div}\left(\tilde{m}_t D_p\tilde{H}_t(\cdot, D\tilde{u}_t)\right)\right]dt, \\ \tilde{m}_{t_0} = m_0, \ \tilde{u}_T = \tilde{G}(\cdot, m_T), \end{cases} \quad (2.10)$$

where (still formally) $d\tilde{M}_t = (\sqrt{2\beta}D_x\tilde{u}_t + v_t(x + \sqrt{2\beta}W_t)) \cdot dW_t$.

Let us explain how we understand the above system. The solution $(\tilde{u}_t)_{t\in[0,T]}$ is seen as an $(\mathcal{F}_t)_{t\in[0,T]}$-adapted process with paths in $\mathcal{C}^0([0,T], \mathcal{C}^{n+2}(\mathbb{T}^d))$, for some fixed $n \geqslant 0$. The process $(\tilde{m}_t)_{t\in[0,T]}$ reads as an $(\mathcal{F}_t)_{t\in[0,T]}$-adapted process with paths in the space $\mathcal{C}^0([0,T], \mathcal{P}(\mathbb{T}^d))$. We shall look for solutions satisfying

$$\sup_{t\in[0,T]} \left(\|\tilde{u}_t\|_{n+2+\alpha}\right) \in L^\infty(\Omega, \mathcal{A}, \mathbb{P}), \quad (2.11)$$

(for some fixed $\alpha \in (0,1)$). The process $(\tilde{M}_t)_{t\in[0,T]}$ is seen as an $(\mathcal{F}_t)_{t\in[0,T]}$-adapted process with paths in the space $\mathcal{C}^0([0,T], \mathcal{C}^n(\mathbb{T}^d))$, such that, for any $x \in \mathbb{T}^d$, $(\tilde{M}_t(x))_{t\in[0,T]}$ is an $(\mathcal{F}_t)_{t\in[0,T]}$ martingale. It is required to satisfy

$$\sup_{t\in[0,T]} \left(\|\tilde{M}_t\|_{n+\alpha}\right) \in L^\infty(\Omega, \mathcal{A}, \mathbb{P}). \quad (2.12)$$

Theorem 2.4.3. *Assume that F, G, and H satisfy (2.4) and (2.5) and that* **(HF1($n+1$))** *and* **(HG1($n+2$))** *hold true for some $n \geqslant 0$ and some $\alpha \in (0,1)$. Then, there exists a unique solution $(\tilde{u}_t, \tilde{m}_t, \tilde{M}_t)_{t\in[0,T]}$ to (2.10), satisfying (2.11) and (2.12).*

Theorem 2.4.3 is proved in Chapter 4 (see Theorem 4.3.1 for more precise estimates). The main difference with the deterministic mean field game system is that the solution $(\tilde{u}_t, \tilde{m}_t)_{0 \leqslant t \leqslant T}$ is sought in a much bigger space, namely $[\mathcal{C}^0([0,T], \mathcal{C}^n(\mathbb{T}^d)) \times \mathcal{C}^0([0,T], \mathcal{P}(\mathbb{T}^d))]^\Omega$, which is not well suited to the use of compactness arguments. Because of that, one can can no longer invoke the Schauder theorem to prove the existence of a solution, which is the standard argument for solving the case $\beta = 0$; see Chapter 3. For this reason, the proof uses instead a continuation method, directly inspired from the literature on finite dimensional forward–backward stochastic systems (see [86]). Notice also that, due to the presence of the noise $(W_t)_{t \in [0,T]}$, the analysis of the time-regularity of the solution becomes a challenging issue and that the continuation method permits bypassing this difficulty.

2.4.3   Second-Order Master Equation

The second main result of this book concerns the analogue of Theorem 2.4.2 when the underlying MFG problem incorporates an additive common noise. Then the master equation involves additional terms, including second-order derivatives in the direction of the measure. It has the form (for some fixed level of common noise $\beta > 0$):

$$
\begin{cases}
-\partial_t U(t,x,m) - (1+\beta)\Delta_x U(t,x,m) + H\big(x, D_x U(t,x,m)\big) \\[2mm]
-F\big(x,m\big) - (1+\beta) \displaystyle\int_{\mathbb{T}^d} \mathrm{div}_y \big[D_m U\big](t,x,m,y)\, dm(y) \\[2mm]
+ \displaystyle\int_{\mathbb{T}^d} D_m U\big(t,x,m,y\big) \cdot D_p H\big(y, D_x U(t,y,m)\big)\, dm(y) \\[2mm]
-2\beta \displaystyle\int_{\mathbb{T}^d} \mathrm{div}_x \big[D_m U\big](t,x,m,y)\, dm(y) \\[2mm]
-\beta \displaystyle\int_{\mathbb{T}^d \times \mathbb{T}^d} \mathrm{Tr}\Big[D_{mm}^2 U\big(t,x,m,y,y'\big)\Big]\, dm(y)\, dm(y') = 0, \\[2mm]
U(T,x,m) = G(x,m), \qquad \text{for } (t,x,m) \in [0,T] \times \mathbb{T}^d \times \mathcal{P}(\mathbb{T}^d).
\end{cases}
\tag{2.13}
$$

Following Definition 2.4.1, we let

Definition 2.4.4.   *We say that a map $U : [0,T] \times \mathbb{T}^d \times \mathcal{P}(\mathbb{T}^d) \to \mathbb{R}$ is a classical solution to the second-order master equation (2.13) if*

- *U is continuous in all its arguments (for the $\mathbf{d}_1$ distance on $\mathcal{P}(\mathbb{T}^d)$), is of class $\mathcal{C}^2$ in x and $\mathcal{C}^1$ in time (the derivatives of order 1 in time and space and of order 2 in space being continuous in all the arguments);*
- *U is of class $\mathcal{C}^2$ with respect to m, the first- and second-order derivatives*

$$
[0,T] \times \mathbb{T}^d \times \mathcal{P}(\mathbb{T}^d) \times \mathbb{T}^d \ni (t,x,m,y) \mapsto \frac{\delta U}{\delta m}(t,x,m,y),
$$

$$
[0,T] \times \mathbb{T}^d \times \mathcal{P}(\mathbb{T}^d) \times \mathbb{T}^d \times \mathbb{T}^d \ni (t,x,m,y,y') \mapsto \frac{\delta^2 U}{\delta m^2}(t,x,m,y),
$$

being continuous in all the arguments, the first-order derivative $\delta U/\delta m$ being twice differentiable in y, the derivatives being continuous in all the arguments, and the second-order derivative $\delta^2 U/\delta m^2$ being also twice differentiable in the pair (y, y'), the derivatives being continuous in all the arguments;

- the function $D_y(\delta U/\delta m) = D_m U$ is differentiable in x, the derivatives being continuous in all the arguments;
- U satisfies the master equation (2.13).

On the model of Theorem 2.4.2, we claim

Theorem 2.4.5. *Assume that F, G and H satisfy (2.4) and (2.5) in Section 2.3 and that $(\mathbf{HF2}(n+1))$ and $(\mathbf{HG2}(n+2))$ hold true for some $n \geqslant 2$ and for some $\alpha \in (0,1)$.*

Then, the second-order master equation (2.13) has a unique solution U.

The solution U enjoys the following regularity: for any $\alpha' \in [0, \alpha)$, $t \in [0, T]$, and $m \in \mathcal{P}(\mathbb{T}^d)$, $U(t, \cdot, m)$, $[\delta U/\delta m](t, \cdot, m, \cdot)$, and $[\delta^2 U/\delta m^2](t, \cdot, m, \cdot, \cdot)$ are in $\mathcal{C}^{n+2+\alpha'}$, $\mathcal{C}^{n+2+\alpha'} \times \mathcal{C}^{n+1+\alpha'}$ and $\mathcal{C}^{n+2+\alpha'} \times \mathcal{C}^{n+\alpha'} \times \mathcal{C}^{n+\alpha'}$ respectively, independently of (t, m). Moreover, the mappings

$$[0, T] \times \mathcal{P}(\mathbb{T}^d) \ni (t, m) \mapsto U(t, \cdot, m) \in \mathcal{C}^{n+2+\alpha'},$$

$$[0, T] \times \mathcal{P}(\mathbb{T}^d) \ni (t, m) \mapsto [\delta U/\delta m](t, \cdot, m, \cdot) \in \mathcal{C}^{n+2+\alpha'} \times \mathcal{C}^{n+1+\alpha'},$$

$$[0, T] \times \mathcal{P}(\mathbb{T}^d) \ni (t, m) \mapsto [\delta^2 U/\delta m^2](t, \cdot, m, \cdot, \cdot) \in \mathcal{C}^{n+2+\alpha'} \times [\mathcal{C}^{n+\alpha'}]^2$$

are continuous. When $\alpha' = 0$, these mappings are Lipschitz continuous in m, uniformly in time.

Chapter 5 is devoted to the proof of Theorem 2.4.5. As for the first-order master equation, the starting point consists in letting, given $(t_0, m_0) \in [0, T] \times \mathcal{P}(\mathbb{T}^d)$,

$$U(t_0, x, m_0) = \tilde{u}_{t_0}(x), \quad x \in \mathbb{T}^d,$$

where $(\tilde{u}_t, \tilde{m}_t, \tilde{M}_t)_{t \in [0,T]}$ is the solution to the MFG system with common noise (2.10), when $(W_t)_{t \in [0,T]}$ in the definition of the coefficients $\tilde{F}$, $\tilde{G}$ and $\tilde{H}$ is replaced by $(W_t - W_{t_0})_{t \in [t_0, T]}$. The key remark (see Lemma 5.1.1), is that, if we let $m_{t_0,t} = [\mathrm{id} + \sqrt{2}(W_t - W_{t_0})] \sharp \tilde{m}_t$, then, for any $h \in [0, T - t_0]$, $\mathbb{P}$ almost surely,

$$\tilde{u}_{t_0+h}(x) = U\big(t_0 + h, x + \sqrt{2}(W_{t_0+h} - W_{t_0}), m_{t_0, t_0+h}\big), \quad x \in \mathbb{T}^d.$$

(Take note that the function entering the push-forward in the definition of $m_{t_0,t}$ is random.) Taking the derivative with respect to h at $h = 0$ on both sides of the equality shows that the map U thus defined satisfies the master equation (up to a tailor-made Itô's formula; see Chapter A.3.2). Of course, the main issue is to prove that U is sufficiently smooth to perform the foregoing computation: for this we need to show that U has first- and second-order derivatives with respect to the measure. As for the deterministic case, this is obtained by linearizing the

MFG system (with common noise). This linearization procedure is complicated by the fact that, because of the noise, the triplet $(\tilde{u}_t, \tilde{m}_t, \tilde{M}_t)_{t \in [0,T]}$ solves an equation in which the coefficients have little time regularity.

As a byproduct of the construction of the master equation, we can come back to the MFG system with common noise.

Definition 2.4.6. *Given $t_0 \in [0,T]$, we call a solution to (2.9) a triplet $(u_t, m_t, v_t)_{t \in [t_0, T]}$ of $(\mathcal{F}_t)_{t \in [t_0, T]}$-adapted processes with trajectories in the space $\mathcal{C}^0([t_0, T], \mathcal{C}^3(\mathbb{T}^d) \times \mathcal{P}(\mathbb{T}^d) \times [\mathcal{C}^2(\mathbb{T}^d)]^d)$ such that the quantity $\sup_{t \in [t_0, T]}(\|u_t\|_3 + \sup_{i=1,\cdots,d} \|v_t^i\|_2)$ belongs to $L^\infty(\Omega, \mathcal{A}, \mathbb{P})$ and (2.9) holds true with probability 1. In (2.9), the forward equation is understood in the sense of distributions, namely, with probability 1, for all test functions $\phi \in \mathcal{C}^2(\mathbb{T}^d)$, for all $t \in [0,T]$,*

$$d \int_{\mathbb{T}^d} \phi \, dm_t = \left((1+\beta) \int_{\mathbb{T}^d} \Delta\phi \, dm_t - \int_{\mathbb{T}^d} D\phi \cdot D_p H(\cdot, Du_t(\cdot)) dm_t \right) dt$$
$$+ \sqrt{2\beta} \left(\int_{\mathbb{T}^d} D\phi \, dm_t \right) dW_t.$$

Observe that, in (2.9), the stochastic integral in the backward equation is a priori well defined up to an exceptional event depending on the spatial position $x \in \mathbb{T}^d$. Using the fact that $\sup_{t \in [t_0, T]} \sup_{i=1,\cdots,d} \|v_t^i\|_2$ is in $L^\infty(\Omega, \mathcal{A}, \mathbb{P})$ and invoking Kolmogorov's continuity theorem, it is standard to find a version of the stochastic integral that is jointly continuous in time and space.

Let U be the solution of the master equation (2.13).

Corollary 2.4.7. *Under the assumptions of Theorem 2.4.5, for any initial data $(t_0, m_0) \in [0,T] \times \mathcal{P}(\mathbb{T}^d)$, the stochastic MFG system (2.9) has a unique solution $(u_t, m_t, v_t)_{t \in [0,T]}$. The process $(u_t, m_t, v_t)_{t \in [0,T]}$ is an $(\mathcal{F}_t)_{t \in [0,T]}$-adapted processes with paths in $\mathcal{C}^0([0,T], \mathcal{C}^{n+2}(\mathbb{T}^d) \times \mathcal{P}(\mathbb{T}^d) \times [\mathcal{C}^{n+1}(\mathbb{T}^d)]^d)$ and the vector field $(v_t)_{t \in [0,T]}$ is given by*

$$v_t(x) = \int_{\mathbb{T}^d} D_m U(t, x, m_t, y) dm_t(y).$$

2.4.4 The Convergence of the Nash System for N Players

We finally study the convergence of Nash equilibria of differential games with N players to the limit system given by the master equation.

We consider the solution $(v^{N,i})_{i \in \{1,\ldots,N\}}$ of the Nash system:

$$\begin{cases} -\partial_t v^{N,i} - \sum_{j=1}^N \Delta_{x_j} v^{N,i} - \beta \sum_{j,k=1}^N \operatorname{Tr} D^2_{x_j, x_k} v^{N,i} + H(x_i, D_{x_i} v^{N,i}) \\ + \sum_{j \neq i} D_p H(x_j, D_{x_j} v^{N,j}) \cdot D_{x_j} v^{N,i} = F(x_i, m_{\boldsymbol{x}}^{N,i}) \quad \text{in } [0,T] \times \mathbb{T}^{Nd} \quad (2.14) \\ v^{N,i}(T, \boldsymbol{x}) = G(x_i, m_{\boldsymbol{x}}^{N,i}) \quad \text{in } \mathbb{T}^{Nd} \end{cases}$$

where we have set, for $\boldsymbol{x} = (x_1, \ldots, x_N) \in (\mathbb{T}^d)^N$, $m_{\boldsymbol{x}}^{N,i} = \dfrac{1}{N-1} \sum\limits_{j \neq i} \delta_{x_j}$.

Let us recall that, under the same assumptions on H, F, and G as in the statement of Theorem 2.4.5, the above system has a unique solution (see, for instance, [70]).

Our main result says that the $(v^{N,i})_{i \in \{1,\ldots,N\}}$ "converge" to the solution of the master equation as $N \to +\infty$. This result, conjectured in Lasry–Lions [74], is somewhat subtle because in the Nash system players observe each other (closed loop form) whereas in the limit system the players just need to observe the theoretical distribution of the population, and not the specific behavior of each player. We first study the convergence of the functions $(v^{N,i})_{i \in \{1,\cdots,N\}}$ and then the convergence of the optimal trajectories.

We have two different ways to express the convergence of the $(v^{N,i})_{i=1,\cdots,N}$, described in the following result:

Theorem 2.4.8. *Let the assumption of Theorem 2.4.5 be in force for some $n \geqslant 2$ and let $(v^{N,i})$ be the solution to (2.14) and U be the classical solution to the second-order master equation. Fix $N \geqslant 1$ and $(t_0, m_0) \in [0, T] \times \mathcal{P}(\mathbb{T}^d)$.*

(i) For any $\boldsymbol{x} \in (\mathbb{T}^d)^N$, let $m_{\boldsymbol{x}}^N := \frac{1}{N} \sum_{i=1}^N \delta_{x_i}$. Then

$$\sup_{i=1,\cdots,N} \left| v^{N,i}(t_0, \boldsymbol{x}) - U(t_0, x_i, m_{\boldsymbol{x}}^N) \right| \leqslant C N^{-1}.$$

(ii) For any $i \in \{1, \ldots, N\}$ and $x_i \in \mathbb{T}^d$, let us set

$$w^{N,i}(t_0, x_i, m_0) := \int_{\mathbb{T}^d} \cdots \int_{\mathbb{T}^d} v^{N,i}(t_0, \boldsymbol{x}) \prod_{j \neq i} m_0(dx_j),$$

where $\boldsymbol{x} = (x_1, \ldots, x_N)$. Then,

$$\left\| w^{N,i}(t_0, \cdot, m_0) - U(t_0, \cdot, m_0) \right\|_{L^1(m_0)} \leqslant \begin{cases} C N^{-1/d} & \text{if } d \geqslant 3 \\ C N^{-1/2} \log(N) & \text{if } d = 2 \\ C N^{-1/2} & \text{if } d = 1 \end{cases}.$$

In (i) and (ii), the constant C does not depend on i, t_0, m_0, or N.

Theorem 2.4.8 says, in two different ways, that the $(v^{N,i})_{i \in \{1,\cdots,N\}}$ are close to U. The first statement explains that, for a fixed $\boldsymbol{x} \in (\mathbb{T}^d)^N$ and a fixed i, the quantity $|v^{N,i}(t_0, \boldsymbol{x}) - U(t_0, x_i, m_{\boldsymbol{x}}^N)|$ is of order N^{-1}. In the second statement, one fixes a measure m_0 and an index i, and one averages in space $v^{N,i}(t_0, \cdot)$ over m_0 for all variables but the i-th one. The resulting map $w^{N,i}$ is at a distance of order $N^{-\min(1/d, 1/2)}$ of $U(t_0, \cdot, m_0)$ (with a logarithmic correction if $d = 2$).

We can also describe the convergence in terms of optimal trajectories. Let $t_0 \in [0, T)$, $m_0 \in \mathcal{P}(\mathbb{T}^d)$ and let $(Z_i)_{i \in \{1,\ldots,N\}}$ be an i.i.d family of N random variables of law m_0. We set $\boldsymbol{Z} = (Z_1, \ldots, Z_N)$. Let also $((B_t^i)_{t \in [0,T]})_{i \in \{1,\ldots,N\}}$ be a family of N independent d-dimensional Brownian motions is also independent of $(Z_i)_{i \in \{1,\ldots,N\}}$ and let $(W_t)_{t \in [0,T]}$ be a d-dimensional Brownian motion independent of the $(B^i)_{i \in \{1,\ldots,N\}}$ and $(Z_i)_{i \in \{1,\ldots,N\}}$. We consider the optimal trajectories $(\boldsymbol{Y}_t = (Y_{1,t}, \ldots, Y_{N,t}))_{t \in [t_0,T]}$ for the N-player game:

$$
\begin{cases}
dY_{i,t} = -D_p H(Y_{i,t}, D_{x_i} v^{N,i}(t, \boldsymbol{Y}_t)) dt + \sqrt{2}\, dB_t^i + \sqrt{2\beta}\, dW_t, & t \in [t_0, T] \\
Y_{i,t_0} = Z_i,
\end{cases}
$$

and the solution $(\tilde{\boldsymbol{X}}_t = (\tilde{X}_{1,t}, \ldots, \tilde{X}_{N,t}))_{t \in [t_0,T]}$ of the stochastic differential equation of (conditional) McKean–Vlasov type:

$$
\begin{cases}
d\tilde{X}_{i,t} = -D_p H\left(\tilde{X}_{i,t}, D_x U\left(t, \tilde{X}_{i,t}, \mathcal{L}(\tilde{X}_{i,t}|W)\right)\right) dt + \sqrt{2}\, dB_t^i + \sqrt{2\beta}\, dW_t, \\
\tilde{X}_{i,t_0} = Z_i,
\end{cases}
$$

Both systems of stochastic differential equations (SDEs) are set on $(\mathbb{R}^d)^N$. Since both are driven by periodic coefficients, solutions generate (canonical) flows of probability measures on $(\mathbb{T}^d)^N$: The flow of probability measures generated in $\mathcal{P}((\mathbb{T}^d)^N)$ by each solution is independent of the representatives in $\mathbb{R}^d$ of the $\mathbb{T}^d$-valued random variables $Z_1, \ldots, Z_N$.

The next result says that the solutions of the two systems are close:

Theorem 2.4.9. *Let the assumption of Theorem 2.4.8 be in force. Then, for any $N \geqslant 1$ and any $i \in \{1, \ldots, N\}$, we have*

$$
\mathbb{E}\left[\sup_{t \in [t_0,T]} \left|Y_{i,t} - \tilde{X}_{i,t}\right|\right] \leqslant
\begin{cases}
CN^{-1/d} & \text{if } d \geqslant 3 \\
CN^{-1/2} \log(N) & \text{if } d = 2 \\
CN^{-1/2} & \text{if } d = 1
\end{cases},
$$

for some constant $C > 0$ independent of t_0, m_0, and N.

In particular, since the $((\tilde{X}_{i,t})_{t \in [t_0,T]})_{i \in \{1,\ldots,N\}}$ are conditionally independent given the realization of W, the above result is a (conditional) propagation of chaos.

The proofs of Theorem 2.4.8 and Theorem 2.4.9 rely on the existence of the solution U of the master equation (2.13) and constitute the aim of Chapter 6. Our starting point is that, for any $N \geqslant 1$, the "projection" of U onto the finite dimensional space $[0, T] \times (\mathbb{T}^d)^N$ is almost a solution to the Nash system (2.14). Namely, if we set, for any $i \in \{1, \ldots, N\}$ and any $\boldsymbol{x} = (x_1, \ldots, x_N) \in (\mathbb{T}^d)^N$,

$$
u^{N,i}(t, \boldsymbol{x}) := U(t, x_i, m_{\boldsymbol{x}}^{N,i}),
$$

then $(u^{N,i})_{i\in\{1,\dots,N\}}$ satisfies (2.14) up to an error term of size $O(1/N)$ for each equation (Proposition 6.1.3). Note that, as the number of equations in (2.14) is N, this could yield to a serious issue because the error terms could add up.

One of the thrusts of our approach is that, somehow, the proofs work under the sole assumption that the master equation (2.13) admits a classical solution. Here the existence of a classical solution is guaranteed under the assumption of Theorem 2.4.5, which includes in particular the monotonicity properties of F and G, but the analysis provided in Chapter 6 shows that monotonicity plays no role in the proofs of Theorems 2.4.8 and 2.4.9. Basically, only the global Lipschitz properties of H and D_pH, together with the nondegeneracy of the diffusions and the various bounds obtained for the solution of the master equation and its derivatives, matter. This is a quite remarkable fact, which demonstrates the efficiency of our strategy.

Chapter Three

A Starter: The First-Order Master Equation

IN THIS CHAPTER we prove Theorem 2.4.2; i.e., we establish the well-posedness of the master equation without common noise:

$$
\begin{cases}
-\partial_t U(t,x,m) - \Delta_x U(t,x,m) + H\big(x, D_x U(t,x,m)\big) \\[2mm]
\quad - \displaystyle\int_{\mathbb{T}^d} \mathrm{div}_y \left[D_m U\right](t,x,m,y)\, dm(y) \\[4mm]
\quad + \displaystyle\int_{\mathbb{T}^d} D_m U(t,x,m,y) \cdot D_p H\big(y, D_x U(t,y,m)\big)\, dm(y) \\[4mm]
= F(x,m) \qquad \text{in } [0,T] \times \mathbb{T}^d \times \mathcal{P}(\mathbb{T}^d), \\[2mm]
U(T,x,m) = G(x,m) \qquad \text{in } \mathbb{T}^d \times \mathcal{P}(\mathbb{T}^d).
\end{cases}
\tag{3.1}
$$

The idea is to represent U by solutions of the mean field game (MFG) system: let us recall that, for any $(t_0, m_0) \in [0,T) \times \mathcal{P}(\mathbb{T}^d)$, the MFG system is the system of forward–backward equations:

$$
\begin{cases}
-\partial_t u - \Delta u + H(x, Du) = F(x, m(t)), & \text{in } (t_0, T) \times \mathbb{T}^d \\[1mm]
\partial_t m - \Delta m - \mathrm{div}(m D_p H(x, Du)) = 0, & \text{in } (t_0, T) \times \mathbb{T}^d \\[1mm]
u(T,x) = G(x, m(T)),\ m(t_0, \cdot) = m_0 & \text{in } \mathbb{T}^d.
\end{cases}
\tag{3.2}
$$

As recalled in the text that follows, under suitable assumptions on the data, there exists a unique solution (u, m) to the foregoing system. Our aim is to show that the map U defined by

$$
U(t_0, \cdot, m_0) := u(t_0, \cdot)
\tag{3.3}
$$

is a solution to (3.1).

Throughout this chapter assumptions (2.4) and (2.5) are in force. Let us, however, underline that the global Lipschitz continuity of H is not absolutely necessary. We just need to know that the solutions of the MFG system are uniformly Lipschitz continuous, independently of the initial conditions: sufficient conditions for this can be found in [74], for instance.

The proof of Theorem 2.4.2 requires several preliminary steps. We first recall the existence of a solution to the MFG system (3.2) (Proposition 3.1.1) and show that this solution depends in a Lipschitz continuous way on the initial measure m_0 (Proposition 3.2.1). Then we show by a linearization procedure that the map U defined in (3.3) is of class $\mathcal{C}^1$ with respect to the measure (Proposition 3.4.3, Corollary 3.4.4). The proof relies on the analysis of a linearized system with a specific structure, for which well-posedness and estimates are given in Lemma 3.3.1. We are then ready to prove Theorem 2.4.2 (Section 3.5). We also show, for later use, that the first-order derivative of U is Lipschitz continuous with respect to m (Proposition 3.6.1). We complete the chapter by explaining how one obtains the solution U as the derivative with respect to the measure m of the value function of an optimal control problem set over flows of probability measures (Theorem 3.7.1).

Some of the proofs given in this chapter consist of a sketch only. One of the reasons is that some of the arguments we use here to investigate the MFG system (3.2) have been already developed in the literature. Another reason is that this chapter constitutes a starter only, specifically devoted to the simpler case without common noise. Arguments will be expanded in detail in the two next chapters, when handling mean field games with a common noise, for which there are fewer available results in the literature.

3.1 SPACE REGULARITY OF U

In this part we investigate the space regularity of U with respect to x. Recall that $U(t_0, \cdot, m_0)$ is defined by

$$U(t_0, x, m_0) = u(t_0, x)$$

where (u, m) is a classical solution to (3.2) with initial condition $m(t_0) = m_0$. By a classical solution to (3.2) we mean a pair $(u, m) \in \mathcal{C}^{1,2} \times \mathcal{C}^0([t_0, T], \mathcal{P}(\mathbb{T}^d))$ such that the equation for u holds in the classical sense while the equation for m holds in the sense of distributions.

Proposition 3.1.1. *Assume that* **(HF1(n))** *and* **(HG1($n+2$))** *hold for some* $n \geqslant 1$. *Then, for any initial condition* $(t_0, m_0) \in [0, T] \times \mathcal{P}(\mathbb{T}^d)$, *the MFG system (3.2) has a unique classical solution* (u, m), *with* $u \in \mathcal{C}^{1+\alpha/2, 2+\alpha}$, *and this solution satisfies*

$$\sup_{t_1 \neq t_2} \frac{\mathbf{d}_1(m(t_1), m(t_2))}{|t_2 - t_1|^{1/2}} + \sum_{|\ell| \leqslant n} \|D^\ell u\|_{1+\alpha/2, 2+\alpha} \leqslant C_n, \tag{3.4}$$

where the constant C_n *does not depend on* (t_0, m_0).

Moreover, m has a continuous, positive density in $(0,T] \times \mathbb{T}^d$ and if, in addition, m_0 is absolutely continuous with a $\mathcal{C}^{2+\alpha}$ positive density, then m is of class $\mathcal{C}^{1+\alpha/2,2+\alpha}$.

Finally, the solution is stable, in the sense that, if m_0^n converges to m_0 in $\mathcal{P}(\mathbb{T}^d)$, then the corresponding solution (u^n, m^n) converges to the solution (u, m) of (3.2) in $\mathcal{C}^{1,2} \times \mathcal{C}^0([0,T], \mathcal{P}(\mathbb{T}^d))$.

The above result is due to Lasry and Lions in [72, 74], in a more general framework. We reproduce the proof here for the sake of completeness.

Note that further regularity of F and G improves the space regularity of u but not its time regularity (as the time regularity of the coefficients depends on that of m, see Proposition 3.1.1). By (3.4), we have, under assumptions **(HF1(n))** and **(HG1($n+2$))**,

$$\sup_{t \in [0,T]} \sup_{m \in \mathcal{P}(\mathbb{T}^d)} \|U(t, \cdot, m)\|_{n+2+\alpha} \leqslant C_n.$$

Proof.   The existence of a solution relies on the Schauder fixed-point argument. Let L be a bound on D_pH and X be the set of time-dependent measures $m \in \mathcal{C}^0([t_0, T], \mathcal{P}(\mathbb{T}^d))$ such that

$$\mathbf{d}_1(m(s), m(t)) \leqslant L|t - s| + \sqrt{2}|t - s|^{1/2} \quad \forall s, t \in [t_0, T]. \tag{3.5}$$

Note that, by Ascoli–Arzela theorem, X is a convex compact space for the uniform distance.

Given $m \in X$, we consider the solution u to the Hamilton–Jacobi equation

$$\begin{cases} -\partial_t u - \Delta u + H(x, Du) = F(x, m(t)) & \text{in } (t_0, T) \times \mathbb{T}^d, \\ u(T, x) = G(x, m(T)) & \text{in } \mathbb{T}^d. \end{cases}$$

Assumption **(HF1(n))** (for $n \geqslant 1$) implies that F is Lipschitz continuous in both variables. Then, by the definition of X, the map $(t, x) \rightarrow F(x, m(t))$ is Hölder continuous in time and space and, more precisely, belongs to $\mathcal{C}^{1/2,1}$ with a Hölder constant independent of m. Moreover, by assumption **(HG1($n+2$))**, the map $x \rightarrow G(x, m(T))$ is of class $\mathcal{C}^{2+\alpha}$ with a constant independent of m. Thus, by the theory of Hamilton–Jacobi equations with Lipschitz continuous Hamiltonian ([70], Theorem V.6.1), there exists a unique classical solution u to the above equation. Beside the L^∞−norm of u (by maximum principle), of Du ([70], Theorem V.3.1) and the $\mathcal{C}^{1+\alpha/2,2+\alpha}$-norm of u ([70], Theorem V.5.4) are bounded independently of m.

Let now $\tilde{m}$ be the weak solution to the Fokker–Planck equation:

$$\begin{cases} \partial_t \tilde{m} - \Delta \tilde{m} - \text{div}(\tilde{m} D_pH(x, Du)) = 0 & \text{in } (t_0, T) \times \mathbb{T}^d, \\ \tilde{m}(t_0, \cdot) = m_0 & \text{in } \mathbb{T}^d. \end{cases}$$

Following [20], the above equation has a unique solution in $\mathcal{C}^0([t_0, T], \mathcal{P}(\mathbb{T}^d))$ in the sense of distributions. Moreover, as $D_p H$ is bounded by L, m satisfies (3.5). In particular, $\tilde{m}$ belongs to X. This defines a map $\Phi : X \to X$ which, to any $m \in X$, associates the map $\tilde{m}$.

Next we claim that Φ is continuous. Indeed let $(m_\ell)_{\ell \geqslant 1}$ be a sequence in X converging to $m \in X$. For each $\ell \geqslant 1$, let u_ℓ and $\tilde{m}_\ell$ be the corresponding solutions to the Hamilton–Jacobi and the Fokker–Planck equations respectively. From our previous estimate, the maps $(u_\ell)_{\ell \geqslant 1}$ are bounded in $\mathcal{C}^{1+\alpha/2,2+\alpha}$. So, by continuity of F and G, any cluster point of the $(u_\ell)_{\ell \geqslant 1}$ is a solution associated with m. By uniqueness of the solution u of this limit problem, the whole sequence $(u_\ell)_{\ell \geqslant 1}$ converges to u. In the same way, $(\tilde{m}_\ell)_{\ell \geqslant 1}$ converges in X to the unique solution $\tilde{m}$ to the Fokker–Planck equation associated with u. This shows the continuity of Φ.

We conclude by the Schauder theorem that Φ has a fixed point (u, m), which is a solution to the MFG system (3.2). Uniqueness of this solution is given by Lasry–Lions monotonicity argument, developped in Lemma 3.1.2.

Note, by uniform parabolicity and strong maximum principle, that m has a continuous positive density on $(0, T] \times \mathbb{T}^d$: see [20].

Let now (u, m) be the solution to (3.2) and assume that m_0 has a smooth density. Then m solves the linear parabolic equation

$$\begin{cases} \partial_t m - \Delta m - Dm \cdot D_p H(x, Du) - m\,\mathrm{div}(D_p H(x, Du)) = 0 \text{ in } (t_0, T) \times \mathbb{T}^d, \\ m(0, \cdot) = m_0 \text{ in } \mathbb{T}^d, \end{cases}$$

with $\mathcal{C}^{\alpha/2,\alpha}$ coefficient and $\mathcal{C}^{2+\alpha}$ initial condition. Thus, by Schauder estimates m is of class $\mathcal{C}^{1+\alpha/2,2+\alpha}$. If, moreover, m_0 is positive, then m remains positive by the strong maximum principle.

Let us now show the improved space regularity for u when F and G are smoother. Fix a direction $e \in \mathbb{R}^d$ and consider $v := Du \cdot e$. Then v solves, in the sense of distributions,

$$\begin{cases} -\partial_t v - \Delta v + D_x H(x, Du) \cdot e + D_p H(x, Du) \cdot Dv = D_x F(x, m(t)) \cdot e \\ \qquad\qquad\qquad\qquad \text{in } (t_0, T) \times \mathbb{T}^d, \\ v(T, x) = D_x G(x, m(T)) \cdot e \qquad \text{in } \mathbb{T}^d. \end{cases}$$

If (HF1(n)) and (HG1(n+2)) hold for $n \geqslant 1$, then, as u is bounded in $\mathcal{C}^{1+\alpha/2,2+\alpha}$, the above equation has uniformly Hölder continuous coefficient and therefore v is also bounded $\mathcal{C}^{1+\alpha/2,2+\alpha}$ by Schauder theory. This proves that

$$\|Du\|_{1+\alpha/2,2+\alpha} \leqslant C_1.$$

By induction on n, one can show that, if (HF1(n)) and (HG1(n+2)) hold, then for any $l \in \mathbb{N}^d$ with $|l| = l_1 + \cdots + l_d = n$, the map $w := D^l u$ is uniformly bounded in $\mathcal{C}^{\alpha/2,1+\alpha}$ and solves, in the sense of distributions, a linear equation of the form

$$
\begin{cases}
-\partial_t w - \Delta w + D_p H(x, Du) \cdot Dw + h_l = D_x^l F(x, m(t)) \\
\qquad\qquad\qquad\qquad\qquad\qquad\qquad\qquad\qquad \text{in } (t_0, T) \times \mathbb{T}^d, \\
w(T, x) = D_x^l G(x, m(T)) \qquad \text{in } \mathbb{T}^d,
\end{cases}
$$

where h_l involves the derivatives of H and u up to order n and is, therefore, uniformly Hölder continuous. We conclude as earlier that w is bounded in $\mathcal{C}^{1+\alpha/2,2+\alpha}$ and therefore that

$$
\sum_{|l|=n} \|D^l u\|_{1+\alpha/2,2+\alpha} \leqslant C_n.
$$

The stability of the solution is a straightforward consequence of the foregoing estimates and of the uniqueness of the solution. $\square$

Lemma 3.1.2 (Lasry–Lions monotonicity argument). *Let $(u^i, m^i)_{i=1,2}$ be two solutions of the MFG system* (3.2) *with possibly different initial conditions $m_i(t_0) = m_0^i \in \mathcal{P}(\mathbb{T}^d)$ for $i = 1, 2$. Then*

$$
\int_{t_0}^{T} \int_{\mathbb{T}^d} (H(x, Du^2) - H(x, Du^1) - D_p H(x, Du^1) \cdot (Du^2 - Du^1)) m^1 dx\, dt
$$
$$
+ \int_{t_0}^{T} \int_{\mathbb{T}^d} (H(x, Du^1) - H(x, Du^2) - D_p H(x, Du^2) \cdot (Du^1 - Du^2)) m^2 dx\, dt
$$
$$
\leqslant - \int_{\mathbb{T}^d} (u^1(t_0, x) - u^2(t_0, x))(m_0^1(dx) - m_0^2(dx)),
$$

where m^1, m^2, u^1, and u^2 and their derivatives are evaluated at (t, x).

Note that, if $m^1(t_0) = m^2(t_0) = m_0$, then the right-hand side of the above inequality vanishes. As H is strictly convex and m^1 and m^2 are positive in $(t_0, T) \times \mathbb{T}^d$, this implies that $Du^1 = Du^2$ in $(t_0, T) \times \mathbb{T}^d$. So m^1 and m^2 solve the same Fokker–Planck equation with a drift of class $\mathcal{C}^{\alpha/2,1+\alpha}$: therefore $m^1 = m^2$ and, as u^1 and u^2 satisfy the same Hamilton–Jacobi equation, they are also equal. This proves the uniqueness of the solution to (3.2).

Another consequence of Lemma 3.1.2 is a stability property: as Du^1 and Du^2 are bounded by a constant C_0, and the positivity of $D_{pp}^2 H$ ensures the existence of a constant C (depending on C_0) such that

$$
\int_{t_0}^{T} \int_{\mathbb{T}^d} |Du^2 - Du^1|^2 (m^1 + m^2) dx\, dt
$$
$$
\leqslant -C \int_{\mathbb{T}^d} (u^1(t_0, x) - u^2(t_0, x))(m_0^1(dx) - m_0^2(dx)).
$$

Proof. We start the proof assuming that $m^1(t_0)$ and $m^2(t_0)$ have a smooth density, so that u and m are in $\mathcal{C}^{1,2}$. The general result can be then obtained by approximation. One computes $\frac{d}{dt}\int_{\mathbb{T}^d}(u^1 - u^2)(m^1 - m^2)dx$ and finds, after integrating by parts and rearranging:

$$\frac{d}{dt}\int_{\mathbb{T}^d}(u^1 - u^2)(m^1 - m^2)dx$$

$$= -\int_{\mathbb{T}^d}(H(x, Du^2) - H(x, Du^1) - D_pH(x, Du^1)\cdot(Du^2 - Du^1))m^1dx$$

$$- \int_{\mathbb{T}^d}(H(x, Du^1) - H(x, Du^2) - D_pH(x, Du^2)\cdot(Du^1 - Du^2))m^2dx$$

$$- \int_{\mathbb{T}^d}(F(x, m^1(t)) - F(x, m^2(t)))(m^1 - m^2).$$

One obtains the claimed inequality by integrating in time and using the monotonicity of F and G. $\qquad\square$

3.2   LIPSCHITZ CONTINUITY OF U

Proposition 3.2.1.   *Assume that* **(HF1(n+1))** *and* **(HG1(n+2))** *hold for some $n \geqslant 1$. Let $m_0^1, m_0^2 \in \mathcal{P}(\mathbb{T}^d)$, $t_0 \in [0, T]$ and (u^1, m^1), (u^2, m^2) be the solutions of the MFG system (3.2) with initial condition (t_0, m_0^1) and (t_0, m_0^2) respectively. Then*

$$\sup_{t\in[0,T]}\left\{\mathbf{d}_1(m^1(t), m^2(t)) + \|u^1(t, \cdot) - u^2(t, \cdot)\|_{n+2+\alpha}\right\} \leqslant C_n\mathbf{d}_1(m_0^1, m_0^2),$$

for a constant C_n independent of t_0, m_0^1 and m_0^2. In particular,

$$\|U(t_0, \cdot, m_0^1) - U(t_0, \cdot, m_0^2)\|_{n+2+\alpha} \leqslant C_n\mathbf{d}_1(m_0^1, m_0^2).$$

Proof.   *First step.* To simplify the notation, we show the result for $t_0 = 0$. We Lasry–Lions monotonicity argument (Lemma 3.1.2): it implies that

$$\int_0^T \int_{\mathbb{T}^d} |Du^1(t, y) - Du^2(t, y)|^2 (m^1(t, y) + m^2(t, y)) \, dy \, dt$$

$$\leqslant C \int_{\mathbb{T}^d} (u^1(0, y) - u^2(0, y)) (m_0^1(dy) - m_0^2(dy)).$$

By definition of $\mathbf{d}_1$, the right-hand side can be estimated by

$$\int_{\mathbb{T}^d} \big(u^1(0,y) - u^2(0,y)\big)\big(m_0^1(dy) - m_0^2(dy)\big) \leqslant C\|D(u^1 - u^2)(0,\cdot)\|_\infty \mathbf{d}_1(m_0^1, m_0^2).$$

Hence

$$\int_0^T \int_{\mathbb{T}^d} \big(m^1(t,y) + m^2(t,y)\big)|Du^1(t,y) - Du^2(t,y)|^2 dy\, dt$$

$$\leqslant C\|D(u^1 - u^2)(0,\cdot)\|_\infty \mathbf{d}_1(m_0^1, m_0^2). \tag{3.6}$$

Second step: Next we estimate $m^1 - m^2$: to do so, let $(\Omega, \mathcal{F}, \mathbb{P})$ be a standard probability space and X_0^1, X_0^2 be random variables on Ω with law m_0^1 and m_0^2 respectively such that $\mathbb{E}[|X_0^1 - X_0^2|] = \mathbf{d}_1(m_0^1, m_0^2)$. Let also $(X_t^1)_{t\in[0,T]}$, $(X_t^2)_{t\in[0,T]}$ be the solutions to

$$dX_t^i = -D_pH(X_t^i, Du^i(t, X_t^i))dt + \sqrt{2}\, dB_t \qquad t \in [0,T], \ i = 1, 2,$$

where $(B_t)_{t\in[0,T]}$ is a d-dimensional Brownian motion. Then the law of X_t^i is $m^i(t)$ for any t. We have

$$\mathbb{E}\big[|X_t^1 - X_t^2|\big] \leqslant \mathbb{E}\big[|X_0^1 - X_0^2|\big]$$

$$+ \mathbb{E}\bigg[\int_0^t \Big(\big|D_pH\big(X_s^1, Du^1(s, X_s^1)\big) - D_pH\big(X_s^2, Du^1(s, X_s^2)\big)\big|$$

$$+ \big|D_pH\big(X_s^2, Du^1(s, X_s^2)\big) - D_pH\big(X_s^2, Du^2(s, X_s^2)\big)\big|\Big)\, ds\bigg].$$

As the maps $x \mapsto D_pH(x, Du^1(s,x))$ and $p \mapsto D_pH(x,p)$ are Lipschitz continuous (see (2.4) and Proposition 3.2.1), we deduce:

$$\mathbb{E}\big[|X_t^1 - X_t^2|\big] \leqslant \mathbb{E}\big[|X_0^1 - X_0^2|\big] + C\int_0^t \mathbb{E}\big[|X_s^1 - X_s^2|\big]\, ds$$

$$+ C\int_0^t \int_{\mathbb{T}^d} |Du^1(s,x) - Du^2(s,x)|m^2(s,x)dx\, ds$$

$$\leqslant \mathbf{d}_1(m_0^1, m_0^2) + C\int_0^t \mathbb{E}\big[|X_s^1 - X_s^2|\big]\, ds$$

$$+ C\bigg(\int_0^t \int_{\mathbb{T}^d} |Du^1(s,x) - Du^2(s,x)|^2 m^2(s,x)dx\, ds\bigg)^{1/2}.$$

In view of (3.6) and Gronwall's lemma, we obtain, for any $t \in [0, T]$

$$\mathbb{E}\big[|X_t^1 - X_t^2|\big] \leqslant C \left[\mathbf{d}_1(m_0^1, m_0^2) + \|D(u^1 - u^2)(0, \cdot)\|_\infty^{1/2} \mathbf{d}_1(m_0^1, m_0^2)^{1/2}\right]. \quad (3.7)$$

As $\mathbf{d}_1(m^1(t), m^2(t)) \leqslant \mathbb{E}[|X_t^1 - X_t^2|]$, we then get

$$\sup_{t \in [0,T]} \mathbf{d}_1(m^1(t), m^2(t))$$

$$\leqslant C \left[\mathbf{d}_1(m_0^1, m_0^2) + \|D(u^1 - u^2)(0, \cdot)\|_\infty^{1/2} \mathbf{d}_1(m_0^1, m_0^2)^{1/2}\right]. \quad (3.8)$$

Third step: We now estimate the difference $w := u^1 - u^2$. We note that w satisfies

$$\begin{cases} -\partial_t w(t, x) - \Delta w(t, x) + V(t, x) \cdot Dw(t, x) = R_1(t, x) & \text{in } [0, T] \times \mathbb{T}^d, \\ w(T, x) = R_T(x) & \text{in } \mathbb{T}^d, \end{cases}$$

where, for $(t, x) \in [0, T] \times \mathbb{T}^d$,

$$V(t, x) = \int_0^1 D_p H\big(x, sDu^1(t, x) + (1 - s)Du^2(t, x)\big)\, ds,$$

$$R_1(t, x) = \int_0^1 \int_{\mathbb{T}^d} \frac{\delta F}{\delta m}\big(x, sm^1(t) + (1 - s)m^2(t), y\big)\big(m^1(t, y) - m^2(t, y)\big) dy\, ds$$

and

$$R_T(x) = \int_0^1 \int_{\mathbb{T}^d} \frac{\delta G}{\delta m}\big(x, sm^1(T) + (1 - s)m^2(T), y\big)\big(m^1(T, y) - m^2(T, y)\big) dy\, ds.$$

By assumption **(HF1(n+1))** and inequality (3.8), we have, for any $t \in [0, T]$,

$$\|R_1(t, \cdot)\|_{n+1+\alpha}$$

$$\leqslant \int_0^1 ds \left\|D_y \frac{\delta F}{\delta m}(\cdot, sm^1(t) + (1 - s)m^2(t), \cdot)\right\|_{C^{n+1+\alpha} \times L^\infty} \mathbf{d}_1(m^1(t), m^2(t))$$

$$\leqslant C \left[\mathbf{d}_1(m_0^1, m_0^2) + \|Dw(0, \cdot)\|_\infty^{1/2} \mathbf{d}_1(m_0^1, m_0^2)^{1/2}\right],$$

and, in the same way (using assumption **(HG1(n+2))**),

$$\|R_T\|_{n+2+\alpha} \leqslant C \left[\mathbf{d}_1(m_0^1, m_0^2) + \|Dw(0, \cdot)\|_\infty^{1/2} \mathbf{d}_1(m_0^1, m_0^2)^{1/2}\right].$$

On the other hand, $V(t, \cdot)$ is bounded in $\mathcal{C}^{n+1+\alpha}$ in view of the regularity of u^1 and u^2 (Proposition 3.1.1). Then Lemma 3.2.2 states that

$$\sup_{t \in [0,T]} \|w(t, \cdot)\|_{n+2+\alpha} \leqslant C \left[\|R_T\|_{n+2+\alpha} + \sup_{t \in [0,T]} \|R_1(t, \cdot)\|_{n+1+\alpha} \right]$$
$$\leqslant C \left[\mathbf{d}_1(m_0^1, m_0^2) + \|Dw(0, \cdot)\|_\infty^{1/2} \mathbf{d}_1(m_0^1, m_0^2)^{1/2} \right].$$

Absorbing the last term of the right-hand side into the left-hand side, we find

$$\sup_{t \in [0,T]} \|w(t, \cdot)\|_{n+2+\alpha} \leqslant C \mathbf{d}_1(m_0^1, m_0^2),$$

and coming back to inequality (3.8), we also obtain

$$\sup_{t \in [0,T]} \mathbf{d}_1(m^1(t), m^2(t)) \leqslant C \mathbf{d}_1(m_0^1, m_0^2). \qquad \square$$

In the proof we used the following estimate:

Lemma 3.2.2. *Let $n \geqslant 1$, V be in $\mathcal{C}^0([0,T], \mathcal{C}^{n-1}(\mathbb{T}^d, \mathbb{R}^d))$ and f be in $\mathcal{C}^0([0,T], \mathcal{C}^{n-1+\alpha})$. Then, for any $z_T \in \mathcal{C}^{n+\alpha}$, the (backward) equation*

$$\begin{cases} -\partial_t z - \Delta z + V(t,x) \cdot Dz = f(t,x), & \text{in } [0,T] \times \mathbb{T}^d \\ z(T,x) = z_T(x) \end{cases} \tag{3.9}$$

has a unique solution in $L^2([0,T], H^1(\mathbb{T}^d))$ and this solution satisfies:

$$\sup_{t \in [0,T]} \|z(t, \cdot)\|_{n+\alpha} \leqslant C \left\{ \|z_T\|_{n+\alpha} + \sup_{t \in [0,T]} \|f(t, \cdot)\|_{n-1+\alpha} \right\},$$

where C depends on $\sup_{t \in [0,T]} \|V(t, \cdot)\|_{n-1}$, d and α only. If, in addition, V belongs to $\mathcal{C}^0([0,T], \mathcal{C}^{n-1+\alpha}(\mathbb{T}^d, \mathbb{R}^d))$, then

$$\sup_{t \neq t'} \frac{\|z(t', \cdot) - z(t, \cdot)\|_{n+\alpha}}{|t' - t|^{\frac{\alpha}{2}}} \leqslant C \left\{ \|z_T\|_{n+\alpha} + \sup_{t \in [0,T]} \|f(t, \cdot)\|_{n-1+\alpha} \right\},$$

where C depends on $\sup_{t \in [0,T]} \|V(t, \cdot)\|_{n-1+\alpha}$, d and α only.

Proof. The existence and the uniqueness of a solution to (3.9) are well known. (see Theorem II.4.2 of [70]). We now prove the estimates.

Step 1. We start the proof with the case $n = 1$ and a homogeneous initial datum $z_T = 0$. Let z solve (3.9) with $z_T = 0$. By the maximum principle, we have

$$\|z\|_\infty \leqslant C\|f\|_\infty,$$

where C depends on T only.

Next we claim that there exists a constant C, depending on $\|V\|_\infty$, α, and d only, such that, if z is the solution to (3.9) with $z_T = 0$, then

$$\|Dz\|_\infty \leqslant C\|f\|_\infty. \tag{3.10}$$

Indeed, assume for a while that there exist V_n, f_n, and z_n, with $\|V_n\|_\infty \leqslant M$ and $\|f_n\|_\infty \leqslant 1$ and where z_n solves

$$\begin{cases} -\partial_t z_n - \Delta z_n + V_n \cdot Dz_n = f_n & \text{in } [0, T] \times \mathbb{T}^d, \\ z_n(T, x) = 0 & \text{in } \mathbb{T}^d, \end{cases}$$

with $k_n := \|Dz_n\|_\infty \to +\infty$. We set $\tilde{z}_n := z_n/k_n$, $\tilde{f}_n := f_n/k_n$. Then $\tilde{z}_n$ solves

$$\begin{cases} \partial_t \tilde{z}_n - \Delta \tilde{z}_n = \tilde{f}_n - V_n \cdot D\tilde{z}_n & \text{in } [0, T] \times \mathbb{T}^d \\ \tilde{z}_n(0, x) = 0 & \text{in } \mathbb{T}^d, \end{cases}$$

with

$$\|\tilde{f}_n - V_n \cdot D\tilde{z}_n\|_\infty \leqslant 1 + M.$$

The L^p estimates on the heat potential (see (3.2) of Chapter 3 in [70]) imply that, for any $p \geqslant 1$,

$$\|(\partial_t \tilde{z}_n, D^2 \tilde{z}_n)\|_{L^p((0,T)\times\mathbb{T}^d)} \leqslant C\|\tilde{f}_n - V_n \cdot D\tilde{z}_n\|_{L^p((0,T)\times\mathbb{T}^d)} \leqslant C(1 + M),$$

for some constant C depending on p and d only. Then by the Sobolev inequality (Lemma II.3.3 in [70]) we have that, for any $\beta \in (0, 1)$,

$$\|D\tilde{z}_n\|_{C^{\beta/2,\beta}} \leqslant C(1 + M),$$

for a constant C depending on β and d only.

On the other hand, multiplying the equation by $\tilde{z}_n$ and integrating in space and on the time interval $[t, T]$ (for $t \in [0, T]$) yields

$$\int_{\mathbb{T}^d} |\tilde{z}_n(t)|^2 dx + \int_t^T \int_{\mathbb{T}^d} |D\tilde{z}_n|^2 dx\, dt = \int_t^T \int_{\mathbb{T}^d} \tilde{z}_n(\tilde{f}_n - V_n \cdot D\tilde{z}_n) dx\, dt$$

$$\leqslant \int_t^T \int_{\mathbb{T}^d} |\tilde{f}_n|^2 dx\, dt + \frac{1}{2} \int_t^T \int_{\mathbb{T}^d} |D\tilde{z}_n|^2 dx\, dt + 3 \int_t^T \int_{\mathbb{T}^d} |\tilde{z}_n|^2 (1 + |V_n|^2) dx\, dt,$$

so that

$$\int_{\mathbb{T}^d} |\tilde{z}_n(t)|^2 dx + \frac{1}{2} \int_t^T \int_{\mathbb{T}^d} |D\tilde{z}_n|^2 dx\, dt \leqslant C\|\tilde{f}_n\|_\infty^2 + C\int_t^T \int_{\mathbb{T}^d} |\tilde{z}_n|^2 dx\, dt,$$

where the constant C depends on T and M. By the Gronwall lemma, we obtain therefore

$$\sup_{t\in[0,T]} \int_{\mathbb{T}^d} |\tilde{z}_n(t,x)|^2 dx + \int_0^T \int_{\mathbb{T}^d} |D\tilde{z}_n|^2 dx dt \leqslant C\|\tilde{f}_n\|_\infty^2.$$

As $(\tilde{f}_n)$ tends to 0 in L^∞, $(D\tilde{z}_n)$ tends to 0 in L^2. This is impossible since $(D\tilde{z}_n)$ is bounded in $\mathcal{C}^{\beta/2,\beta}$ and satisfies $\|D\tilde{z}_n\|_\infty = 1$. This proves our claim (3.10).

We now improve inequality (3.10) into a Hölder bound. Let z be a solution to (3.9) with $z(T,\cdot) = 0$. Then z solves the heat equation with right-hand side given by $f - V \cdot Dz$. So, arguing as above, we have, thanks to (3.10) and for any $\beta \in (0,1)$,

$$\|Dz\|_{\mathcal{C}^{\beta/2,\beta}} \leqslant C\|f - V \cdot Dz\|_\infty \leqslant C\|f\|_\infty, \tag{3.11}$$

where the constant C depends on $\|V\|_\infty$, d, T, and β only.

Step 2. Next we prove the inequality for the derivatives of the solution z of (3.9), still with an homogeneous terminal condition $z(T,\cdot) = 0$. For $l \in \mathbb{N}^d$ with $|l| \leqslant n - 1$, the map $w := D^l z$ solves

$$\begin{cases} -\partial_t w - \Delta w + V(t,x) \cdot Dw = D^l f(t,x) + g_l(t,x), & \text{in } [0,T] \times \mathbb{T}^d \\ z(T,x) = z_T(x) \end{cases}$$

where $g_l(t,x)$ is a linear combination of the space derivatives of z up to order $|l|$ with coefficients proportional to space derivatives of V up to order $|l|$. So, by (3.11), we have, for any $\beta \in (0,1)$,

$$\|Dw\|_{\mathcal{C}^{\beta/2,\beta}} \leqslant C\|D^l f + g_l\|_\infty \leqslant C\Big(\sup_{t\in[0,T]} \|f(t,\cdot)\|_{|l|} + \sup_{t\in[0,T]} \|z(t,\cdot)\|_{|l|}\Big),$$

where C depends on l, β, d, and $\sup_{t\in[0,T]} \|V(t,\cdot)\|_{|l|}$ only. By induction on $|l|$, the above inequality implies for the choice $\beta = \alpha$ that

$$\sup_{|l|\leqslant n} \|D^l z(t,\cdot)\|_{\mathcal{C}^{\alpha/2,\alpha}} \leqslant C \sup_{t\in[0,T]} \|f(t,\cdot)\|_{n-1}, \tag{3.12}$$

where C depends on n, α, d and $\sup_{t\in[0,T]} \|V(t,\cdot)\|_{n-1}$ only.

Step 3. We now remove the assumption that $z_T = 0$. We write z as the sum $z = z_1 + z_2$, where z_1 solves the heat equation with terminal condition z_T and z_2

solves equation (3.9) with the right-hand side $f - V \cdot Dz_1$ and homogeneous terminal condition $z_2(T, \cdot) = 0$. For any $l \in \mathbb{N}^d$ with $|l| \leqslant n$, the derivative $D^l z$ solves the heat equation with terminal condition $D^l z_T$. So, by the classical Hölder regularity of the solution of the heat equation (inequality III.2.2 of [70]), we have

$$\|D^l z_1\|_{\mathcal{C}^{\alpha/2,\alpha}} \leqslant C\|D^l z_T\|_\alpha,$$

where C depends only on α, d, and $|l|$. By (3.12) and the estimate on z_1, we also have

$$
\begin{aligned}
\sup_{|l|\leqslant n} \|D^l z_2\|_{\mathcal{C}^{\alpha/2,\alpha}} &\leqslant\ C \sup_{t\in[0,T]} \|f(t,\cdot) - V(t,\cdot)\cdot Dz_1(t,\cdot)\|_{n-1} \\
&\leqslant\ C(\sup_{t\in[0,T]} \|f(t,\cdot)\|_{n-1} + \|z_T\|_{n+\alpha}),
\end{aligned}
$$

where C depends on n, α, d, and $\sup_{t\in[0,T]} \|V(t,\cdot)\|_{n-1}$ only. Putting together the estimates for z_1 and z_2 gives

$$\sup_{|l|\leqslant n} \|D^l z\|_{\mathcal{C}^{\alpha/2,\alpha}} \leqslant C(\sup_{t\in[0,T]} \|f(t,\cdot)\|_{n-1} + \|z_T\|_{n+\alpha}). \qquad (3.13)$$

In particular,

$$\sup_{t\in[0,T]} \|z(t,\cdot)\|_{n+\alpha} \leqslant C(\sup_{t\in[0,T]} \|f(t,\cdot)\|_{n-1} + \|z_T\|_{n+\alpha}), \qquad (3.14)$$

where C is as above.

Step 4. We finally check the time regularity. For any fixed $y \in \mathbb{R}^d$, the difference $w(t,x) := z(t, x+y) - z(t,x)$ satisfies

$$
\begin{cases}
-\partial_t w - \Delta w + V(t,x)\cdot Dw = (f(t, x+y) - f(t,x)) \\
\qquad -(V(t, x+y) - V(t,x))\cdot Dz(t, x+y) \qquad \text{in } [0,T]\times\mathbb{T}^d, \\
z(T,x) = z_T(x+y) - z_T(x) \text{ in } \mathbb{T}^d.
\end{cases}
$$

From (3.13) we therefore have

$$
\begin{aligned}
\sup_{|l|\leqslant n} \|D^l w\|_{\mathcal{C}^{\alpha/2,\alpha}} &\leqslant\ C(\|z_T(\cdot + y) - z_T(\cdot)\|_{n+\alpha} \\
&\qquad + \sup_{t\in[0,T]} \|f(t, \cdot + y) - f(t,\cdot)\|_{n-1} \\
&\qquad + \sup_{t\in[0,T]} \|(V(t, \cdot + y) - V(t,\cdot))\cdot Dz(t, \cdot + y)\|_{n-1}) \\
&\leqslant\ C(\sup_{t\in[0,T]} \|f(t,\cdot)\|_{n-1+\alpha} + \|z_T\|_{n+\alpha})|y|^\alpha,
\end{aligned}
$$

where we used the Hölder regularity of V in space and the bound on Dz in (3.14) for the last inequality. Note that the constant C now also depends on $\sup_{t \in [0,T]} \|V(t,\cdot)\|_{n-1+\alpha}$. Using the time Hölder regularity of w, one then obtains

$$\sup_{t' \neq t} \frac{\|z(t',\cdot) - z(t,\cdot)\|_{n+\alpha}}{|t'-t|^{\alpha/2}} \leqslant C(\sup_{t \in [0,T]} \|f(t,\cdot)\|_{n-1+\alpha} + \|z_T\|_{n+\alpha}). \qquad \square$$

3.3   ESTIMATES ON A LINEAR SYSTEM

In the sequel we will repetitively encounter forward–backward systems of linear equations. To minimize the computation, we collect in this section estimates on such systems, which have the generic form:

$$\begin{cases} (i) \quad -\partial_t z - \Delta z + V(t,x) \cdot Dz = \dfrac{\delta F}{\delta m}(x, m(t))(\rho(t)) + b(t,x) \\[2mm] (ii) \quad \partial_t \rho - \Delta \rho - \operatorname{div}(\rho V) - \operatorname{div}(m\Gamma Dz + c) = 0 \quad \text{in } [t_0, T] \times \mathbb{T}^d, \\[2mm] (iii) \quad z(T,x) = \dfrac{\delta G}{\delta m}(x, m(T))(\rho(T)) + z_T(x), \ \rho(t_0) = \rho_0 \quad \text{in } \mathbb{T}^d, \end{cases} \qquad (3.15)$$

where V is a given vector field in $\mathcal{C}^0([0,T], \mathcal{C}^{n+1+\alpha}(\mathbb{T}^d; \mathbb{R}^d))$ (for some $n \geqslant 0$), m is a continuous time-dependent family of probability measures; i.e., $m \in \mathcal{C}^0([t_0, T], \mathcal{P}(\mathbb{T}^d))$, $\Gamma : [t_0, T] \times \mathbb{T}^d \to \mathbb{R}^{d \times d}$ is a continuous map with values into the family of symmetric matrices and where the maps $b : [t_0, T] \times \mathbb{T}^d \to \mathbb{R}$, $c : [t_0, T] \times \mathbb{T}^d \to \mathbb{R}^d$ and $z_T : \mathbb{T}^d \to \mathbb{R}$ are given. Above, we used the shortened notation $\frac{\delta F}{\delta m}(x, m(t))(\rho(t))$ for the duality bracket between $\frac{\delta F}{\delta m}(x, m(t), \cdot)$ and $\rho(t)$, and similarly for $\frac{\delta G}{\delta m}(x, m(T))(\rho(T))$. In (i), z and its derivatives are evaluated at $(t,x) \in [t_0, T] \times \mathbb{T}^d$. Moreover, we always assume that there is a constant $\bar{C} > 0$ such that

$$\begin{aligned} \forall t, t' \in [t_0, T], \qquad & \mathbf{d}_1(m(t), m(t')) \leqslant \bar{C}|t-t'|^{1/2}, \\ \forall (t,x) \in [t_0, T] \times \mathbb{T}^d, \qquad & \bar{C}^{-1} I_d \leqslant \Gamma(t,x) \leqslant \bar{C} I_d. \end{aligned} \qquad (3.16)$$

Typically, $V(t,x) = D_p H(x, Du(t,x))$, $\Gamma(t,x) = D_{pp}^2 H(x, Du(t,x))$ for some solution (u, m) of the MFG system (3.2) starting from some initial data $m(t_0) = m_0$. Recall that the derivative Du is globally Lipschitz continuous with a constant independent of (t_0, m_0) (Proposition 3.1.1), so that assumption (2.4) gives the existence of a constant $\bar{C}$ for which (3.16) holds. We also note for later use that this constant does not depend on (t_0, m_0).

We establish the existence of a solution and its regularity under the assumption that b and c belong to $\mathcal{C}^0([0,T], \mathcal{C}^{n+1+\alpha})$ and $\mathcal{C}([0,T], (\mathcal{C}^{n+\alpha}(\mathbb{T}^d; \mathbb{R}^d))')$ respectively and that z_T and ρ_0 belong respectively to $\mathcal{C}^{n+2+\alpha}$ and to $(\mathcal{C}^{n+1+\alpha})'$.

Lemma 3.3.1. *Under assumption* **(HF1($n+1$))** *and* **(HG1($n+2$))**, *system* (3.15) *has a unique solution* (z, ρ) *in* $\mathcal{C}^0([0,T], \mathcal{C}^{n+2+\alpha} \times (\mathcal{C}^{n+1+\alpha})')$, *with*

$$\sup_{t\in[t_0,T]} \|z(t,\cdot)\|_{n+2+\alpha} + \sup_{t\neq t'} \frac{\|z(t',\cdot) - z(t,\cdot)\|_{n+2+\alpha}}{|t'-t|^{\frac{\alpha}{2}}} \leqslant C_n M. \qquad (3.17)$$

and

$$\sup_{t\in[t_0,T]} \|\rho(t)\|_{-(n+1+\alpha)} + \sup_{t\neq t'} \frac{\|\rho(t') - \rho(t)\|_{-(n+1+\alpha)}}{|t-t'|^{\frac{1}{2}}} \leqslant C_n M, \qquad (3.18)$$

where the constant C_n depends on n, T, α, $\sup_{t\in[t_0,T]} \|V(t,\cdot)\|_{n+1+\alpha}$, the constant $\bar{C}$ in (3.16), F, and G and where M is given by

$$M := \|z_T\|_{n+2+\alpha} + \|\rho_0\|_{-(n+1+\alpha)} + \sup_{t\in[t_0,T]} \left(\|b(t,\cdot)\|_{n+1+\alpha} \right.$$

$$\left. + \|c(t)\|_{-(n+\alpha)} \right), \qquad (3.19)$$

where $\|c(t)\|_{-(n+\alpha)}$ stands for the supremum of $\|(c(t))_i\|_{-(n+\alpha)}$ over $i \in \{1,\ldots,d\}$, $(c(t))_i$ denoting the ith coordinate of $c(t)$.

Proof. Without loss of generality we assume $t_0 = 0$. We prove the existence of a solution to (3.15) by Leray–Schauder argument. The proof requires several steps, the key point being precisely to prove the estimates (3.17) and (3.18). In Step 1, we suppose that m, c, and ρ_0 are smooth and prove the existence of the solution under this additional condition. We remove this assumption in the very last step.

Step 1: *Definition of a map* **T** *satisfying the Leray–Schauder theorem.* We set $X := \mathcal{C}^0([0,T], (\mathcal{C}^{n+1+\alpha})')$. For $\rho \in X$, we define $\mathbf{T}(\rho)$ as follows. First, we call z the solution to

$$\begin{cases} -\partial_t z - \Delta z + V(t,x) \cdot Dz = \dfrac{\delta F}{\delta m}(x, m(t))(\rho(t)) + b(t,x) \\ \qquad\qquad\qquad\qquad\qquad\qquad\text{in } [0,T] \times \mathbb{T}^d, \\ z(T) = \dfrac{\delta G}{\delta m}(x, m(T))(\rho(T)) + z_T \quad \text{in } \mathbb{T}^d. \end{cases} \qquad (3.20)$$

From Lemma 3.2.2, there exists a unique solution z in $\mathcal{C}^{\alpha/2}([0,T], \mathcal{C}^{n+2+\alpha})$ to the above equation and it satisfies

$$\sup_{t\in[0,T]} \|z(t,\cdot)\|_{n+2+\alpha} + \sup_{t\neq t'} \frac{\|z(t',\cdot) - z(t,\cdot)\|_{n+2+\alpha}}{|t'-t|^{\frac{\alpha}{2}}}$$

$$\leqslant C\left\{\|\frac{\delta G}{\delta m}(x,m(T))(\rho(T))\|_{n+2+\alpha} + \|z_T\|_{n+2+\alpha}\right.$$

$$\left. + \sup_{t\in[0,T]} \|\frac{\delta F}{\delta m}(\cdot,m(t))(\rho(t))\|_{n+1+\alpha} + \sup_{t\in[0,T]} \|b(t,\cdot)\|_{n+1+\alpha}\right\} \quad (3.21)$$

$$\leqslant C\left\{\|z_T\|_{n+2+\alpha} + \sup_{t\in[0,T]} \|\rho(t))\|_{-(n+1+\alpha)} + \|b\|\right\},$$

where the constant C depends on the constant in $(\mathbf{HF1}(n+1))$ and $(\mathbf{HG1}(n+2))$, on $\sup_{t\in[0,T]}\|V(t,\cdot)\|_{n+1+\alpha}$, d, and α, and where we have set, for simplicity, $\|b\| := \sup_{t\in[0,T]}\|b(t,\cdot)\|_{n+1+\alpha}$.

Next we define $\tilde{\rho}$ as the solution in the sense of distributions to

$$\begin{cases} \partial_t\tilde{\rho} - \Delta\tilde{\rho} - \operatorname{div}(\tilde{\rho}V) - \operatorname{div}(m\Gamma Dz + c) = 0 & \text{in } [0,T]\times\mathbb{T}^d, \\ \tilde{\rho}(0) = \rho_0 & \text{in } \mathbb{T}^d. \end{cases} \quad (3.22)$$

As ρ_0, m, and c are assumed to be smooth in this step and as V and Γ satisfy (3.16), there exists a unique solution $\tilde{\rho}$ in $L^2([0,T],H^1(\mathbb{T}^d))$ to the above parabolic equation (Theorem II.4.1 of [70]). Furthermore, as the coefficients of this parabolic equation are bounded, $\tilde{\rho}$ is bounded in $\mathcal{C}^{\beta/2,\beta}$ for some $\beta \in (0,1)$, where the bounds depend on ρ_0, c, and m (but not on ρ since $\|Dz\|_\infty$ does not depend on ρ).

We set $\mathbf{T}(\rho) := \tilde{\rho}$ and note that $\mathbf{T}$ maps X into X since $\mathcal{C}^{\beta/2,\beta}$ is compactly embedded in X. We now check that $\mathbf{T}$ is continuous and compact on X. Let (ρ_n) converge to ρ in X. We denote by z_n (respectively z) the solution to (3.20) associated with ρ_n (and, respectively ρ). We also set $\tilde{\rho}_n := \mathbf{T}(\rho_n)$ and $\tilde{\rho} := \mathbf{T}(\rho)$. Note that the maps $(t,x) \to \frac{\delta F}{\delta m}(x,m(t))(\rho_n(t))$ and $(t,x) \to \frac{\delta G}{\delta m}(x,m(T))(\rho_n(T))$ converge uniformly to the maps $(t,x) \to \frac{\delta F}{\delta m}(x,m(t))(\rho(t))$ and $(t,x) \to \frac{\delta G}{\delta m}(x,m(T))(\rho(T))$ respectively. Hence (z_n) converges uniformly to z. By the bounds on the derivatives of z_n, (Dz_n) also converges uniformly to Dz. The stability property of equation (3.22) then implies that the sequence $(\tilde{\rho}_n)$ converges to $\tilde{\rho} := \mathbf{T}(\rho)$ in $L^2([0,T],H^1(\mathbb{T}^d))$, and thus in X by uniform Hölder estimate on the solution to equation (3.22). The same uniform Hölder estimate implies that the map $\mathbf{T}$ is compact.

In Steps 2 and 3, we show that, if $\rho = \sigma\mathbf{T}(\rho)$ for some $(\rho,\sigma) \in X\times[0,1]$, then ρ satisfies (3.18). This estimate proves that the norm in X of ρ is bounded independently of σ. Then we can conclude by Leray–Schauder theorem the existence of a fixed point for $\mathbf{T}$, which, by construction, yields a solution to (3.15).

From now on we fix $(\rho, \sigma) \in X \times [0,1]$ such that $\rho = \sigma \mathbf{T}(\rho)$ and let z be the solution to (3.20). Note that the pair (z, ρ) satisfies

$$
\begin{cases}
-\partial_t z - \Delta z + V(t,x) \cdot Dz = \left(\dfrac{\delta F}{\delta m}(x, m(t))(\rho(t)) + b(t,x) \right) \\
\qquad\qquad\qquad\qquad\qquad\qquad\qquad\qquad \text{in } [0,T] \times \mathbb{T}^d, \\
\partial_t \rho - \Delta \rho - \operatorname{div}(\rho V) - \sigma \operatorname{div}(m\Gamma Dz + c) = 0 \qquad \text{in } [0,T] \times \mathbb{T}^d, \\
\rho(0) = \sigma \rho_0, \qquad z(T) = \left(\dfrac{\delta G}{\delta m}(x, m(T))(\rho(T)) + z_T \right) \qquad \text{in } \mathbb{T}^d.
\end{cases}
$$

Our goal is to show that (3.17) and (3.18) hold for z and ρ respectively. In Steps 2 and 3, we will not use the smoothness of m, c, and ρ_0, so that these estimates also hold for any solution (z, ρ) to the above system, and, in particular, for solutions of (3.15).

Step 2: *Estimate of ρ.* Let us recall that, for any integer n, we abbreviate the notation for the duality product $\langle \cdot, \cdot \rangle_{(C^{n+\alpha})' \times C^{n+\alpha}}$ as $\langle \cdot, \cdot \rangle_{n+\alpha}$. By duality, we have

$$
\langle \rho(T), z(T, \cdot) \rangle_{n+1+\alpha} - \sigma \langle \rho_0, z(0, \cdot) \rangle_{n+1+\alpha}
$$

$$
= - \int_0^T \left\langle \rho(t), \frac{\delta F}{\delta m}(\cdot, m(t))(\rho(t)) + b(t, \cdot) \right\rangle_{n+1+\alpha} dt
$$

$$
- \sigma \int_0^T \int_{\mathbb{T}^d} Dz(t,x) \cdot \left[\Gamma(t,x) Dz(t,x) \right] m(t, dx) dt
$$

$$
- \sigma \int_0^T \langle c(t), Dz(t, \cdot) \rangle_{n+\alpha} dt.
$$

Using the terminal condition of z and the regularity and the monotonicity of F and G, we therefore obtain

$$
\sigma \int_0^T \int_{\mathbb{T}^d} \Gamma(t,x) Dz(t,x) \cdot Dz(t,x) m(t, dx) dt
$$

$$
\leqslant \sup_{t \in [0,T]} \|\rho(t)\|_{-(n+1+\alpha)} \left(\|z_T\|_{n+1+\alpha} + \|b\| \right) \tag{3.23}
$$

$$
+ \sigma \sup_{t \in [0,T]} \|z(t, \cdot)\|_{n+1+\alpha} \left(\|\rho_0\|_{-(n+1+\alpha)} + \|c\| \right),
$$

where we have set $\|b\| := \sup_{t \in [0,T]} \|b(t, \cdot)\|_{n+1+\alpha}$ and $\|c\| := \sup_{t \in [0,T]} \|c(t)\|_{-(n+\alpha)}$.

We now estimate ρ by duality. Let $\tau \in (0, T]$, $\xi \in C^{n+1+\alpha}$ and w be the solution to the backward equation

$$
-\partial_t w - \Delta w + V(t,x) \cdot Dw = 0 \text{ in } [0, \tau] \times \mathbb{T}^d, \qquad w(\tau) = \xi \text{ in } \mathbb{T}^d. \tag{3.24}
$$

Lemma 3.2.2 states that

$$\sup_{t \in [0,T]} \|w(t, \cdot)\|_{n+1+\alpha} + \sup_{t \neq t'} \frac{\|w(t', \cdot) - w(t, \cdot)\|_{n+1+\alpha}}{|t' - t|^{\frac{\alpha}{2}}} \leqslant C \|\xi\|_{n+1+\alpha}, \quad (3.25)$$

where C depends on $\sup_{t \in [0,T]} \|V(t, \cdot)\|_{n+\alpha}$. We have

$$\langle \rho(\tau), \xi \rangle_{n+1+\alpha} = \sigma \langle \rho_0, w(0) \rangle_{n+1+\alpha}$$

$$- \sigma \int_0^\tau \int_{\mathbb{T}^d} Dw(t,x) \cdot \big(\Gamma(t,x)Dz(t,x)\big) m(t,dx)\, dt \qquad (3.26)$$

$$- \sigma \int_0^\tau \langle c(t), Dz(t) \rangle_{n+\alpha}\, dt,$$

where the right-hand side is bounded above by

$$\sigma \|w(0)\|_{n+1+\alpha} \|\rho_0\|_{-(n+1+\alpha)} + \sigma \sup_{t \in [0,T]} \|Dw(t, \cdot)\|_{n+\alpha} \|c\|$$

$$+ \sigma \left(\int_0^T \int_{\mathbb{T}^d} \Gamma(t,x)Dw(t,x) \cdot Dw(t,x) m(t,dx)\, dt \right)^{1/2}$$

$$\times \left(\int_0^T \int_{\mathbb{T}^d} \Gamma(t,x)Dz(t,x) \cdot Dz(t,x) m(t,dx)\, dt \right)^{1/2}$$

$$\leqslant C \left[\sigma \|\xi\|_{n+1+\alpha} \|\rho_0\|_{-(n+1+\alpha)} + \sigma \sup_{t \in [0,T]} \|Dw(t, \cdot)\|_{n+\alpha} \|c\| \right.$$

$$\left. + \sigma \|Dw\|_\infty \left(\int_0^T \int_{\mathbb{T}^d} \Gamma(t,x)Dz(t,x) \cdot Dz(t,x) m(t,dx)\, dt \right)^{1/2} \right].$$

Using (3.23) and (3.25) we therefore obtain

$$\langle \rho(\tau), \xi \rangle_{n+1+\alpha} \leqslant C \|\xi\|_{n+1+\alpha} \left[\sigma \|\rho_0\|_{-(n+1+\alpha)} + \sigma \|c\| \right.$$

$$+ \sigma^{1/2} \sup_{t \in [0,T]} \|\rho(t)\|_{-(n+1+\alpha)}^{1/2} \big(\|z_T\|_{n+1+\alpha}^{1/2} + \|b\|^{1/2} \big)$$

$$\left. + \sigma \sup_{t \in [0,T]} \|z(t, \cdot)\|_{n+1+\alpha}^{1/2} \big(\|\rho_0\|_{-(n+1+\alpha)}^{1/2} + \|c\|^{1/2} \big) \right].$$

Taking the supremum over ξ with $\|\xi\|_{C^{n+\alpha+1}} \leqslant 1$ and over $\tau \in [0,T]$ yields

$$\sup_{t \in [0,T]} \|\rho(t)\|_{-(n+1+\alpha)} \leqslant C \left[\sigma \|\rho_0\|_{-(n+1+\alpha)} + \sigma \|c\| \right.$$

$$+ \sigma^{1/2} \sup_{t \in [0,T]} \|\rho(t)\|_{-(n+1+\alpha)}^{1/2} \big(\|z_T\|_{n+1+\alpha}^{1/2} + \|b\|^{1/2} \big)$$

$$\left. + \sigma \sup_{t \in [0,T]} \|z(t, \cdot)\|_{n+1+\alpha}^{1/2} \big(\|\rho_0\|_{-(n+1+\alpha)}^{1/2} + \|c\|^{1/2} \big) \right].$$

Rearranging and using the definition of M in (3.19), we obtain

$$
\sup_{t\in[0,T]} \|\rho(t)\|_{-(n+1+\alpha)}
$$

$$
\leqslant C\left[\sigma M + \sigma \sup_{t\in[0,T]} \|z(t,\cdot)\|_{n+1+\alpha}^{1/2}\left(\|\rho_0\|_{-(n+1+\alpha)}^{1/2} + \|c\|^{1/2}\right)\right]. \tag{3.27}
$$

We can use the same kind of argument to obtain the regularity of ρ with respect to the time variable: by (3.26), we have, for any $\tau \in [0,T]$ and $t \in [0,\tau]$ and using the Hölder estimate in time in (3.25):

$$
\langle \rho(\tau) - \rho(t), \xi \rangle_{n+1+\alpha}
$$

$$
= \langle \rho(t), w(t) - w(\tau) \rangle_{n+1+\alpha}
$$

$$
- \sigma \int_t^\tau \int_{\mathbb{T}^d} Dw(s,x) \cdot \left[\Gamma(s,x)Dz(s,x)\right] m(s,dx)\,ds
$$

$$
- \sigma \int_t^\tau \langle c(s), Dw(s,\cdot) \rangle_{n+\alpha}\,ds
$$

$$
\leqslant C(\tau-t)^{\frac{\alpha}{2}} \|\xi\|_{n+1+\alpha} \sup_{t\in[0,T]} \|\rho(t)\|_{-(n+1+\alpha)}
$$

$$
+ (\tau-t)^{\frac{1}{2}} \|Dw\|_\infty \left(\int_0^T \int_{\mathbb{T}^d} \Gamma(s,x)Dz(s,x) \cdot Dz(s,x)m(s,dx)\,ds\right)^{1/2}
$$

$$
+ (\tau-t)\|c\| \sup_{t\in[0,T]} \|w(t,\cdot)\|_{n+1+\alpha}.
$$

Plugging (3.27) into (3.23), we get that the root of the left-hand side in (3.23) satisfies the same bound as the left-hand side in (3.27). Therefore, together with (3.25), we obtain

$$
\langle (\rho(\tau) - \rho(t)), \xi \rangle_{n+1+\alpha}
$$

$$
\leqslant C(\tau-t)^{\frac{\alpha}{2}} \|\xi\|_{n+1+\alpha} \left[M + \sup_{t\in[0,T]} \|z(t,\cdot)\|_{n+1+\alpha}^{1/2}\left(\|\rho_0\|_{-(n+1+\alpha)}^{1/2} + \|c\|^{1/2}\right)\right].
$$

Dividing by $(\tau-t)^{\alpha/2}$ and taking the supremum over ξ yields

$$
\sup_{t\neq t'} \frac{\|\rho(t') - \rho(t)\|_{-(n+1+\alpha)}}{|t-t'|^{\frac{\alpha}{2}}}
$$

$$
\leqslant C\left[M + \sup_{t\in[0,T]} \|z(t,\cdot)\|_{n+1+\alpha}^{1/2}\left(\|\rho_0\|_{-(n+1+\alpha)}^{1/2} + \|c\|^{1/2}\right)\right]. \tag{3.28}
$$

Step 3: *Estimate of z*. In view of the equation satisfied by z, we have, by Lemma 3.2.2,

$$\sup_{t\in[0,T]} \|z(t,\cdot)\|_{n+2+\alpha} + \sup_{t\neq t'} \frac{\|z(t',\cdot)-z(t,\cdot)\|_{n+2+\alpha}}{|t'-t|^{\frac{\alpha}{2}}}$$

$$\leqslant C\left[\sup_{t\in[0,T]} \left\|\frac{\delta F}{\delta m}(x,m(t))(\rho(t))+b(t,\cdot)\right\|_{n+1+\alpha}\right.$$

$$\left. + \left\|\frac{\delta G}{\delta m}(x,m(T))(\rho(T))+z_T\right\|_{n+2+\alpha}\right],$$

where C depends on $\sup_{t\in[0,T]}\|V(t,\cdot)\|_{n+1+\alpha}$. Assumptions **(HF1($n$+1))** and **(HG1($n$+2))** on F and G imply that the right-hand side of the previous inequality is bounded above by

$$C\left[\sup_{t\in[0,T]} \|\rho(t)\|_{-(n+1+\alpha)} + \|b\| + \|\rho(T)\|_{-(n+2+\alpha)} + \|z_T\|_{n+2+\alpha}\right].$$

Estimate (3.27) on ρ then implies (since $\|\rho(T)\|_{-(n+2+\alpha)} \leqslant \|\rho(T)\|_{-(n+1+\alpha)}$):

$$\sup_{t\in[0,T]} \|z(t,\cdot)\|_{n+2+\alpha} + \sup_{t\neq t'} \frac{\|z(t',\cdot)-z(t,\cdot)\|_{n+2+\alpha}}{|t'-t|^{\frac{\alpha}{2}}}$$

$$\leqslant C\left[M + \sup_{t\in[0,T]} \|z(t,\cdot)\|_{n+1+\alpha}^{1/2}\left(\|\rho_0\|_{-(n+1+\alpha)}^{1/2} + \|c\|^{1/2}\right)\right].$$

Rearranging we obtain (3.17). Plugging this estimate into (3.27) and (3.28) then gives (3.18).

Step 4: *The case of general data*. We now prove that the results are also valid for general data. Let (m_k), (c_k), and (ρ_0^k) be smooth and converge to m, c, and ρ_0 in $\mathcal{C}^0([0,T],\mathcal{P}(\mathbb{T}^d))$, $\mathcal{C}^0([0,T],(\mathcal{C}^{n+\alpha})')$ and $(\mathcal{C}^{n+1+\alpha})'$ respectively. Let (z_k,ρ_k) be the solution to (3.15) associated with m_k, c_k, and ρ_0^k. Note that z_k and ρ_k satisfy inequalities of the form (3.17) and (3.18) with a right-hand side $C_n M_k$ bounded uniformly with respect to k.

For $k \neq k'$, the difference $(z_{k,k'},\rho_{k,k'}) := (z_k - z_{k'}, \rho_k - \rho_{k'})$ is a solution to (3.15) associated with

$$b_{k,k'}(t,x) := \frac{\delta F}{\delta m}(x,m_k(t))(\rho_{k'}(t)) - \frac{\delta F}{\delta m}(x,m_{k'}(t))(\rho_{k'}(t)),$$

$$c_{k,k'} := (m_{k'}-m_k)\Gamma(t,x)Dz_{k'} + c_k - c_{k'},$$

$$z_T^{k,k'}(x) := \frac{\delta G}{\delta m}(x,m_k(T))(\rho_{k'}(T)) - \frac{\delta G}{\delta m}(x,m_{k'}(T))(\rho_{k'}(T)),$$

and

$$\rho_0^{k,k'} := \rho_0^k - \rho_0^{k'}.$$

By (3.17), we have

$$\sup_{t \in [0,T]} \|z_{k,k'}(t,\cdot)\|_{n+2+\alpha} + \sup_{t \neq t'} \frac{\|z_{k,k'}(t',\cdot) - z_{k,k'})(t,\cdot)\|_{n+2+\alpha}}{|t'-t|^{\frac{\alpha}{2}}} \leqslant C_n M_{k,k'}$$

where $M_{k,k'}$ is given by

$$\|z_T^{k,k'}\|_{n+2+\alpha} + \|\rho_0^{k,k'}\|_{-(n+1+\alpha)} + \sup_{t \in [0,T]} \left(\|b_{k,k'}(t,\cdot)\|_{n+1+\alpha} + \|c_{k,k'}(t)\|_{-(n+\alpha)} \right).$$

So, by assumptions **(HF1($n+1$))** and **(HG1($n+2$))**, $M_{k,k'}$ is bounded above by

$$\mathrm{Lip}_{n+2}(\frac{\delta G}{\delta m}) \mathbf{d}_1(m_k(T), m_{k'}(T)) \|\rho_{k'}(T)\|_{-(n+2+\alpha)} + \|\rho_0^k - \rho_0^{k'}\|_{-(n+1+\alpha)}$$
$$+ \mathrm{Lip}_{n+1}(\frac{\delta F}{\delta m}) \sup_{t \in [0,T]} \mathbf{d}_1(m_k(t), m_{k'}(t)) \sup_{t \in [0,T]} \|\rho_{k'}(t)\|_{-(n+1+\alpha)}$$
$$+ \sup_{t \in [0,T]} \mathbf{d}_1(m_k(t), m_{k'}(t)) \|\Gamma\|_\infty \|Dz_k\|_\infty + \sup_{t \in [0,T]} \|c_k(t) - c_{k'}(t)\|_{-(n+\alpha)},$$

and thus by

$$C(1 + \sup_k M_k) \Big(\|\rho_0^k - \rho_0^{k'}\|_{-(n+1+\alpha)}$$
$$+ \sup_{t \in [0,T]} (\mathbf{d}_1(m_k(t), m_{k'}(t)) + \|c_k(t) - c_{k'}(t)\|_{-(n+\alpha)}) \Big).$$

Therefore (z_k) is a Cauchy sequence in $\mathcal{C}^{\alpha/2}([0,T], \mathcal{C}^{n+2+\alpha})$. Using in the same way estimate (3.18), we find that (ρ_n) is a Cauchy sequence in the space $\mathcal{C}^{\alpha/2}([0,T], (\mathcal{C}^{n+1+\alpha})')$. One then easily checks that the limit (z, ρ) of (z_k, ρ_k) solves system (3.15) associated with m, b, c, z_T, and ρ_0, which yields the existence of a solution. Uniqueness comes from the fact that system (3.15) is affine and that estimates (3.17) and (3.18) hold. $\qquad\square$

3.4 DIFFERENTIABILITY OF U WITH RESPECT TO THE MEASURE

In this section we show that the map U has a derivative with respect to m. To do so, we linearize the MFG system (3.2). Let us fix $(t_0, m_0) \in [0,T] \times \mathcal{P}(\mathbb{T}^d)$ and let (m, u) be the solution to the MFG system (3.2) with initial condition $m(t_0) = m_0$. Recall that, by definition, $U(t_0, x, m_0) = u(t_0, x)$.

For any μ_0 in a suitable space, we consider the solution (v, μ) to the *linearized system*

$$
\begin{cases}
-\partial_t v - \Delta v + D_p H(x, Du) \cdot Dv = \dfrac{\delta F}{\delta m}(x, m(t))(\mu(t)), \\[2mm]
\partial_t \mu - \Delta \mu - \operatorname{div}\big(\mu D_p H(x, Du)\big) - \operatorname{div}\big(m D_{pp}^2 H(x, Du)Dv\big) = 0, \quad (3.29) \\[2mm]
v(T, x) = \dfrac{\delta G}{\delta m}(x, m(T))(\mu(T)), \ \mu(t_0, \cdot) = \mu_0.
\end{cases}
$$

Our aim is to prove that U is of class $\mathcal{C}^1$ with respect to m with

$$
v(t_0, x) = \int_{\mathbb{T}^d} \frac{\delta U}{\delta m}(t_0, x, m_0, y)\mu_0(y)dy.
$$

Let us start by showing that the linearized system (3.29) has a solution and give estimates on this solution.

Proposition 3.4.1. *Assume that* **(HF1(n+1))** *and* **(HG1(n+2))** *hold for some $n \geqslant 0$. If $m_0 \in \mathcal{P}(\mathbb{T}^d)$ and $\mu_0 \in (\mathcal{C}^{n+1+\alpha})'$, there is a unique solution (v, μ) of (3.29) and this solution satisfies*

$$
\sup_{t \in [t_0, T]} \big\{ \|v(t, \cdot)\|_{n+2+\alpha} + \|\mu(t)\|_{-(n+1+\alpha)} \big\} \leqslant C\|\mu_0\|_{-(n+1+\alpha)},
$$

where the constant C depends on n, T, H, F, and G (but not on (t_0, m_0)).

Note that the map $\mu_0 \to (v, \mu)$ is linear and continuous from $(\mathcal{C}^{n+1+\alpha})'$ into $C^0([t_0, T], \mathcal{C}^{n+2+\alpha} \times (\mathcal{C}^{n+1+\alpha})')$.

Proof. It is a straightforward application of Lemma 3.3.1, with the choice $V(t, x) = D_p H(x, Du(t, x))$, $\Gamma(t, x) = D_{pp}^2 H(x, Du(t, x))$ and $z_T = b = c = 0$. Note that V belongs to $\mathcal{C}^0([0, T], \mathcal{C}^{n+1+\alpha})$ in view of Proposition 3.1.1. $\qquad\square$

Let us recall that we use for simplicity the abbreviated notation $\langle \cdot, \cdot \rangle_{n+1+\alpha}$ for $\langle \cdot, \cdot \rangle_{(\mathcal{C}^{n+1+\alpha})', \mathcal{C}^{n+1+\alpha}}$.

Corollary 3.4.2. *Under the assumptions of Proposition 3.4.1, there exists, for any (t_0, m_0), a $\mathcal{C}^{n+2+\alpha} \times \mathcal{C}^{n+1+\alpha}$ map $(x, y) \mapsto K(t_0, x, m_0, y)$ such that, for any $\mu_0 \in (\mathcal{C}^{n+1+\alpha})'$, the v component of the solution of (3.29) is given by*

$$
v(t_0, x) = \langle \mu_0, K(t_0, x, m_0, \cdot) \rangle_{n+1+\alpha}. \tag{3.30}
$$

Moreover

$$
\|K(t_0, \cdot, m_0, \cdot)\|_{(n+2+\alpha, n+1+\alpha)} \leqslant C_n,
$$

and K and its derivatives in (x, y) are continuous on $[0, T] \times \mathbb{T}^d \times \mathcal{P}(\mathbb{T}^d) \times \mathbb{T}^d$.

Proof. For $\ell \in \mathbb{N}^d$ with $|\ell| \leqslant n+1$ and $y \in \mathbb{T}^d$, let $(v^{(\ell)}(\cdot, \cdot, y), \mu^{(\ell)}(\cdot, \cdot, y))$ be the solution to (3.29) with initial condition $\mu_0 = D^\ell \delta_y$ (the ℓ-th derivative of the Dirac mass at y). Note that $\mu_0 \in (\mathcal{C}^{n+1+\alpha})'$. We set $K(t_0, x, m_0, y) := v^{(0)}(t_0, x, y)$.

Let us check that $\partial_{y_1} K(t_0, x, m_0, y) = -v^{(e_1)}(t_0, x, y)$, where $e_1 = (1, 0, \ldots, 0)$. Indeed, since $\epsilon^{-1}(\delta_{y+\epsilon e_1} - \delta_y)$ converges to $-D^{e_1} \delta_y$ in $(\mathcal{C}^{n+1+\alpha})'$ while, by linearity, the map $\epsilon^{-1}(K(\cdot, \cdot, m_0, y + \epsilon e_1) - K(\cdot, \cdot, m_0, y))$ is the first component of the solution of (3.29) with initial condition $\epsilon^{-1}(\delta_{y+\epsilon e_1} - \delta_y)$, this map converges to the first component of the solution with initial condition $-D^{e_1} \delta_y$, which is $-v^{(e_1)}(\cdot, \cdot, y)$. This proves our claim.

One can then check in the same way by induction that, for $|\ell| \leqslant n+1$,

$$D_y^\ell K(t_0, x, m_0, y) = (-1)^{|\ell|} v^{(\ell)}(t_0, x, y).$$

Finally, if $|\ell| \leqslant n+1$, Proposition 3.4.1 combined with the linearity of system (3.29) implies that

$$\left\| D_y^\ell K(t_0, \cdot, m_0, y) - D_y^\ell K(t_0, \cdot, m_0, y') \right\|_{n+2+\alpha}$$
$$\leqslant C \| D^\ell \delta_y - D^\ell \delta_{y'} \|_{-(n+1+\alpha)} \leqslant C \| \delta_y - \delta_{y'} \|_{-\alpha} \leqslant C |y - y'|^\alpha.$$

Therefore $K(t_0, \cdot, m_0, \cdot)$ belongs to $\mathcal{C}^{n+2+\alpha} \times \mathcal{C}^{n+1+\alpha}$. Continuity of K and its derivatives in (t_0, m_0) easily follows from the estimates on (u, m) and on (v, μ), which are independent of the initial measure m_0. $\square$

We now show that K is indeed the derivative of U with respect to m.

Proposition 3.4.3. *Assume that* **(HF1(n+1))** *and* **(HG1(n+2))** *hold for some $n \geqslant 0$. Fix $t_0 \in [0, T]$ and $m_0, \hat{m}_0 \in \mathcal{P}(\mathbb{T}^d)$. Let (u, m) and $(\hat{u}, \hat{m})$ be the solutions of the MFG system (3.2) starting from (t_0, m_0) and $(t_0, \hat{m}_0)$ respectively and let (v, μ) be the solution to (3.29) with initial condition $(t_0, \hat{m}_0 - m_0)$. Then*

$$\sup_{t \in [t_0, T]} \left\{ \|\hat{u}(t, \cdot) - u(t, \cdot) - v(t, \cdot)\|_{n+2+\alpha} + \|\hat{m}(t, \cdot) - m(t, \cdot) \right.$$
$$\left. - \mu(t, \cdot)\|_{-(n+1+\alpha)} \right\} \leqslant C \mathbf{d}_1^2(m_0, \hat{m}_0).$$

As a straightforward consequence, we obtain the differentiability of U with respect to the measure:

Corollary 3.4.4. *Under the assumption of Proposition 3.4.3, the map U is of class $\mathcal{C}^1$ (in the sense of Definition 2.2.1) with*

$$\frac{\delta U}{\delta m}(t_0, x, m_0, y) = K(t_0, x, m_0, y),$$

whose regularity is given by Corollary 3.4.2. Moreover,

$$\left\| U(t_0, \cdot, \hat{m}_0) - U(t_0, \cdot, m_0) - \int_{\mathbb{T}^d} \frac{\delta U}{\delta m}(t_0, \cdot, m_0, y) d(\hat{m}_0 - m_0)(y) \right\|_{n+2+\alpha}$$

$$\leqslant C\mathbf{d}_1^2(m_0, \hat{m}_0).$$

Remark 3.4.5. Let us recall that the derivative $\delta U/\delta m$ is defined up to an additive constant and that our normalization condition is

$$\int_{\mathbb{T}^d} \frac{\delta U}{\delta m}(t_0, x, m_0, y) dm_0(y) = 0.$$

Let us check that we have indeed

$$\int_{\mathbb{T}^d} K(t, x, m_0, y) dm_0(y) = 0. \tag{3.31}$$

For this let us choose $\mu_0 = m_0$ in (3.29). Since, by normalization condition, $\frac{\delta F}{\delta m}(t, m(t))(m(t)) = 0$, for any $t \in [0, T]$, and $\frac{\delta G}{\delta m}(t, m(T))(m(T)) = 0$, it is clear that the solution to (3.29) is just $(v, \mu) = (0, m)$. So, by (3.30), (3.31) holds.

Proof of Proposition 3.4.3. Let us set $z := \hat{u} - u - v$ and $\rho := \hat{m} - m - \mu$. The proof consists in estimating the pair (z, ρ), which satisfies

$$\begin{cases} -\partial_t z - \Delta z + D_p H(x, Du) \cdot Dz = \dfrac{\delta F}{\delta m}(x, m(t))(\rho(t)) + b(t, x), \\[2mm] \partial_t \rho - \Delta \rho - \text{div}\big(\rho D_p H(x, Du)\big) - \text{div}\big(m D_{pp}^2 H(x, Du)Dz\big) - \text{div}(c) = 0, \\[2mm] z(T, x) = \dfrac{\delta G}{\delta m}(x, m(T))(\rho(T)) + z_T(x), \ \rho(t_0, \cdot) = 0, \end{cases}$$

where

$$b(t, x) = A(t, x) + B(t, x),$$

with

$$A(t, x) = -\int_0^1 \big(D_p H(x, sD\hat{u} + (1-s)Du) - D_p H(x, Du)\big) \cdot D(\hat{u} - u) \, ds,$$

and

$$B(t,x) = \int_0^1 \int_{\mathbb{T}^d} \left(\frac{\delta F}{\delta m}(x, s\hat{m}(t) + (1-s)m(t), y) \right.$$

$$\left. - \frac{\delta F}{\delta m}(x, m(t), y) \right) d(\hat{m}(t) - m(t))(y)ds,$$

$$c(t) = (\hat{m} - m)(t)D_{pp}^2 H(\cdot, Du(t, \cdot))(D\hat{u} - Du)(t, \cdot)$$

$$+ \hat{m} \int_0^1 \left(D_{pp}^2 H(\cdot, sD\hat{u}(t, \cdot) + (1-s)Du(t, \cdot)) \right.$$

$$\left. - D_{pp}^2 H(\cdot, Du(t, \cdot)) \right)(D\hat{u} - Du)(t, \cdot)ds,$$

(note that $c(t)$ is a signed measure) and

$$z_T(x) = \int_0^1 \int_{\mathbb{T}^d} \left(\frac{\delta G}{\delta m}(x, s\hat{m}(T) + (1-s)m(T), y) \right.$$

$$\left. - \frac{\delta G}{\delta m}(x, m(T), y) \right) d(\hat{m}(T) - m(T))(y)ds.$$

We apply Lemma 3.3.1 to get

$$\sup_{t \in [t_0, T]} \|(z(t), \rho(t))\|_{\mathcal{C}^{n+2+\alpha} \times (\mathcal{C}^{n+1+\alpha})'}$$

$$\leqslant C \left[\|z_T\|_{n+2+\alpha} + \|\rho_0\|_{-(n+1+\alpha)} + \sup_{t \in [t_0, T]} (\|b(t)\|_{n+1+\alpha} + \|c(t)\|_{-(n+\alpha)}) \right].$$

It remains to estimate the various quantities in the right-hand side. We have

$$\sup_{t \in [t_0, T]} \|b(t, \cdot)\|_{n+1+\alpha} \leqslant \sup_{t \in [t_0, T]} \|A(t, \cdot)\|_{n+1+\alpha} + \sup_{t \in [t_0, T]} \|B(t, \cdot)\|_{n+1+\alpha},$$

where

$$\sup_{t \in [t_0, T]} \|A(t, \cdot)\|_{n+1+\alpha} \leqslant C \sup_{t \in [t_0, T]} \|(\hat{u} - u)(t, \cdot)\|_{n+2+\alpha}^2 \leqslant C\mathbf{d}_1^2(m_0, \hat{m}_0),$$

according to Proposition 3.2.1 (using also the bounds in Proposition 3.1.1). To estimate B and $\|z_T\|_{\mathcal{C}^{n+2+\alpha}}$, we argue in the same way:

$$\|z_T\|_{\mathcal{C}^{n+2+\alpha}} + \sup_{t \in [t_0, T]} \|B(t, \cdot)\|_{n+1+\alpha} \leqslant C\mathbf{d}_1^2(m_0, \hat{m}_0),$$

where we have used as above Proposition 3.2.1, now combined with assumptions (**HF1(n+1)**) and (**HG1(n+2)**), which imply (e.g., for F) that, for any

$m_1, m_2 \in \mathcal{P}(\mathbb{T}^d)$,

$$\left\| \int_{\mathbb{T}^d} \left(\frac{\delta F}{\delta m}(\cdot, m_1, y) - \frac{\delta F}{\delta m}(\cdot, m_2, y) \right) d(m_1 - m_2)(y) \right\|_{n+1+\alpha}$$

$$\leqslant \mathbf{d}_1(m_1, m_2) \left\| D_y \frac{\delta F}{\delta m}(\cdot, m_1, \cdot) - D_y \frac{\delta F}{\delta m}(\cdot, m_2, \cdot) \right\|_{\mathcal{C}^{n+1+\alpha} \times L^\infty}$$

$$\leqslant C \mathbf{d}_1^2(m_1, m_2).$$

Finally,

$$\sup_{t \in [t_0, T]} \|c(t)\|_{-(n+\alpha)} \leqslant \sup_{t \in [t_0, T]} \sup_{\|\xi\|_{n+\alpha} \leqslant 1} \langle c(t), \xi \rangle_{n+\alpha},$$

where, for $\|\xi\|_{n+\alpha} \leqslant 1$,

$$\langle c(t), \xi \rangle_{n+\alpha} = \Big\langle \xi, (\hat{m} - m) D_{pp}^2 H(\cdot, Du(t, \cdot))(D\hat{u} - Du)(t, \cdot)$$

$$+ \hat{m}(t) \int_0^1 \big(D_{pp}^2 H(\cdot, [sD\hat{u} + (1-s)Du](t, \cdot))$$

$$- D_{pp}^2 H(\cdot, Du(t, \cdot)) \big) (D\hat{u} - Du)(t, \cdot) ds \Big\rangle_{n+\alpha}$$

$$\leqslant C \Big(\|\xi\|_1 \|u - \hat{u}\|_2 \mathbf{d}_1(m_0, \hat{m}_0) + \|\xi\|_{L^\infty} \|u - \hat{u}\|_1^2 \Big).$$

So again by Proposition 3.2.1 we get

$$\sup_{t \in [t_0, T]} \|c(t)\|_{-(n+\alpha)} \leqslant C \mathbf{d}_1^2(m_0, \hat{m}_0).$$

This completes the proof. $\square$

3.5   PROOF OF THE SOLVABILITY OF THE FIRST-ORDER MASTER EQUATION

Proof of Theorem 2.4.2 (existence).   We check in a first step that the map U defined by (3.3) is a solution of the first-order master equation. Let us first assume that $m_0 \in \mathcal{C}^\infty(\mathbb{T}^d)$, with $m_0 > 0$. For $t_0 \in [0, T)$, let (u, m) be the classical solution of the MFG system (3.2) starting from m_0 at time t_0. Then

$$\frac{U(t_0 + h, x, m_0) - U(t_0, x, m_0)}{h} = \frac{U(t_0 + h, x, m_0) - U(t_0 + h, x, m(t_0 + h))}{h}$$

$$+ \frac{U(t_0 + h, x, m(t_0 + h)) - U(t_0, x, m_0)}{h}.$$

Let us set $m_s := (1-s)m(t_0) + sm(t_0 + h)$. Note from the equation satisfied by m and from the regularity of U, as given by Corollary 3.4.4, that

$$U\big(t_0 + h, x, m(t_0 + h)\big) - U\big(t_0 + h, x, m(t_0)\big)$$

$$= \int_0^1 \int_{\mathbb{T}^d} \frac{\delta U}{\delta m}(t_0 + h, x, m_s, y)(m(t_0 + h, y) - m(t_0, y))\, dy\, ds$$

$$= \int_0^1 \int_{\mathbb{T}^d} \int_{t_0}^{t_0+h} \left[\frac{\delta U}{\delta m}(t_0 + h, x, m_s, y) \right.$$

$$\times \left. \Big(\Delta m(t,y) + \operatorname{div}\big(m(t,y)D_p H(y, Du(t,y))\big)\Big) \right] dt\, dy\, ds$$

$$= \int_0^1 \int_{\mathbb{T}^d} \int_{t_0}^{t_0+h} \Delta_y \frac{\delta U}{\delta m}(t_0 + h, x, m_s, y)m(t,y)dt\, dy\, ds$$

$$- \int_0^1 \int_{\mathbb{T}^d} \int_{t_0}^{t_0+h} D_y \frac{\delta U}{\delta m}(t_0 + h, x, m_s, y) \cdot D_p H\big(y, Du(t,y)\big)m(t,y)\, dt\, dy\, ds.$$

Dividing by h and using the continuity of $D_m U$ and its smoothness with respect to the space variables, we obtain

$$\lim_{h \to 0} \frac{U(t_0 + h, x, m(t_0 + h)) - U(t_0 + h, x, m_0)}{h}$$

$$= \int_{\mathbb{T}^d} \Big(\operatorname{div}_y \left[D_m U \right](t_0, x, m_0, y)$$

$$- D_m U(t_0, x, m_0, y) \cdot D_p H\big(y, Du(t_0, y)\big)\Big)m_0(y)\, dy.$$

On the other hand, for $h > 0$,

$$U(t_0 + h, x, m(t_0 + h)) - U(t_0, x, m_0) = u(t_0 + h, x) - u(t_0, x)$$

$$= h\partial_t u(t_0, x) + o(h),$$

since u is smooth, so that

$$\lim_{h \to 0+} \frac{U(t_0 + h, x, m(t_0 + h)) - U(t_0, x, m_0)}{h} = \partial_t u(t_0, x).$$

Therefore $\partial_t U(t_0, x, m_0)$ exists and, using the equation satisfied by u, is equal to

$$\partial_t U(t_0, x, m_0)$$

$$= - \int_{\mathbb{T}^d} \text{div}_y \left[D_m U \right] (t_0, x, m_0, y) m_0(y) \, dy$$

$$+ \int_{\mathbb{T}^d} D_m U(t_0, x, m_0, y) \cdot D_p H \left(x, D_x U(t_0, y, m_0) \right) m_0(y) \, dy \tag{3.32}$$

$$- \Delta_x U(t_0, x, m_0) + H \left(x, D_x U(t_0, x, m_0) \right) - F(x, m_0).$$

This means that U has a continuous time derivative at any point (t_0, x, m_0) where $m_0 \in C^\infty(\mathbb{T}^d)$ with $m_0 > 0$ and satisfies (2.6) at such a point (when m_0 is regarded as an element of $\mathcal{P}(\mathbb{T}^d)$). By continuity of the right-hand side of (3.32), U has a time derivative everywhere and (2.6) holds at any point. $\qquad\square$

Next we turn to the uniqueness part of the theorem:

Proof of Theorem 2.4.2 (uniqueness). To prove uniqueness of the solution of the master equation, we explicitly show that the solutions of the MFG system (3.2) coincide with the characteristics of the master equation. Let V be another solution to the master equation. The main point is that, by definition of a solution, $D^2_{x,y} \frac{\delta V}{\delta m}$ is bounded, and therefore $D_x V$ is Lipschitz continuous with respect to the measure variable.

Let us fix (t_0, m_0) with $m_0 \in C^\infty(\mathbb{T}^d)$. In view of the Lipschitz continuity of $D_x V$, one can easily uniquely solve in $C^0([0, T], \mathcal{P}(\mathbb{T}^d))$ the Fokker–Planck equation:

$$\begin{cases} \partial_t \tilde{m} - \Delta \tilde{m} - \text{div} \left(\tilde{m} D_p H(x, D_x V(t, x, \tilde{m})) \right) = 0 & \text{in } [t_0, T] \times \mathbb{T}^d, \\ \tilde{m}(t_0) = m_0 & \text{in } \mathbb{T}^d. \end{cases}$$

Then let us set $\tilde{u}(t, x) = V(t, x, \tilde{m}(t))$. By the regularity properties of V, $\tilde{u}$ is at least of class $C^{1,2}$ with

$$\partial_t \tilde{u}(t, x)$$

$$= \partial_t V(t, x, \tilde{m}(t)) + \left\langle \frac{\delta V}{\delta m}(t, x, \tilde{m}(t), \cdot), \partial_t \tilde{m}(t) \right\rangle_{C^2, (C^2)'}$$

$$= \partial_t V(t, x, \tilde{m}(t)) + \left\langle \frac{\delta V}{\delta m}(t, x, \tilde{m}(t), \cdot), \Delta \tilde{m} \right.$$

$$\left. + \text{div}(\tilde{m} D_p H \left(x, D_x V(t, x, \tilde{m}) \right)) \right\rangle_{C^2, (C^2)'}$$

$$= \partial_t V(t, x, \tilde{m}(t)) + \int_{\mathbb{T}^d} \text{div}_y \left[D_m V \right] (t, x, \tilde{m}(t), y) d\tilde{m}(t)(y)$$

$$- \int_{\mathbb{T}^d} D_m V(t, x, \tilde{m}(t), y) \cdot D_p H(y, D_x V(t, y, \tilde{m})) d\tilde{m}(t)(y).$$

Recalling that V satisfies the master equation, we obtain

$$\partial_t \tilde{u}(t, x) = -\Delta_x V(t, x, \tilde{m}(t)) + H\big(x, D_x V(t, x, \tilde{m}(t))\big) - F(x, \tilde{m}(t))$$

$$= -\Delta \tilde{u}(t, x) + H(x, D\tilde{u}(t, x)) - F(x, \tilde{m}(t))$$

with terminal condition $\tilde{u}(T, x) = V(T, x, \tilde{m}(T)) = G(x, \tilde{m}(T))$. Therefore the pair $(\tilde{u}, \tilde{m})$ is a solution of the MFG system (3.2). As the solution of this system is unique, we get that $V(t_0, x, m_0) = U(t_0, x, m_0)$ if m_0 has a smooth density. The equality $V - U$ holds then everywhere by continuity of V and U. $\qquad\square$

3.6   LIPSCHITZ CONTINUITY OF $\frac{\delta U}{\delta m}$ WITH RESPECT TO m

We later need the Lipschitz continuity of the derivative of U with respect to the measure.

Proposition 3.6.1.   *Let us assume that* **(HF1($n+1$))** *and* **(HG1($n+2$))** *hold for some $n \geqslant 2$. Then*

$$\sup_{t \in [0,T]} \sup_{m_1 \neq m_2} (\mathbf{d}_1(m_1, m_2))^{-1} \left\| \frac{\delta U}{\delta m}(t, \cdot, m_1, \cdot) - \frac{\delta U}{\delta m}(t, \cdot, m_2, \cdot) \right\|_{(n+2+\alpha, n+\alpha)} \leqslant C,$$

where C depends on n, F, G, H, and T.

Proof.   Let us set $(z, \mu) := (v^1 - v^2, \mu^1 - \mu^2)$. We first write an equation for (z, μ). To avoid too heavy notation, we set $H_1'(t, x) = D_p H(x, Du^1(t, x))$, $H_1''(t, x) = D_{pp}^2 H(x, Du^1(t, x))$, $F_1'(x, \mu) = \int_{\mathbb{T}^d} \frac{\delta F}{\delta m}(x, m^1, y)\mu(y)\, dy$, etc... Then (z, μ) satisfies

$$\begin{cases} -\partial_t z - \Delta z + H_1' \cdot Dz = F_1'(\cdot, \mu) + b & \text{in } [t_0, T] \times \mathbb{T}^d, \\ \partial_t \mu - \Delta \mu - \text{div}(\mu H_1') - \text{div}(m^1 H_1'' Dz) - \text{div}(c) = 0 & \text{in } [t_0, T] \times \mathbb{T}^d, \\ z(T) = G_1'(\mu(T)) + z_T,\ \mu(t_0) = 0 & \text{in } \mathbb{T}^d, \end{cases}$$

where

$$b(t, x) := F_1'\big(x, \mu^2(t)\big) - F_2'\big(x, \mu^2(t)\big) - \big[(H_1' - H_2') \cdot Dv^2\big](t, x),$$

$$c(t, x) := \mu^2(t, x)(H_1' - H_2')(t, x) + \big[(m^1 H_1'' - m^2 H_2'')Dv^2\big](t, x),$$

$$z_T(x) := G_1'(\mu^2(T)) - G_2'(\mu^2(T)).$$

We apply Lemma 3.3.1 with $V = H_1'$ and $\Gamma = H_1''$: It says that, under assumptions **(HF1($n+1$))** and **(HG1($n+2$))**,

$$\sup_{t \in [t_0, T]} \|z(t, \cdot)\|_{n+2+\alpha} \leqslant C\Big[\|z_T\|_{n+2+\alpha} + \sup_{t \in [t_0, T]} \big(\|b(t, \cdot)\|_{n+1+\alpha} + \|c(t, \cdot)\|_{-(n+\alpha)}\big)\Big].$$

Let us estimate the various terms in the right-hand side:

$$\|z_T\|_{n+2+\alpha}$$

$$\leqslant \left\| \int_{\mathbb{T}^d} \left(\frac{\delta G}{\delta m}(0,\cdot,m^1(T),y) - \frac{\delta G}{\delta m}(0,\cdot,m^2(T),y) \right) \mu^2(T,y)dy \right\|_{n+2+\alpha}$$

$$\leqslant \left\| \frac{\delta G}{\delta m}(0,\cdot,m^1(T),\cdot) - \frac{\delta G}{\delta m}(0,\cdot,m^2(T),\cdot) \right\|_{(n+2+\alpha,n+1+\alpha)} \|\mu^2(T)\|_{-(n+1+\alpha)}$$

$$\leqslant C\mathbf{d}_1(m_0^1,m_0^2)\,\|\mu_0\|_{-(n+1+\alpha)},$$

where we have used Propositions 3.2.1 and 3.4.1 in the last inequality. Moreover, we have

$$\|b(t,\cdot)\|_{n+1+\alpha} \leqslant \left\| F_1'(\cdot,\mu^2(t)) - F_2'(\cdot,\mu^2(t)) \right\|_{n+1+\alpha}$$
$$+ \left\| (H_1' - H_2')(t,\cdot)Dv^2(t,\cdot) \right\|_{n+1+\alpha},$$

where the first term can be estimated as z_T:

$$\left\| F_1'(\cdot,\mu^2(t)) - F_2'(\cdot,\mu^2(t)) \right\|_{n+1+\alpha} \leqslant C\mathbf{d}_1(m_0^1,m_0^2)\|\mu_0\|_{-(n+1+\alpha)}.$$

The second one is bounded by

$$\left\| (H_1' - H_2')(t,\cdot)Dv^2(t,\cdot) \right\|_{n+1+\alpha}$$
$$= \left\| \left(D_pH(\cdot,Du^1(t,\cdot)) - D_pH(\cdot,Du^2(t,\cdot)) \right)Dv^2(t,\cdot) \right\|_{n+1+\alpha}$$
$$\leqslant \left\| (u^1 - u^2)(t,\cdot) \right\|_{n+2+\alpha} \|v^2(t,\cdot)\|_{n+2+\alpha}$$
$$\leqslant C\mathbf{d}_1(m_0^1,m_0^2)\,\|\mu_0\|_{-(n+1+\alpha)},$$

where the last inequality comes from Proposition 3.2.1 and Proposition 3.4.1 thanks to assumptions **(HF1($n+1$))** and **(HG1($n+2$))**. Finally, by a similar argument,

$$\|c(t)\|_{-(n+\alpha)} = \sup_{\|\phi\|_{n+\alpha}\leqslant 1} \left\langle \phi, \left[\mu^2(H_1' - H_2') \right. \right.$$

$$\left. \left. + \left((m^1 - m^2)H_1'' + m^2(H_1'' - H_2'') \right)Dv^2 \right] \right\rangle_{n+\alpha}$$

$$\leqslant \sup_{\|\phi\|_{n+\alpha}\leqslant 1} \left\| \phi(H_1' - H_2')(t,\cdot) \right\|_{n+\alpha} \|\mu^2(t,\cdot)\|_{-(n+\alpha)}$$

$$+ \mathbf{d}_1\left(m^1(t),m^2(t) \right) \sup_{\|\phi\|_1\leqslant 1} \left\| \phi(H_1''Dv^2)(t,\cdot) \right\|_1$$

$$+ \sup_{\|\phi\|_0\leqslant 1} \left\| \phi(H_1'' - H_2'')(t,\cdot)Dv^2(t,\cdot) \right\|_0$$

$$\leqslant C\big\|(u^1 - u^2)(t, \cdot)\big\|_{n+1+\alpha}\|\mu_0\|_{-(n+\alpha)}$$
$$+ C\mathbf{d}_1\big(m^1(t), m^2(t)\big)\|v^2(t, \cdot)\|_2 + C\big\|(u^1 - u^2)(t, \cdot)\big\|_1\|v^2(t, \cdot)\|_1$$
$$\leqslant C\mathbf{d}_1(m_0^1, m_0^2)\|\mu_0\|_{-(n+\alpha)}.$$

This shows that

$$\sup_{t \in [t_0, T]} \|z(t, \cdot)\|_{n+2+\alpha} \leqslant C\mathbf{d}_1(m_0^1, m_0^2)\|\mu_0\|_{-(n+\alpha)}.$$

As

$$z(t_0, x) = \int_{\mathbb{T}^d} \left(\frac{\delta U}{\delta m}(t_0, x, m_0^1, y) - \frac{\delta U}{\delta m}(t_0, x, m_0^2, y) \right) \mu_0(y)dy,$$

we have proved

$$\sup_{m_1 \neq m_2} (\mathbf{d}_1(m_1, m_2))^{-1} \left\| \frac{\delta U}{\delta m}(t_0, \cdot, m_1, \cdot) - \frac{\delta U}{\delta m}(t_0, \cdot, m_2, \cdot) \right\|_{(n+2+\alpha, n+\alpha)} \leqslant C.$$

$\square$

3.7 LINK WITH THE OPTIMAL CONTROL OF THE FOKKER–PLANCK EQUATION

We now explain that, when F and G derive from potentials $\mathcal{F}$ and $\mathcal{G}$, the space derivative $D_x U$ is nothing but the derivative with respect to the measure of the solution $\mathcal{U}$ of a Hamilton–Jacobi equation stated in the space of measures. The fact that the mean field game system can be viewed as a necessary condition for an optimal transport of the Kolmogorov equation goes back to Lasry and Lions [74]. As explained by Lions [76], one can also write the value function of this optimal control problem, which turns out to be a Hamilton–Jacobi equation in the space of measure. The (directional) derivative with respect to the measure of the value function is then (at least formally) the solution of the master equation. This is rigorously derived, for the short horizon and first-order (in space and measure) master equation, by Gangbo and Swiech [45]. We show here that this holds true for the master equation without common noise.

Let us assume that F and G derive from $\mathcal{C}^1$ potential maps $\mathcal{F} : \mathcal{P}(\mathbb{T}^d) \to \mathbb{R}$ and $\mathcal{G} : \mathcal{P}(\mathbb{T}^d) \to \mathbb{R}$:

$$F(x, m) = \frac{\delta \mathcal{F}}{\delta m}(x, m), \qquad G(x, m) = \frac{\delta \mathcal{G}}{\delta m}(x, m). \tag{3.33}$$

Note for later use that the monotonicity of F and G implies the convexity of $\mathcal{F}$ and $\mathcal{G}$.

Theorem 3.7.1. *Under the assumptions of Theorem 2.4.2, let U be the solution to the master equation (3.1) and suppose that (3.33) holds. Then the Hamilton–Jacobi equation*

$$
\begin{cases}
-\partial_t \mathcal{U}(t,m) + \displaystyle\int_{\mathbb{T}^d} H\left(y, D_m\mathcal{U}(t,m,y)\right) dm(y) \\
\quad - \displaystyle\int_{\mathbb{T}^d} \mathrm{div}\left[D_m\mathcal{U}\right](t,m,y) dm(y) = \mathcal{F}(m) \qquad \text{in } [0,T] \times \mathcal{P}(\mathbb{T}^d), \\
\mathcal{U}(T,m) = \mathcal{G}(m) \qquad \text{in } \mathcal{P}(\mathbb{T}^d)
\end{cases}
\tag{3.34}
$$

has a unique classical solution $\mathcal{U}$ and

$$
D_m\mathcal{U}(t,x,m) = D_x U(t,x,m) \qquad \forall (t,x,m) \in [0,T] \times \mathbb{T}^d \times \mathcal{P}(\mathbb{T}^d). \tag{3.35}
$$

We represent the solution $\mathcal{U}$ of (3.34) as the value function of an optimal control problem: for an initial condition $(t_0, m_0) \in [0,T] \times \mathcal{P}(\mathbb{T}^d)$, let

$$
\begin{aligned}
\mathcal{U}(t_0, m_0) := \inf_{(m,\alpha)} \int_{t_0}^{T} &\left[\int_{\mathbb{T}^d} H^*\left(x, \alpha(t,x)\right) m(t,dx) \right] dt \\
&+ \int_{t_0}^{T} \mathcal{F}(m(t)) dt + \mathcal{G}(m(T)),
\end{aligned}
\tag{3.36}
$$

(where H^* is the convex conjugate of H with respect to the second variable) under the constraint that $m \in C^0([0,T], \mathcal{P}(\mathbb{T}^d))$, α is a bounded and Borel measurable function from $[0,T] \times \mathbb{T}^d$ into $\mathbb{R}^d$, and the pair (m,α) satisfies in the sense of distributions:

$$
\partial_t m - \Delta m - \mathrm{div}(\alpha m) = 0 \text{ in } [0,T] \times \mathbb{T}^d, \qquad m(t_0) = m_0 \text{ in } \mathbb{T}^d. \tag{3.37}
$$

Of course, (3.37) is understood as the Fokker–Planck equation describing the flow of measures generated on the torus by the stochastic differential equation (SDE)

$$
dZ_t = -\alpha(t, Z_t) dt + dB_t, \quad t \in [0,T],
$$

which is known to be uniquely solvable in the weak sense. Notice that, throughout the section, we shall use, as in (3.36), the notation $m(t,dx)$ to denote the integral on the torus with respect to the (time-dependent) measure $m(t)$.

The following characterization of the optimal path for $\mathcal{U}$ is due to Lasry and Lions [74]:

Proposition 3.7.2. *For an initial position $(t_0, m_0) \in [0,T] \times \mathcal{P}(\mathbb{T}^d)$, let (u,m) be the solution of the MFG system (3.2). Then, the pair $(m,\alpha) = (m, D_p H(\cdot, Du(\cdot, \cdot)))$ is a minimizer for $\mathcal{U}(t_0, m_0)$.*

Proof. For a function $\hat{m} \in \mathcal{C}^0([0,T], \mathcal{P}(\mathbb{T}^d))$ and a bounded and measurable function $\hat{\alpha}$ from $[0,T] \times \mathbb{T}^d$ into $\mathbb{R}^d$, we let

$$J(\hat{m}, \hat{\alpha}) := \int_{t_0}^T \int_{\mathbb{T}^d} H^*(x, \hat{\alpha}(t,x)) d\hat{m}(t) + \int_{t_0}^T \mathcal{F}(\hat{m}(t)) dt + \mathcal{G}(\hat{m}(T)),$$

where $\hat{m}$ solves

$$\partial_t \hat{m} - \Delta \hat{m} - \operatorname{div}(\hat{\alpha}\hat{m}) = 0 \text{ in } [0,T] \times \mathbb{T}^d, \qquad \hat{m}(t_0) = m_0 \text{ in } \mathbb{T}^d.$$

Since, for any $m' \in \mathcal{P}(\mathbb{T}^d)$ and $\alpha' \in \mathbb{R}^d$,

$$H^*(x, \alpha') = \sup_{p \in \mathbb{R}^d} (\alpha' \cdot p - H(x,p)),$$

we have, by convexity of $\mathcal{F}$ and $\mathcal{G}$,

$$
\begin{aligned}
J(\hat{m}, \hat{\alpha}) \geqslant & \int_{t_0}^T \left[\int_{\mathbb{T}^d} \left[\hat{\alpha}(t,x) \cdot Du(t,x) - H(x, Du(t,x)) \right] \hat{m}(t, dx) \right] dt \\
& + \int_{t_0}^T \left[\mathcal{F}(m(t)) + F(\cdot, m(t))(\hat{m}(t) - m(t)) \right] dt \\
& + \mathcal{G}(m(T)) + G(\cdot, m(T))(\hat{m}(T) - m(T)) \\
= & \, J(m, \alpha) \\
& + \int_{t_0}^T \left[\int_{\mathbb{T}^d} \left[Du(t,x) \cdot (\hat{\alpha}(t,x)\hat{m}(t, dx) - \alpha(t,x)m(t, dx)) \right. \right. \\
& \left. \left. - H(x, Du(t,x))(\hat{m} - m)(t, dx) \right] \right] dt \\
& + \int_{t_0}^T F(\cdot, m(t))(\hat{m} - m)(t) dt + G(\cdot, m(T))(\hat{m}(T) - m(T)).
\end{aligned}
$$

because

$$\alpha(t,x) \cdot Du(t,x) - H(x, Du(t,x)) = H^*(x, \alpha(t,x)).$$

Using the equation satisfied by (m, w) and $(\hat{m}, \hat{w})$ we have

$$\int_{t_0}^{T} \left[\int_{\mathbb{T}^d} Du(t, x) \cdot \left(\hat{\alpha}(t, x) \hat{m}(t, dx) - \alpha(t, x) m(t, dx) \right) \right] dt$$

$$= - \left[\int_{\mathbb{T}^d} u(t, x) (\hat{m} - m)(t, dx) \right]_0^T$$

$$+ \int_0^T \left[\int_{\mathbb{T}^d} (\partial_t u + \Delta u)(t, x) (\hat{m} - m)(t, dx) \right] dt$$

$$= -G(\cdot, m(T)) \left(\hat{m}(T) - m(T) \right)$$

$$+ \int_0^T \left[\int_{\mathbb{T}^d} \left(H(x, Du(t, x)) - F(x, m(t)) \right) (\hat{m} - m)(t, dx) \right] dt.$$

This proves that $J(\hat{m}, \hat{\alpha}) \geq J(m, \alpha)$ and shows the optimality of (m, α).   □

Proof of Theorem 3.7.1.   *First step.* Let us first check that $\mathcal{U}$, defined by (3.36), is $\mathcal{C}^1$ with respect to m and satisfies

$$\frac{\delta \mathcal{U}}{\delta m}(t, x, m) = U(t, x, m) - \int_{\mathbb{T}^d} U(t, y, m) dm(y),$$

$$(t, x, m) \in [0, T] \times \mathbb{T}^d \times \mathcal{P}(\mathbb{T}^d). \tag{3.38}$$

Note that taking the derivative with respect to x on both sides shows (3.35).

We now prove (3.38). Let $m_0, \hat{m}_0$ be two initial measures and (u, m) and $(\hat{u}, \hat{m})$ the solutions of the MFG system (3.2) with initial conditions (t_0, m_0) and $(t_0, \hat{m}_0)$ respectively. Let also (v, μ) be the solution of the linearized system (3.29) with initial condition $(t_0, \hat{m}_0 - m_0)$. Let us recall that, according to Proposition 3.4.3, we have

$$\sup_{t \in [t_0, T]} \left\{ \| \hat{u} - u - v \|_{n+2+\alpha} + \| \hat{m} - m - \mu \|_{-(n+1+\alpha)} \right\} \leq C \mathbf{d}_1^2(m_0, \hat{m}_0) \tag{3.39}$$

while Propositions 3.2.1 and 3.4.1 imply that

$$\sup_{t \in [0, T]} \left\{ \| \hat{u} - u \|_{n+2+\alpha} + \| \mu \|_{-(n+1+\alpha)} \right\} \leq C \mathbf{d}_1(m_0, \hat{m}_0).$$

Our aim is to show that

$$\mathcal{U}(t_0, \hat{m}_0) - \mathcal{U}(t_0, m_0) - \int_{\mathbb{T}^d} U(t_0, x, m_0) d(\hat{m}_0 - m_0)(x)$$

$$= O\left(\mathbf{d}_1^2(m_0, \hat{m}_0) \right). \tag{3.40}$$

Indeed, if (3.40) holds true, then U is a derivative of $\mathcal{U}$ and, by convention (2.1), proves (3.38).

Second step. We now turn to the proof of (3.40). Since (u, m) and $(\hat{u}, \hat{m})$ are optimal in $\mathcal{U}(t_0, m_0)$ and $\mathcal{U}(t_0, \hat{m}_0)$ respectively, we have

$$\mathcal{U}(t_0, \hat{m}_0) - \mathcal{U}(t_0, m_0) = \int_{t_0}^T \left(\int_{\mathbb{T}^d} H^*(x, D_p H(x, D\hat{u}(t, x))) \hat{m}(t, dx) \right.$$

$$\left. - \int_{\mathbb{T}^d} H^*(x, D_p H(x, Du(t, x))) m(t, dx) \right) dt$$

$$+ \int_{t_0}^T \left(\mathcal{F}(\hat{m}(t)) - \mathcal{F}(m(t)) \right) dt + \mathcal{G}(\hat{m}(T)) - \mathcal{G}(m(T)).$$

Note that, by (3.39),

$$\int_{t_0}^T \left(\int_{\mathbb{T}^d} H^*\left(x, D_p H(x, D\hat{u}(t, x))\right) \hat{m}(t, dx) \right.$$

$$\left. - \int_{\mathbb{T}^d} H^*\left(x, D_p H(x, Du(t, x))\right) m(t, dx) \right) dt$$

$$= \int_{t_0}^T \left(\int_{\mathbb{T}^d} H^*\left(x, D_p H(x, Du(t, x))\right) \mu(t, dx) \right.$$

$$+ \int_{\mathbb{T}^d} D_q H^*\left(x, D_p H(x, Du(t, x))\right)$$

$$\cdot \left[D_{pp}^2 H(x, Du(t, x)) Dv(t, x) \right] m(t, dx) \bigg) dt + O(\mathbf{d}_1^2(m_0, \hat{m}_0))$$

$$= \int_{t_0}^T \left(\int_{\mathbb{T}^d} \left(Du(t, x) \cdot D_p H(x, Du(t, x)) - H(x, Du(t, x)) \right) \mu(t, dx) \right.$$

$$+ \int_{\mathbb{T}^d} Du(t, x) \cdot \left[D_{pp}^2 H(x, Du(t, x)) Dv(t, x) \right] m(t, dx) \bigg) dt + O(\mathbf{d}_1^2(m_0, \hat{m}_0)),$$

where we have used the properties of the Fenchel conjugate in the last equality, while

$$\int_{t_0}^T \left[\mathcal{F}(\hat{m}(t)) - \mathcal{F}(m(t)) \right] dt + \mathcal{G}(\hat{m}(T)) - \mathcal{G}(m(T))$$

$$= \int_{t_0}^T \left(\int_{\mathbb{T}^d} F(x, m(t)) \mu(t, dx) \right) dt + \int_{\mathbb{T}^d} G(x, m(T)) \mu(T, dx)$$

$$+ O(\mathbf{d}_1^2(m_0, \hat{m}_0)).$$

Recalling the equation satisfied by u and μ, we have

$$\frac{d}{dt}\int_{\mathbb{T}^d} u(t,x)\mu(t,dx) = \int_{\mathbb{T}^d}\Big[H\big(x,Du(t,x)\big) - F\big(x,m(t)\big)\Big]\mu(t,dx)$$

$$- \int_{\mathbb{T}^d} Du(t,x)\cdot D_pH\big(x,Du(t,x)\big)\mu(t,dx)$$

$$- \int_{\mathbb{T}^d} Du(t,x)\cdot\Big[D^2_{pp}H\big(x,Du(t,x)\big)Dv(t,x)\Big]m(t,dx).$$

Putting the last three identities together, we obtain

$$\mathcal{U}(t_0,\hat{m}_0) - \mathcal{U}(t_0,m_0)$$

$$= -\int_{t_0}^{T}\Big(\frac{d}{dt}\int_{\mathbb{T}^d} u(t,x)\mu(t,dx)\Big)dt + \int_{\mathbb{T}^d} G\big(x,m(T)\big)\mu(T,dx)$$

$$+ O\big(\mathbf{d}_1^2(m_0,\hat{m}_0)\big)$$

$$= \int_{\mathbb{T}^d} u(t_0,x)\mu(t_0,dx) + O\big(\mathbf{d}_1^2(m_0,\hat{m}_0)\big)$$

$$= \int_{\mathbb{T}^d} U(t_0,x,m_0)d(\hat{m}_0 - m_0)(x) + O\big(\mathbf{d}_1^2(m_0,\hat{m}_0)\big).$$

This completes the proof of (3.38).

Third step. Next we show that $\mathcal{U}$ is a classical solution to the Hamilton–Jacobi equation (3.34). Let us fix $(t_0,m_0) \in [0,T) \times \mathcal{P}(\mathbb{T}^d)$, where m_0 has a smooth, positive density. Let also (u,m) be the solution of the MFG system (3.2) with initial condition (t_0,m_0). Proposition 3.7.2 states that $(m,D_pH(\cdot,Du(\cdot,\cdot)))$ is a minimizer for $\mathcal{U}(t_0,m_0)$. By standard dynamic programming principle, we have therefore, for any $h \in (0,T-t_0)$,

$$\mathcal{U}(t_0,m_0) = \int_{t_0}^{t_0+h}\int_{\mathbb{T}^d} H^*\Big(x,D_pH\big(x,Du(t,x)\big)\Big)m(t,x)\,dx\,dt$$

$$+ \int_{t_0}^{t_0+h}\mathcal{F}\big(m(t)\big)dt + \mathcal{U}\big(t_0+h,m(t_0+h)\big).$$

(3.41)

Now we note that

$$\frac{\mathcal{U}(t_0+h,m_0) - \mathcal{U}(t_0,m_0)}{h} = \frac{\mathcal{U}(t_0+h,m_0) - \mathcal{U}(t_0+h,m(t_0+h))}{h}$$

$$+ \frac{\mathcal{U}(t_0+h,m(t_0+h)) - \mathcal{U}(t_0,m_0)}{h}.$$

(3.42)

We can handle the first term in the right-hand side of (3.42) by using the fact that $\mathcal{U}$ is $\mathcal{C}^1$ with respect to m. Letting $m_{s,h} := (1-s)m_0 + sm(t_0+h)$, we have

$$\mathcal{U}(t_0 + h, m(t_0 + h)) - \mathcal{U}(t_0 + h, m_0)$$

$$= \int_0^1 \int_{\mathbb{T}^d} \frac{\delta \mathcal{U}}{\delta m}(t_0 + h, m_{s,h}, y) d(m(t_0 + h) - m_0)(y) \, ds$$

$$= - \int_0^1 \int_{\mathbb{T}^d} \int_{t_0}^{t_0 + h} D_m \mathcal{U}(t_0 + h, m_{s,h}, y)$$

$$\cdot \Big(Dm(t,y) + D_p H(y, Du(t,y))m(t,y) \Big) \, dt \, dy \, ds.$$

Dividing by h, letting $h \to 0^+$ and rearranging gives

$$\lim_{h \to 0^+} \frac{\mathcal{U}(t_0 + h, m(t_0 + h)) - \mathcal{U}(t_0 + h, m_0)}{h}$$

$$= \int_{\mathbb{T}^d} \operatorname{div} [D_m \mathcal{U}] (t_0, m_0, y) dm_0(y)$$

$$- \int_{\mathbb{T}^d} D_m \mathcal{U}(t_0, m_0, y) \cdot D_p H(y, Du(t_0, y)) dm_0(y).$$

To handle the second term in the right-hand side of (3.42), we use (3.41) and get

$$\lim_{h \to 0^+} \frac{\mathcal{U}(t_0 + h, m(t_0 + h)) - \mathcal{U}(t_0, m_0)}{h}$$

$$= - \int_{\mathbb{T}^d} H^*(x, D_p H(x, Du(t_0, x))) dm_0(x) - \mathcal{F}(m_0).$$

As $Du(t_0, x) = D_x U(t_0, x, m_0) = D_m \mathcal{U}(t_0, m_0, x)$, we have

$$- H^*(x, D_p H(x, Du(t, x))) + D_m \mathcal{U}(t_0, m_0, x) \cdot D_p H(y, Du(t_0, y))$$

$$= -H^*(x, D_p H(x, D_m \mathcal{U}(t_0, m_0, x)))$$

$$+ D_m \mathcal{U}(t_0, m_0, x) \cdot D_p H(x, D_m \mathcal{U}(t_0, m_0, x))$$

$$= H(x, D_m \mathcal{U}(t_0, m_0, x)).$$

Collecting the above equalities, we therefore obtain

$$\lim_{h \to 0^+} \frac{\mathcal{U}(t_0 + h, m_0) - \mathcal{U}(t_0, m_0)}{h} = - \int_{\mathbb{T}^d} \operatorname{div} [D_m \mathcal{U}] (t_0, m_0, y) dm_0(y)$$

$$+ \int_{\mathbb{T}^d} H(x, D_m \mathcal{U}(t_0, m_0, x)) dm_0(x) - \mathcal{F}(m_0).$$

As the right-hand side of the above equality is continuous in all variables, this shows that $\mathcal{U}$ is continuously differentiable with respect to t and satisfies (3.34).

Last step. We finally check that $\mathcal{U}$ is the unique classical solution to (3.34). For this we use the standard comparison argument. Let $\mathcal{V}$ be another classical solution and assume that $\mathcal{V} \neq \mathcal{U}$. To fix the ideas, let us suppose that $\sup(\mathcal{V}-\mathcal{U})$ is positive. Then, for any $\epsilon > 0$ small enough,

$$\sup_{(t,m)\in(0,T]\times\mathcal{P}(\mathbb{T}^d)} \left[\mathcal{V}(t,m) - \mathcal{U}(t,m) + \epsilon \log(\frac{t}{T}) \right]$$

is positive. Let $(\hat{t},\hat{m})$ be a maximum point. Note that $\hat{t} < T$ because $\mathcal{V}(T,\cdot) = \mathcal{U}(T,\cdot)$. By optimality of $(\hat{t},\hat{m})$ and regularity of $\mathcal{V}$ and $\mathcal{U}$, we have

$$\partial_t \mathcal{V}(\hat{t},\hat{m}) - \partial_t \mathcal{U}(\hat{t},\hat{m}) + \frac{\epsilon}{\hat{t}} = 0 \qquad \text{and} \qquad \frac{\delta \mathcal{V}}{\delta m}(\hat{t},\hat{m},\cdot) = \frac{\delta \mathcal{U}}{\delta m}(\hat{t},\hat{m},\cdot),$$

so that

$$D_m \mathcal{V}(\hat{t},\hat{m},\cdot) = D_m \mathcal{U}(\hat{t},\hat{m},\cdot) \text{ and div} [D_m \mathcal{V}](\hat{t},\hat{m},\cdot) = \text{div} [D_m \mathcal{U}](\hat{t},\hat{m},\cdot).$$

Using the equation satisfied by $\mathcal{U}$ and $\mathcal{V}$ yields $\frac{\epsilon}{\hat{t}} = 0$, a contradiction.   □

Chapter Four

Mean Field Game System with a Common Noise

THE MAIN PURPOSE OF THIS CHAPTER and Chapter 5 is to show that the same approach as the one developed in Chapter 3 may be applied to the case when the whole system is forced by a so-called "common noise." Such a common noise is sometimes referred to as a "systemic noise," see, for instance, Lions' lectures at the *Collège de France*.

In terms of a game with a finite number of players, the common noise describes some noise that affects all the players in the same way, so that the dynamics of one given particle reads[1]

$$dX_t = -D_p H(X_t, Du_t(X_t)) \, dt + \sqrt{2} dB_t + \sqrt{2\beta} dW_t, \quad t \in [0, T], \qquad (4.1)$$

where β is a nonnegative parameter, B and W are two independent d-dimensional Wiener processes, B standing for the same idiosyncratic noise as in the previous chapter and W now standing for the so-called common noise. Throughout the chapter, we use the standard convention from the theory of stochastic processes that consists in indicating the time parameter as an index in random functions.

As we shall see next, the effect of the common noise is to randomize the mean field game (MFG) equilibria so that, with the same notations as earlier $(m_t)_{t \geq 0}$ becomes a random flow of measures. Precisely, it reads as the flow of conditional marginal measures of $(X_t)_{t \in [0,T]}$ given the realization of W. To distinguish things properly, we shall refer to the situation discussed in the previous chapter as the "deterministic" or "first-order" case. In this way, we point out that, without common noise, equilibria are completely deterministic. Compared to the notation of the introductory Chapter 1 or of Chapter 2, we let the level of common noise β be equal to 1 throughout the chapter: this is without loss of generality and simplifies (a little) the notation.

This chapter is specifically devoted to the analysis of the MFG system in the presence of the common noise (see (1.9)). Using a *continuation like* argument (instead of the classical strategy based on the Schauder fixed-point theorem), we investigate the existence and uniqueness of a solution. On the model of the first-

---

[1] Equation (4.1) is set on $\mathbb{R}^d$ but the solution may be canonically mapped onto $\mathbb{T}^d$ since the coefficients are periodic: when the process $(X_t)_{t \in [0,T]}$ is initialized with a probability measure on $\mathbb{T}^d$, the dynamics on the torus are independent of the representative in $\mathbb{R}^d$ of the initial condition.

order case, we also investigate the linearized system. The derivation of the master equation is deferred to the next chapter. The use of the continuation method in the analysis of MFG systems has been widely investigated by Gomes and his coauthors in the absence of common noise (see, for instance, [37, 46, 47] together with Chapter 11 in [48]), but it seems to be a new point in the presence of common noise. The analysis provided in the text that follows is directly inspired by the study of finite dimensional forward–backward systems: its application is here made possible thanks to the monotonicity assumption required on F and G.

As already mentioned, we assume without loss of generality that $\beta = 1$ throughout this chapter.

4.1   STOCHASTIC FOKKER–PLANCK/ HAMILTON–JACOBI–BELLMAN SYSTEM

The major difficulty in handling MFG with a common noise is that the system made of the Fokker–Planck and Hamilton–Jacobi–Bellman equations in (3.2) becomes stochastic. Its general form has already been discussed in [27]. Both the forward and the backward equations become stochastic as both the equilibrium $(m_t)_{0 \leqslant t \leqslant T}$ and the value function $(u_t)_{0 \leqslant t \leqslant T}$ depend on the realization of the common noise W. Unfortunately, the stochastic system does not consist of a simple randomization of the coefficients: to ensure that the value function u_t at time t depends only on the past before t in the realization of $(W_s)_{0 \leqslant s \leqslant T}$, the backward equation incorporates an additional correction term that is reminiscent of the theory of finite-dimensional backward stochastic differential equations.

The Fokker–Planck equation satisfied by $(m_t)_{t \in [0,T]}$ reads

$$d_t m_t = \left[2\Delta m_t + \mathrm{div}\big(m_t D_p H(m_t, Du_t)\big) \right] dt - \sqrt{2}\mathrm{div}(m_t dW_t),$$
$$t \in [0,T]. \tag{4.2}$$

The value function u is sought as the solution of the stochastic Hamilton–Jacobi–Bellman equation:

$$d_t u_t = \left\{ -2\Delta u_t + H(x, Du_t) - F(x, m_t) - \sqrt{2}\mathrm{div}(v_t) \right\} dt + v_t \cdot dW_t, \tag{4.3}$$

where, at any time t, v_t is a random function of x with values in $\mathbb{R}^d$. Once again, we emphasize that the term $v_t \cdot dW_t = \sum_{i=1}^{d} v_t^i dW_t^i$ allows us to guarantee that $(u_t)_{0 \leqslant t \leqslant T}$ is adapted with respect to the filtration generated by the common noise. The extra term $-\sqrt{2}\mathrm{div}(v_t)$ may be explained by the so-called Itô–Wentzell formula, which is the chain rule for random fields applied to random processes; see, for instance, [68]. It allows us to cancel out the bracket that arises in the application of the Itô–Wentzell formula[2] to $(u_t(X_t))_{t \in [0,T]}$,

---

[2] In the application of Itô–Wentzell formula, u_t is seen as a (random) periodic function from $\mathbb{R}^d$ to $\mathbb{R}$.

with $(X_t)_{0 \leqslant t \leqslant T}$ as in (4.1); see Subsection A.3.1 in the Appendix for a complete overview. Indeed, when expanding the infinitesimal variation of $(u_t(X_t))_{t \in [0,T]}$, the martingale term contained in u_t works with the martingale term contained in X and generates an additional bracket term. This additional bracket term is precisely $\sqrt{2}\mathrm{div}(v_t)(X_t)$; it thus cancels out with the term $-\sqrt{2}\mathrm{div}(v_t)(X_t)$ that appears in the dynamics of u_t. For the sake of completeness, we here provide a rough version of the computations that enter the definition of this additional bracket (we refer to the Appendix for a more complete account). When expanding the difference $u_{t+dt}(X_{t+dt}) - u_t(X_t)$, for $t \in [0,T]$ and an infinitesimal variation dt, the martingale structure in (4.3) brings about a term of the form $v_t(X_{t+dt}) \cdot (W_{t+dt} - W_t)$. By the standard Itô formula, it looks like

$$
v_t(X_{t+dt}) \cdot \left(W_{t+dt} - W_t\right) = \sum_{i=1}^{d} v_t^i(X_{t+dt})\left(W_{t+dt}^i - W_t^i\right)
$$

$$
= \sum_{i=1}^{d} v_t^i(X_t)dW_t^i + \sqrt{2}\sum_{i=1}^{d} \frac{\partial v_t^i}{\partial x_i}(X_t)\, dt, \qquad (4.4)
$$

the last term matching precisely the divergence term (up to the sign) that appears in (4.3).

As in the deterministic case, our aim is to define U by means of the same formula as in (3.3), that is, $U(0, x, m_0)$ is the value at point x of the value function taken at time 0 when the population is initialized with the distribution m_0.

To proceed, the idea is to reduce the equations by taking advantage of the additive structure of the common noise. The point is to make the (formal) change of variable

$$
\tilde{u}_t(x) := u_t(x + \sqrt{2}W_t), \quad \tilde{m}_t(x) := m_t(x + \sqrt{2}W_t), \quad x \in \mathbb{T}^d, \quad t \in [0, T].
$$

The second definition makes sense when m_t is a density, which is the case in the analysis because of the smoothing effect of the noise. A more rigorous way to define $\tilde{m}_t$ is to let it be the push-forward of m_t by the shift $\mathbb{T}^d \ni x \mapsto x - \sqrt{2}W_t \in \mathbb{T}^d$. Take note that such a definition is completely licit, as m_t reads as a conditional measure given the common noise. As the conditioning consists in freezing the common noise, the shift $x \mapsto x - \sqrt{2}W_t$ may be seen as a "deterministic" mapping.

As a result, $\tilde{m}_t$ is the conditional law of the process $(X_t - \sqrt{2}W_t)_{t \in [0,T]}$ given the common noise. Since

$$
d\left(X_t - \sqrt{2}W_t\right) = -D_pH\big(X_t - \sqrt{2}W_t
$$
$$
+ \sqrt{2}W_t, Du_t(X_t - \sqrt{2}W_t + \sqrt{2}W_t)\big)\, dt + \sqrt{2}dB_t,
$$

for $t \in [0, T]$, we get that $(\tilde{m}_t)_{t \in [0,T]}$ should satisfy

$$
d_t\tilde{m}_t = \left\{\Delta\tilde{m}_t + \mathrm{div}\big(\tilde{m}_t D_pH(\cdot + \sqrt{2}W_t, D\tilde{u}_t)\big)\right\} dt
$$
$$
= \left\{\Delta\tilde{m}_t + \mathrm{div}\big(\tilde{m}_t D_p\tilde{H}_t(\cdot, D\tilde{u}_t)\big)\right\} dt,
\qquad (4.5)
$$

where we have denoted $\tilde{H}_t(x, p) = H(x + \sqrt{2}W_t, p)$. This reads as the standard Fokker–Planck equation but in a random medium. Such a computation may be recovered by applying the Itô–Wentzell formula to $(m_t(x + \sqrt{2}W_t))_{t \in [0,T]}$, provided that each m_t is smooth enough in space. Quite remarkably, $(\tilde{m}_t)_{t \in [0,T]}$ is of absolutely continuous variation in time, which has a clear meaning when $(\tilde{m}_t)_{t \in [0,T]}$ is seen as a process with values in a set of smooth functions; when $(\tilde{m}_t)_{t \in [0,T]}$ is seen as a process with values in $\mathcal{P}(\mathbb{T}^d)$, the process $(\langle \varphi, \tilde{m}_t \rangle)_{t \in [0,T]}$ ($\langle \cdot, \cdot \rangle$ standing for the duality bracket) is indeed of absolutely continuous variation when φ is a smooth function on $\mathbb{T}^d$.

Similarly, we can apply (at least formally) Itô–Wentzell formula to $(u_t(x + \sqrt{2}W_t))_{t \in [0,T]}$ in order to express the dynamics of $(\tilde{u}_t)_{t \in [0,T]}$.

$$
\begin{aligned}
d_t \tilde{u}_t &= \left\{ -\Delta \tilde{u}_t + H\left(\cdot + \sqrt{2}W_t, D\tilde{u}_t \right) - F\left(\cdot + \sqrt{2}W_t, m_t \right) \right\} dt + \tilde{v}_t \cdot dW_t, \\
&= \left\{ -\Delta \tilde{u}_t + \tilde{H}_t(\cdot, D\tilde{u}_t) - \tilde{F}_t(\cdot, m_t) \right\} dt + \tilde{v}_t \cdot dW_t, \quad t \in [0, T],
\end{aligned}
\tag{4.6}
$$

where $\tilde{F}_t(x, m) = F(x + \sqrt{2}W_t, m)$; for a new representation term $\tilde{v}_t(x) = \sqrt{2}D_x \tilde{u}_t(x) + v_t(x + \sqrt{2}W_t)$, the boundary condition can be written $\tilde{u}_T(\cdot) = \tilde{G}(\cdot, m_T)$ with $\tilde{G}(x, m) = G(x + \sqrt{2}W_T, m)$. In such a way, we completely avoid any discussion about the smoothness of $\tilde{v}$. Take note that there is no way to get rid of the stochastic integral, as it permits us to ensure that $\tilde{u}_t$ remains adapted with respect to the observation up until time t.

Below, we shall investigate the system (4.5)–(4.6) directly. It is only in the next chapter (see Section 5.5) that we make the connection with the original formulation (4.2)–(4.3) and then complete the proof of Corollary 2.4.7. The reason is that it suffices to define the solution of the master equation by letting $U(0, x, m_0)$ be the value of $\tilde{u}_0(x)$ with m_0 as initial distribution. Notice indeed that $\tilde{u}_0(x)$ is expected to match $\tilde{u}_0(x) = u_0(x - \sqrt{2}W_0) = u_0(x)$. Of course, the same strategy may be applied at any time $t \in [0, T]$ by investigating $(\tilde{u}_s(x + \sqrt{2}(W_s - W_t)))_{s \in [t,T]}$.

With these notations, the monotonicity assumption takes the form:

Lemma 4.1.1. *Let m and m' be two elements of $\mathcal{P}(\mathbb{T}^d)$. For some $t \in [0, T]$ and for some realization of the noise, denote by $\tilde{m}$ and $\tilde{m}'$ the push-forwards of m and m' by the mapping $\mathbb{T}^d \ni x \mapsto x - \sqrt{2}W_t \in \mathbb{T}^d$. Then, for the given realization of $(W_s)_{s \in [0,T]}$,*

$$
\int_{\mathbb{T}^d} \left(\tilde{F}_t(x, m) - \tilde{F}_t(x, m') \right) d(\tilde{m} - \tilde{m}') \geqslant 0, \quad \int_{\mathbb{T}^d} \left(\tilde{G}(x, m) - \tilde{G}(x, m') \right) d(\tilde{m} - \tilde{m}') \geqslant 0.
$$

Proof. The proof consists of a straightforward change of variable.   □

Remark 4.1.2. *Below, we shall use quite systematically, without recalling it, the notation tilde ($\sim$) in order to denote the new coefficients and the new solutions after the random change of variable $x \mapsto x + \sqrt{2}W_t$.*

4.2 PROBABILISTIC SETUP

Throughout the chapter, we shall use the probability space $(\Omega, \mathcal{A}, \mathbb{P})$ equipped with two independent d-dimensional Brownian motions $(B_t)_{t \geqslant 0}$ and $(W_t)_{t \geqslant 0}$. The probability space is assumed to be complete. We then denote by $(\mathcal{F}_t)_{t \geqslant 0}$ the completion of the filtration generated by $(W_t)_{t \geqslant 0}$. When needed, we shall also use the filtration generated by $(B_t)_{t \geqslant 0}$.

Given an initial distribution $m_0 \in \mathcal{P}(\mathbb{T}^d)$, we consider the system

$$
\begin{aligned}
d_t \tilde{m}_t &= \left\{ \Delta \tilde{m}_t + \operatorname{div}\left(\tilde{m}_t D_p \tilde{H}_t(\cdot, D\tilde{u}_t) \right) \right\} dt, \\
d_t \tilde{u}_t &= \left\{ -\Delta \tilde{u}_t + \tilde{H}_t(\cdot, D\tilde{u}_t) - \tilde{F}_t(\cdot, m_t) \right\} dt + d\tilde{M}_t,
\end{aligned}
\tag{4.7}
$$

with the initial condition $\tilde{m}_0 = m_0$ and the terminal boundary condition $\tilde{u}_T = \tilde{G}(\cdot, m_T)$, with $\tilde{G}(x, m_T) = G(x + \sqrt{2}W_T, m_T)$.

The solution $(\tilde{u}_t)_{t \in [0,T]}$ is seen as an $(\mathcal{F}_t)_{t \in [0,T]}$-adapted process with paths in the space $\mathcal{C}^0([0,T], \mathcal{C}^n(\mathbb{T}^d))$, where n is a large enough integer (see the precise statements in the text that follows). The process $(\tilde{m}_t)_{t \in [0,T]}$ reads as an $(\mathcal{F}_t)_{t \in [0,T]}$-adapted process with paths in the space $\mathcal{C}^0([0,T], \mathcal{P}(\mathbb{T}^d))$, $\mathcal{P}(\mathbb{T}^d)$ being equipped with the Monge–Kantorovich distance $\mathbf{d}_1$. We shall look for solutions satisfying

$$
\sup_{t \in [0,T]} \left(\|\tilde{u}_t\|_{n+\alpha} \right) \in L^\infty(\Omega, \mathcal{A}, \mathbb{P}),
\tag{4.8}
$$

for some $\alpha \in (0,1)$.

The process $(\tilde{M}_t)_{t \in [0,T]}$ is seen as an $(\mathcal{F}_t)_{t \in [0,T]}$-adapted process with paths in the space $\mathcal{C}^0([0,T], \mathcal{C}^{n-2}(\mathbb{T}^d))$, such that, for any $x \in \mathbb{T}^d$, $(\tilde{M}_t(x))_{t \in [0,T]}$ is an $(\mathcal{F}_t)_{t \in [0,T]}$ martingale. It is required to satisfy

$$
\sup_{t \in [0,T]} \left(\|\tilde{M}_t\|_{n-2+\alpha} \right) \in L^\infty(\Omega, \mathcal{A}, \mathbb{P}).
\tag{4.9}
$$

Notice that, for our purpose, there is no need to discuss the representation of the martingale as a stochastic integral.

4.3 SOLVABILITY OF THE STOCHASTIC FOKKER–PLANCK/ HAMILTON–JACOBI–BELLMAN SYSTEM

The objective is to discuss the existence and uniqueness of a classical solution to the system (4.7) under the same assumptions as in the deterministic case. Theorem 4.3.1 covers Theorem 2.4.3 in Chapter 2:

Theorem 4.3.1. *Assume that F, G, and H satisfy* (2.4) *and* (2.5) *in Section 2.3. Assume moreover that, for some integer $n \geqslant 2$ and some[3] $\alpha \in [0,1)$,* **(HF1(n-1))** *and* **(HG1(n))** *hold true.*

Then, there exists a unique solution $(\tilde{m}_t, \tilde{u}_t, \tilde{M}_t)_{t \in [0,T]}$ to (4.7)*, with the prescribed initial condition $\tilde{m}_0 = m_0$, satisfying* (4.8) *and* (4.9)*. It satisfies* $\sup_{t \in [0,T]}(\|\tilde{u}_t\|_{n+\alpha} + \|\tilde{M}_t\|_{n+\alpha-2}) \in L^\infty(\Omega, \mathcal{A}, \mathbb{P})$.

Moreover, we can find a constant C such that, for any two initial conditions m_0 and m_0' in $\mathcal{P}(\mathbb{T}^d)$, we have

$$\sup_{t \in [0,T]}\left(\mathbf{d}_1^2(\tilde{m}_t, \tilde{m}_t') + \|\tilde{u}_t - \tilde{u}_t'\|_{n+\alpha}^2\right) \leqslant C\mathbf{d}_1^2(m_0, m_0') \qquad \mathbb{P} - a.e.,$$

where $(\tilde{m}, \tilde{u}, \tilde{M})$ and $(\tilde{m}', \tilde{u}', \tilde{M}')$ denote the solutions to (4.7) *with m_0 and m_0' as initial conditions.*

Theorem 4.3.1 is the analogue of Propositions 3.1.1 and 3.2.1 in the deterministic setting, except that we do not discuss the time regularity of the solutions (which, as well known in the theory of finite-dimensional backward stochastic differential equations, may be a rather difficult question).

The strategy of proof relies on the so-called continuation method. We emphasize that, differently from the standard argument that is used in the deterministic case, we will not make use of the Schauder theorem to establish the existence of a solution. The reason is that, in order to apply the Schauder theorem, we would need a compactness criterion on the space on which the equilibrium is defined, namely $L^\infty(\Omega, \mathcal{A}, \mathbb{P}; \mathcal{C}^0([0,T], \mathcal{P}(\mathbb{T}^d)))$. As already noticed in an earlier paper [29], this would ask for a careful (and certainly complicated) discussion on the choice of Ω and then on the behavior of the solution to (4.7) with respect to the topology put on Ω.

Here the idea is as follows. Given two parameters $(\vartheta, \varpi) \in [0,1]^2$, we shall first have a look at the parameterized system:

$$\begin{aligned} d_t\tilde{m}_t &= \left\{\Delta\tilde{m}_t + \text{div}\big[\tilde{m}_t\big(\vartheta D_p\tilde{H}_t(\cdot, D\tilde{u}_t) + b_t\big)\big]\right\}dt, \\ d_t\tilde{u}_t &= \left\{-\Delta\tilde{u}_t + \vartheta\tilde{H}_t(\cdot, D\tilde{u}_t) - \varpi\tilde{F}_t(\cdot, m_t) + f_t\right\}dt + d\tilde{M}_t, \end{aligned} \qquad (4.10)$$

with the initial condition $\tilde{m}_0 = m_0$ and the terminal boundary condition $\tilde{u}_T = \varpi\tilde{G}(\cdot, m_T) + g_T$, where $((b_t, f_t)_{t \in [0,T]}, g_T)$ is some input.

In (4.10), there are two extreme regimes: when $\vartheta = \varpi = 0$ and the input is arbitrary, the equation is known to be explicitly solvable; when $\vartheta = \varpi = 1$ and the input is set equal to 0, (4.10) fits the original one. This is our precise purpose, to prove first, by a standard contraction argument, that the equation is solvable when $\vartheta = 1$ and $\varpi = 0$ and then to propagate existence and uniqueness from

---

[3]In most of the analysis, α is assumed to be (strictly) positive, except in this statement, where it may be 0. Including the case $\alpha = 0$ allows for a larger range of application of the uniqueness property.

the case $(\vartheta, \varpi) = (1, 0)$ to the case $(\vartheta, \varpi) = (1, 1)$ by means of a continuation argument.

Throughout the analysis, the assumption of Theorem 4.3.1 is in force. Generally speaking, the inputs $(b_t)_{t \in [0,T]}$ and $(f_t)_{t \in [0,T]}$ are $(\mathcal{F}_t)_{t \in [0,T]}$ adapted processes with paths in the space $\mathcal{C}^0([0,T], [\mathcal{C}^1(\mathbb{T}^d)]^d)$ and $\mathcal{C}^0([0,T], \mathcal{C}^{n-1}(\mathbb{T}^d))$ respectively. Similarly, g_T is an $\mathcal{F}_T$-measurable random variable with realizations in $\mathcal{C}^{n+\alpha}(\mathbb{T}^d)$. We shall require that

$$\sup_{t \in [0,T]} \|b_t\|_1, \quad \sup_{t \in [0,T]} \|f_t\|_{n-1+\alpha}, \quad \|g_T\|_{n+\alpha}$$

are bounded (in $L^\infty(\Omega, \mathcal{A}, \mathbb{P})$).

It is worth mentioning that, whenever $\varphi : [0, T] \times \mathbb{T}^d \to \mathbb{R}$ is a continuous mapping such that $\varphi(t, \cdot) \in \mathcal{C}^\alpha(\mathbb{T}^d)$ for any $t \in [0, T]$, the mapping $[0, T] \ni t \mapsto \|\varphi(t, \cdot)\|_\alpha$ is lower semicontinuous and, thus, the mapping $[0, T] \ni t \mapsto \sup_{s \in [0,t]} \|\varphi(s, \cdot)\|_\alpha$ is left-continuous. In particular, whenever $(f_t)_{t \in [0,T]}$ is a process with paths in $\mathcal{C}^0([0,T], \mathcal{C}^k(\mathbb{T}^d))$, for some $k \geqslant 0$, the quantity $\sup_{t \in [0,T]} \|f_t\|_{k+\alpha}$ is a random variable, equal to $\sup_{t \in [0,T] \cap \mathbb{Q}} \|f_t\|_{k+\alpha}$, and the nondecreasing process $(\sup_{s \in [0,t]} \|f_s\|_{k+\alpha})_{t \in [0,T]}$ has continuous paths. As a byproduct,

$$\text{essup}_{\omega \in \Omega} \sup_{t \in [0,T]} \|f_t\|_{k+\alpha} = \sup_{t \in [0,T]} \text{essup}_{\omega \in \Omega} \|f_t\|_{k+\alpha}.$$

4.3.1   Case $\vartheta = \varpi = 0$

We start with the following simple lemma:

Lemma 4.3.2.   *Assume that $\vartheta = \varpi = 0$. Then, with the same type of inputs as above, (4.10) has a unique solution $(\tilde{m}_t, \tilde{u}_t, \tilde{M}_t)_{t \in [0,T]}$, with the prescribed initial condition. As required, it satisfies (4.8) and (4.9). Moreover, there exists a constant C, depending only on n and T, such that*

$$\text{essup}_{\omega \in \Omega} \sup_{t \in [0,T]} \|\tilde{u}_t\|_{n+\alpha} \leqslant C \big(\text{essup}_{\omega \in \Omega} \|g_T\|_{n+\alpha} + \text{essup}_{\omega \in \Omega} \sup_{t \in [0,T]} \|f_t\|_{n-1+\alpha} \big).$$
$$(4.11)$$

Proof of Lemma 4.3.2.   When $\vartheta = \varpi = 0$, the forward equation simply reads

$$d_t \tilde{m}_t = \big\{ \Delta \tilde{m}_t + \text{div}[\tilde{m}_t b_t] \big\} \, dt, \quad t \in [0, T]$$

with initial condition m_0. This is a standard Kolmogorov equation (with random coefficient) that is pathwise solvable. Its solution can be interpreted as the marginal law of the solution to the stochastic differential equation (compare with (4.1)): $dX_t = -b_t(X_t) \, dt + \sqrt{2} dB_t$, $t \in [0, T]$, X_0 having m_0 as statistical distribution. Recalling the Kantorovich–Rubinstein duality formula for $\mathbf{d}_1$ (see, for instance, Subsection A.1.1), we have

$$\mathrm{essup}_{\omega \in \Omega} \sup_{s \neq t} \frac{\mathbf{d}_1(\tilde{m}_s, \tilde{m}_t)}{|s - t|^{\frac{1}{2}}} \leqslant C\big(1 + \mathrm{essup}_{\omega \in \Omega} \|b\|_\infty\big),$$

for a constant C depending only on d and T. As $\vartheta = \varpi = 0$, the backward equation in (4.10) has the form

$$d_t \tilde{u}_t = \big\{ -\Delta \tilde{u}_t + f_t \big\} \, dt + d\tilde{M}_t, \quad t \in [0, T],$$

with the terminal boundary condition $\tilde{u}_T = g_T$. Although the equation is infinite-dimensional, it may be solved in a quite straightforward way. Taking the conditional expectation given $s \in [0, T]$ in the above equation, we indeed get that any solution should satisfy (provided we can exchange differentiation and conditional expectation)

$$d_t \mathbb{E}\big[\tilde{u}_t | \mathcal{F}_s\big] = \big\{ -\Delta \mathbb{E}\big[\tilde{u}_t | \mathcal{F}_s\big] + \mathbb{E}\big[f_t | \mathcal{F}_s\big] \big\} \, dt, \quad t \in [s, T],$$

which suggests letting as a candidate for a solution:

$$\begin{aligned}
\tilde{u}_s(x) &= \mathbb{E}\big[\bar{u}_s(x) | \mathcal{F}_s\big], \\
\bar{u}_s(x) &= P_{T-s} g_T(x) - \int_s^T P_{t-s} f_t(x) \, dt, \quad s \in [0, T], \ x \in \mathbb{T}^d,
\end{aligned} \tag{4.12}$$

where P denotes the heat semigroup (but associated with the Laplace operator Δ instead of $(1/2)\Delta$). For any $s \in [0, T]$ and $x \in \mathbb{T}^d$, the conditional expectation is uniquely defined up to a negligible event under $\mathbb{P}$. We claim that, for any $s \in [0, T]$, we can find a version of the conditional expectation in such a way that the process $[0, T] \ni s \mapsto (\mathbb{T}^d \ni x \mapsto \tilde{u}_s(x))$ reads as a progressively measurable random variable with values in $\mathcal{C}^0([0, T], \mathcal{C}^0(\mathbb{T}^d))$. By the representation formula (4.12), we indeed have that, $\mathbb{P}$ almost surely, $\bar{u}$ is jointly continuous in time and space. Making use of Lemma 4.3.4, we deduce that the realizations of $[0, T] \ni s \mapsto (\mathbb{T}^d \ni x \mapsto \tilde{u}_s(x))$ belong to $\mathcal{C}^0([0, T], \mathcal{C}^0(\mathbb{T}^d))$, the mapping $[0, T] \times \Omega \ni (s, \omega) \mapsto (\mathbb{T}^d \ni x \mapsto (\tilde{u}_s(\omega))(x))$ being measurable with respect to the progressive σ-field

$$\mathscr{P} = \big\{ A \in \mathcal{B}([0, T]) \otimes \mathcal{A} : \ \forall t \in [0, T], \ A \cap ([0, t] \times \Omega) \in \mathcal{B}([0, t]) \otimes \mathcal{F}_t \big\}. \tag{4.13}$$

By the maximum principle, we can find a constant C, depending only on T and d, such that

$$\begin{aligned}
\mathrm{essup}_{\omega \in \Omega} \sup_{s \in [0, T]} \|\tilde{u}_s\|_0 &\leqslant \mathrm{essup}_{\omega \in \Omega} \sup_{s \in [0, T]} \|\bar{u}_s\|_0 \\
&\leqslant C\big(\mathrm{essup}_{\omega \in \Omega} \|g_T\|_0 + \mathrm{essup}_{\omega \in \Omega} \sup_{0 \leqslant s \leqslant T} \|f_s\|_0\big).
\end{aligned}$$

More generally, taking the representation formula (4.12) at two different $x, x' \in \mathbb{T}^d$ and then taking the difference, we get

$$\operatorname{essup}_{\omega \in \Omega} \sup_{s \in [0,T]} \|\tilde{u}_s\|_\alpha \leqslant C\big(\operatorname{essup}_{\omega \in \Omega} \|g_T\|_\alpha + \operatorname{essup}_{\omega \in \Omega} \sup_{s \in [0,T]} \|f_s\|_\alpha\big).$$

We now proceed with the derivatives of higher order. Generally speaking, there are two ways to differentiate the representation formula (4.12). The first one is to say that, for any $k \in \{1, \ldots, n-1\}$,

$$D_x^k \bar{u}_s(x) = P_{T-s}\big(D^k g_T\big)(x) - \int_s^T P_{t-s}\big(D_x^k f_t\big)(x)\, dt,$$
$$(s, x) \in [0, T] \times \mathbb{T}^d, \tag{4.14}$$

which may be established by a standard induction argument. The second way is to make use of the regularization property of the heat kernel in order to go one step further, namely, for any $k \in \{1, \ldots, n\}$,

$$D_x^k \bar{u}_s(x) = P_{T-s}\big(D^k g_T\big)(x) - \int_s^T DP_{t-s}\big(D_x^{k-1} f_t\big)(x)\, dt,$$
$$= P_{T-s}\big(D^k g_T\big)(x) - \int_0^{T-s} DP_t\big(D_x^{k-1} f_{t+s}\big)(x)\, dt, \tag{4.15}$$

for $(s, x) \in [0, T] \times \mathbb{T}^d$, where DP_{t-s} stands for the derivative of the heat semigroup. Equation (4.15) is easily derived from (4.14). It permits handling the fact that f is $(n-1)$-times differentiable only.

Recalling that $|DP_t \varphi| \leqslant ct^{-1/2} \|\varphi\|_\infty$ for any bounded Borel function $\varphi : \mathbb{T}^d \to \mathbb{R}$ and for some $c \geqslant 1$ independent of φ and of $t \in [0, T]$, we deduce that, for any $k \in \{1, \ldots, n\}$, the mapping $[0, T] \times \mathbb{T}^d \ni (s, x) \mapsto D_x^k \bar{u}_s(x)$ is continuous. Moreover, we can find a constant C such that, for any $s \in [0, T]$,

$$\operatorname{essup}_{\omega \in \Omega} \|\bar{u}_s\|_{k+\alpha} \leqslant \operatorname{essup}_{\omega \in \Omega} \|g_T\|_{k+\alpha}$$
$$+ C \int_s^T \frac{1}{\sqrt{t-s}} \operatorname{essup}_{\omega \in \Omega} \|f_t\|_{k+\alpha-1}\, dt. \tag{4.16}$$

In particular, invoking once again Lemma 4.3.4, we can find a version of the conditional expectation in the representation formula $\tilde{u}_s(x) = \mathbb{E}[\bar{u}_s(x)|\mathcal{F}_s]$ such that $\tilde{u}$ has paths in $\mathcal{C}^0([0, T], \mathcal{C}^n(\mathbb{T}^d))$. For any $k \in \{1, \ldots, n\}$, $D_x^k \tilde{u}$ is progressively measurable and, for all $(s, x) \in [0, T] \times \mathbb{T}^d$, it holds that $D_x^k \tilde{u}_s(x) = \mathbb{E}[D_x^k \bar{u}_s(x)|\mathcal{F}_s]$.

Using (4.16), we have, for any $k \in \{1, \ldots, n\}$,

$$\operatorname{essup}_{\omega \in \Omega} \sup_{s \in [0,T]} \|\tilde{u}_s\|_{k+\alpha} \leqslant C\big(\operatorname{essup}_{\omega \in \Omega} \|g_T\|_{k+\alpha} + \operatorname{essup}_{\omega \in \Omega} \sup_{s \in [0,T]} \|f_s\|_{k+\alpha-1}\big).$$

Now that $\tilde{u}$ has been constructed, it remains to reconstruct the martingale part $(\tilde{M}_t)_{0 \leqslant t \leqslant T}$ in the backward equation of the system (4.10) (with $\vartheta = \varpi = 0$ therein). Since $\tilde{u}$ has trajectories in $\mathcal{C}^0([0,T], \mathcal{C}^{n+\alpha}(\mathbb{T}^d))$, $n \geqslant 2$, we can let

$$\tilde{M}_t(x) = \tilde{u}_t(x) - \tilde{u}_0(x) + \int_0^t \Delta \tilde{u}_s(x)\, ds - \int_0^t f_s(x)\, ds, \quad t \in [0,T], \ x \in \mathbb{T}^d.$$

It is then clear that $\tilde{M}$ has trajectories in $\mathcal{C}^0([0,T], \mathcal{C}^{n-2}(\mathbb{T}^d))$ and that

$$\mathrm{essup}_{\omega \in \Omega} \sup_{t \in [0,T]} \|\tilde{M}_t\|_{n+\alpha-2} < \infty.$$

It thus remains to prove that, for each $x \in \mathbb{T}^d$, the process $(\tilde{M}_t(x))_{0 \leqslant t \leqslant T}$ is a martingale (starting from 0). Clearly, it has continuous and $(\mathcal{F}_t)_{0 \leqslant t \leqslant T}$-adapted paths. Moreover,

$$\tilde{M}_T(x) - \tilde{M}_t(x) = g_T(x) - \tilde{u}_t(x) + \int_t^T \Delta \tilde{u}_s(x)\, ds$$

$$- \int_t^T f_s(x)\, ds, \quad t \in [0,T], \ x \in \mathbb{T}^d.$$

Now, recalling the relationship $\mathbb{E}[\Delta \tilde{u}_s(x)|\mathcal{F}_t] = \mathbb{E}[\Delta \bar{u}_s(x)|\mathcal{F}_t]$, we get

$$\mathbb{E}\left[\int_t^T \Delta \tilde{u}_s(x)\, ds \Big| \mathcal{F}_t\right] = \mathbb{E}\left[\int_t^T \Delta \bar{u}_s(x)\, ds \Big| \mathcal{F}_t\right].$$

Taking the conditional expectation given $\mathcal{F}_t$, we deduce that

$$\mathbb{E}\big[\tilde{M}_T(x) - \tilde{M}_t(x)|\mathcal{F}_t\big] = \mathbb{E}\left[g_T(x) - \bar{u}_t(x) - \int_t^T f_s(x)\, ds\right.$$

$$\left. + \int_t^T \Delta \bar{u}_s(x)\, ds \Big| \mathcal{F}_t\right] = 0,$$

the second equality following from (4.12). This shows that $\tilde{M}_t(x) = \mathbb{E}[\tilde{M}_T(x)|\mathcal{F}_t]$, so that the process $(\tilde{M}_t(x))_{0 \leqslant t \leqslant T}$ is a martingale, as required. $\square$

Remark 4.3.3. *Notice that, alternatively to (4.11), we also have, by Doob's inequality,*

$$\mathbb{E}\big[\sup_{t \in [0,T]} \|\tilde{u}_t\|_{n+\alpha}^2\big] \leqslant C\mathbb{E}\big[\|g_T\|_{n+\alpha}^2 + \sup_{t \in [0,T]} \|f_t\|_{n+\alpha-1}^2\big]. \tag{4.17}$$

Lemma 4.3.4. *Consider a random field $\mathcal{U} : [0,T] \times \mathbb{T}^d \to \mathbb{R}$, with continuous paths (in the variable $(t,x) \in [0,T] \times \mathbb{T}^d$), such that*

$$\mathrm{essup}_{\omega \in \Omega} \|\mathcal{U}\|_0 < \infty.$$

Then, we can find a version of the random field $[0,T] \times \mathbb{T}^d \ni (t,x) \mapsto \mathbb{E}[\mathcal{U}(t,x)|\mathcal{F}_t]$ such that $[0,T] \ni t \mapsto (\mathbb{T}^d \ni x \mapsto \mathbb{E}[\mathcal{U}(t,x)|\mathcal{F}_t])$ is a progressively measurable random variable with values in $\mathcal{C}^0([0,T], \mathcal{C}^0(\mathbb{T}^d))$, the progressive σ-field $\mathscr{P}$ being defined in (4.13).

More generally, if, for some $k \geqslant 1$, the paths of $\mathcal{U}$ are k-times differentiable in the space variable, the derivatives up to the order k having jointly continuous (in (t,x)) paths and satisfying

$$\operatorname{essup}_{\omega \in \Omega} \sup_{t \in [0,T]} \|\mathcal{U}(t,\cdot)\|_k < \infty,$$

then we can find a version of the random field $[0,T] \times \mathbb{T}^d \ni (t,x) \mapsto \mathbb{E}[\mathcal{U}(t,x)|\mathcal{F}_t]$ that is progressively measurable and that has paths in $\mathcal{C}^0([0,T], \mathcal{C}^k(\mathbb{T}^d))$, the derivative of order i written $[0,T] \times \mathbb{T}^d \ni (t,x) \mapsto \mathbb{E}[D_x^i \mathcal{U}(t,x)|\mathcal{F}_t]$.

Proof. *First step.* We first prove the first part of the statement (existence of a progressively measurable version with continuous paths). Existence of a differentiable version will be handled next. A key fact in the proof is that, the filtration $(\mathcal{F}_t)_{t \in [0,T]}$ being generated by $(W_t)_{t \in [0,T]}$, any martingale with respect to $(\mathcal{F}_t)_{t \in [0,T]}$ admits a continuous version.

Throughout the proof, we denote by w the (pathwise) modulus of continuity of $\mathcal{U}$ on the compact set $[0,T] \times \mathbb{T}^d$, namely

$$w(\delta) = \sup_{x,y \in \mathbb{T}^d : |x-y| \leqslant \delta} \sup_{s,t \in [0,T] : |t-s| \leqslant \delta} |\mathcal{U}(s,x) - \mathcal{U}(t,y)|, \quad \delta > 0.$$

Since $\operatorname{essup}_{\omega \in \Omega} \|\mathcal{U}\|_0 < \infty$, we have, for any $\delta > 0$,

$$\operatorname{essup}_{\omega \in \Omega} w(\delta) < \infty.$$

By Doob's inequality, we have that, for any integer $p \geqslant 1$,

$$\forall \varepsilon > 0, \quad \mathbb{P}\left(\sup_{s \in [0,T]} \mathbb{E}\left[w\left(\frac{1}{p}\right) \Big| \mathcal{F}_s \right] \geqslant \varepsilon \right) \leqslant \varepsilon^{-1} \mathbb{E}\left[w\left(\frac{1}{p}\right) \right],$$

the right-hand side converging to 0 as p tends to ∞, thanks to Lebesgue's dominated convergence theorem. Therefore, by a standard application of the Borel–Cantelli lemma, we can find an increasing sequence of integers $(a_p)_{p \geqslant 1}$ such that the sequence $(\sup_{s \in [0,T]} \mathbb{E}[w(1/a_p)|\mathcal{F}_s])_{p \geqslant 1}$ converges to 0 with probability 1.

We now come back to the original problem. For any $(t,x) \in [0,T] \times \mathbb{T}^d$, we let

$$\mathcal{V}(t,x) = \mathbb{E}[\mathcal{U}(t,x)|\mathcal{F}_t].$$

The difficulty comes from the fact that each $\mathcal{V}(t,x)$ is uniquely defined up to a negligible set. The objective is thus to choose each of these negligible sets in a relevant way.

Denoting by $\mathcal{T}$ a dense countable subset of $[0,T]$ and by $\mathcal{X}$ a dense countable subset of $\mathbb{T}^d$, we can find a negligible event $N \subset \mathcal{A}$ such that, outside N, the process $[0,T] \ni s \mapsto \mathbb{E}[\mathcal{U}(t,x)|\mathcal{F}_s]$ has a continuous version for any $t \in \mathcal{T}$ and $x \in \mathcal{X}$. Modifying the set N if necessary, we have, outside N, for any integer $p \geqslant 1$, any $t,t' \in \mathcal{T}$ and $x,x' \in \mathcal{X}$, with $|t - t'| + |x - x'| \leqslant 1/a_p$,

$$\sup_{s \in [0,T]} \left| \mathbb{E}[\mathcal{U}(t,x)|\mathcal{F}_s] - \mathbb{E}[\mathcal{U}(t',x')|\mathcal{F}_s] \right| \leqslant \sup_{s \in [0,T]} \mathbb{E}\big[w(\frac{1}{a_p})|\mathcal{F}_s\big],$$

the right-hand side converging to 0 as p tends to ∞. Therefore, by a uniform continuity extension argument, it is thus possible to extend continuously, outside N, the mapping $\mathcal{T} \times \mathcal{X} \ni (t,x) \mapsto ([0,T] \ni s \mapsto \mathbb{E}[\mathcal{U}(t,x)|\mathcal{F}_s]) \in \mathcal{C}^0([0,T],\mathbb{R})$ to the entire $[0,T] \times \mathbb{T}^d$. For any $(t,x) \in [0,T] \times \mathbb{T}^d$, the value of the extension is a version of the conditional expectation $\mathbb{E}[\mathcal{U}(t,x)|\mathcal{F}_s]$. Outside N, the slice $(s,x) \mapsto \mathbb{E}[\mathcal{U}(s,x)|\mathcal{F}_s]$ is obviously continuous. Moreover, it satisfies, for all $p \geqslant 1$,

$$\forall x,x' \in \mathbb{T}^d, \quad |x - x'| \leqslant \frac{1}{a_p} \Rightarrow$$

$$\sup_{s \in [0,T]} \left| \mathbb{E}[\mathcal{U}(s,x)|\mathcal{F}_s] - \mathbb{E}[\mathcal{U}(s,x')|\mathcal{F}_s] \right| \leqslant \sup_{s \in [0,T]} \mathbb{E}\big[w(\frac{1}{a_p})|\mathcal{F}_s\big],$$

which says that, for each realization outside N, the functions $(\mathbb{T}^d \ni x \mapsto \mathbb{E}[\mathcal{U}(s,x)|\mathcal{F}_s])_{s \in [0,T]}$ are equicontinuous. Together with the continuity in s, we deduce that, outside N, the function $[0,T] \ni s \mapsto (\mathbb{T}^d \ni x \mapsto \mathbb{E}[\mathcal{U}(s,x)|\mathcal{F}_s]) \in \mathcal{C}^0(\mathbb{T}^d)$ is continuous. On N, we can arbitrarily let $\mathcal{V} \equiv 0$, which is licit since N has 0 probability. Progressive measurability is then easily checked (the fact that $\mathcal{V}$ is arbitrarily defined on N does not matter because the filtration is complete).

Second step. We now handle the second part of the statement (existence of a $\mathcal{C}^k$ version). By a straightforward induction argument, it suffices to treat the case $k = 1$. By the first step, we already know that the random field $[0,T] \times \mathbb{T}^d \ni (t,x) \mapsto \mathbb{E}[D_x \mathcal{U}(t,x)|\mathcal{F}_t]$ has a continuous version. In particular, for any unit vector $e \in \mathbb{R}^d$, it makes sense to consider the mapping

$$\mathbb{T}^d \times \mathbb{R}^* \ni (x,h) \mapsto \frac{1}{h}\Big(\mathbb{E}[\mathcal{U}(t,x+he)|\mathcal{F}_t] - \mathbb{E}[\mathcal{U}(t,x)|\mathcal{F}_t]\Big) - \mathbb{E}\big[\langle D_x\mathcal{U}(t,x),e\rangle|\mathcal{F}_t\big].$$

Notice that we can find an event of probability 1, on which, for all $t \in [0,T]$, $x \in \mathbb{T}^d$ and $h \in \mathbb{R}^d$,

$$\left| \frac{1}{h}\Big(\mathbb{E}[\mathcal{U}(t,x+he)|\mathcal{F}_t] - \mathbb{E}[\mathcal{U}(t,x)|\mathcal{F}_t]\Big) - \mathbb{E}\big[\langle D_x\mathcal{U}(t,x),e\rangle|\mathcal{F}_t\big] \right|$$

$$= \left| \mathbb{E}\bigg[\int_0^1 \Big\langle D_x\mathcal{U}(t,x+\lambda he) - D_x\mathcal{U}(t,x), e \Big\rangle d\lambda | \mathcal{F}_t \bigg] \right| \qquad (4.18)$$

$$= \left| \int_0^1 \Big(\mathbb{E}\big[\langle D_x\mathcal{U}(t,x+\lambda he),e\rangle|\mathcal{F}_t\big] - \mathbb{E}\big[\langle D_x\mathcal{U}(t,x),e\rangle|\mathcal{F}_t\big] \Big) d\lambda \right|,$$

where we used the fact the mapping $[0,T] \times \mathbb{T}^d \ni (t,x) \mapsto \mathbb{E}[D_x\mathcal{U}(t,x)|\mathcal{F}_t]$ has continuous paths in order to guarantee the integrability of the integrand in the third line. By continuity of the paths again, the right-hand side tends to 0 with h (uniformly in t and x). $\qquad\square$

Instead of (4.11), we will sometimes make use of the following:

Lemma 4.3.5. *We can find a constant C such that, whenever $\vartheta = \varpi = 0$, any solution to (4.10) satisfies*

$$\forall k \in \{1,\ldots,n\}, \quad \int_t^T \frac{\operatorname{essup}_{\omega \in \Omega}\|\tilde{u}_s\|_{k+\alpha}}{\sqrt{s-t}}\,ds$$

$$\leqslant C\left(\operatorname{essup}_{\omega \in \Omega}\|g_T\|_{k+\alpha} + \int_t^T \operatorname{essup}_{\omega \in \Omega}\|f_s\|_{k+\alpha-1}\,ds\right).$$

Proof. Assume that we have a solution to (4.10). Then, making use of (4.16) in the proof of Lemma 4.3.2, we have that, for all $k \in \{1,\ldots,n\}$ and all $s \in [0,T]$,

$$\operatorname{essup}_{\omega \in \Omega}\|\tilde{u}_s\|_{k+\alpha} \leqslant C\left(\operatorname{essup}_{\omega \in \Omega}\|g_T\|_{k+\alpha} + \int_s^T \frac{\operatorname{essup}_{\omega \in \Omega}\|f_r\|_{k+\alpha-1}}{\sqrt{r-s}}\,dr\right).$$

$$(4.19)$$

Dividing by $\sqrt{s-t}$ for a given $t \in [0,T]$, integrating from t to T, and modifying the value of C if necessary, we deduce that

$$\int_t^T \frac{\operatorname{essup}_{\omega \in \Omega}\|\tilde{u}_s\|_{k+\alpha}}{\sqrt{s-t}}\,ds$$

$$\leqslant C\left(\operatorname{essup}_{\omega \in \Omega}\|g_T\|_{k+\alpha} + \int_t^T ds \int_s^T \frac{\operatorname{essup}_{\omega \in \Omega}\|f_r\|_{k+\alpha-1}}{\sqrt{s-t}\sqrt{r-s}}\,dr\right)$$

$$= C\left[\operatorname{essup}_{\omega \in \Omega}\|g_T\|_{k+\alpha} + \int_t^T \operatorname{essup}_{\omega \in \Omega}\|f_r\|_{k+\alpha-1}\left(\int_t^r \frac{1}{\sqrt{s-t}\sqrt{r-s}}\,ds\right)dr\right],$$

the last line following from the Fubini theorem. The result easily follows. $\qquad\square$

Following (4.17), we shall use the following variant of Lemma 4.3.5:

Lemma 4.3.6. *For $p \in \{1,2\}$, we can find a constant C such that, whenever $\vartheta = \varpi = 0$, any solution to (4.10) satisfies, for all $t \in [0,T]$:*

$$\forall k \in \{1,\ldots,n\}, \quad \mathbb{E}\left[\int_t^T \frac{\|\tilde{u}_s\|_{k+\alpha}^p}{\sqrt{s-t}}\,ds\Big|\mathcal{F}_t\right]$$

$$\leqslant C\mathbb{E}\left[\|g_T\|_{k+\alpha}^p + \int_t^T \|f_r\|_{k+\alpha-1}^p\,dr \,\Big|\, \mathcal{F}_t\right].$$

Proof. The proof goes along the same lines as that of Lemma 4.3.5. We start with the following variant of (4.16), that holds, for any $s \in [0, T]$,

$$\|\tilde{u}_s\|_{k+\alpha}^p \leqslant C\mathbb{E}\left[\|g_T\|_{k+\alpha}^p + \int_s^T \frac{\|f_r\|_{k+\alpha-1}^p}{\sqrt{r-s}} dr \,\Big|\, \mathcal{F}_s\right]. \qquad (4.20)$$

Therefore, for any $0 \leqslant t \leqslant s \leqslant T$, we get

$$\mathbb{E}\left[\|\tilde{u}_s\|_{k+\alpha}^p | \mathcal{F}_t\right] \leqslant C\mathbb{E}\left[\|g_T\|_{k+\alpha}^p + \int_s^T \frac{\|f_r\|_{k+\alpha-1}^p}{\sqrt{r-s}} dr \,\Big|\, \mathcal{F}_t\right].$$

Dividing by $\sqrt{s-t}$ and integrating in s, we get

$$\mathbb{E}\left[\left(\int_t^T \frac{\|\tilde{u}_s\|_{k+\alpha}^p}{\sqrt{s-t}} ds\right) | \mathcal{F}_t\right]$$
$$\leqslant C\mathbb{E}\left[\|g_T\|_{k+\alpha}^p + \int_t^T \|f_r\|_{k+\alpha-1}^p \left(\int_t^r \frac{1}{\sqrt{r-s}\sqrt{s-t}} ds\right) dr | \mathcal{F}_t\right].$$

Therefore,

$$\mathbb{E}\left[\left(\int_t^T \frac{\|\tilde{u}_s\|_{k+\alpha}^p}{\sqrt{s-t}} ds\right) | \mathcal{F}_t\right] \leqslant C\mathbb{E}\left[\|g_T\|_{k+\alpha}^p + \int_t^T \|f_r\|_{k+\alpha-1}^p dr | \mathcal{F}_t\right],$$

which completes the proof. $\qquad\qquad\qquad\qquad\qquad\qquad\qquad\qquad\qquad\qquad\square$

4.3.2   A Priori Estimates

In the previous subsection, we handled the case $\vartheta = \varpi = 0$. To handle the more general case when $(\vartheta, \varpi) \in [0, 1]^2$, we shall use the following a priori regularity estimate:

Lemma 4.3.7. *Let* $(b_t^0)_{t\in[0,T]}$ *and* $(f_t^0)_{t\in[0,T]}$ *be* $(\mathcal{F}_t)_{t\in[0,T]}$ *adapted processes with paths in the space* $\mathcal{C}^0([0,T], \mathcal{C}^1(\mathbb{T}^d, \mathbb{R}^d))$ *and* $\mathcal{C}^0([0,T], \mathcal{C}^{n-1}(\mathbb{T}^d))$ *and* g_T *be an* $\mathcal{F}_T$-*measurable random variable with values in* $\mathcal{C}^n(\mathbb{T}^d)$, *such that*

$$\text{essup}_{\omega \in \Omega} \sup_{t\in[0,T]} \|b_t^0\|_1, \ \text{essup}_{\omega \in \Omega} \sup_{t\in[0,T]} \|f_t^0\|_{n+\alpha-1}, \ \text{essup}_{\omega \in \Omega}\|g_T^0\|_{n+\alpha} \leqslant C,$$

for some constant $C \geqslant 0$. *Then, for any* $k \in \{0, \ldots, n\}$, *we can find two constants* λ_k *and* Λ_k, *depending on* C, *such that, denoting by* $\mathcal{B}$ *the set*

$$\mathcal{B} := \Big\{ w \in \mathcal{C}^0([0,T], \mathcal{C}^n(\mathbb{T}^d)) : \forall k \in \{1, \ldots, n\}, \ \forall t \in [0, T],$$
$$\|w_t\|_{k+\alpha} \leqslant \Lambda_k \exp\big(\lambda_k(T-t)\big) \Big\},$$

it holds that, for any integer $N \geqslant 1$, any family of adapted processes $(\tilde{m}^i, \tilde{u}^i)_{i=1,\dots,N}$ with paths in $\mathcal{C}^0([0,T], \mathcal{P}(\mathbb{T}^d)) \times \mathcal{B}$, any families $(a^i)_{i=1,\dots,N} \in [0,1]^N$ and $(b^i)_{i=1,\dots,N} \in [0,1]^N$ with $a^1 + \cdots + a^N \leqslant 2$ and $b^1 + \cdots + b^N \leqslant 2$, and any input $(f_t)_{t \in [0,T]}$ and g_T of the form

$$f_t = \sum_{i=1}^N [a^i \tilde{H}_t(\cdot, D\tilde{u}_t^i) - b^i \tilde{F}_t(\cdot, \tilde{m}_t^i)] + f_t^0, \quad g_T = \sum_{i=1}^N b^i \tilde{G}(\cdot, \tilde{m}_T^i) + g_T^0,$$

any solution $(\tilde{m}, \tilde{u})$ to (4.10) for some $\vartheta, \varpi \in [0,1]$ has paths in $\mathcal{C}^0([0,T], \mathcal{P}(\mathbb{T}^d)) \times \mathcal{B}$, that is,

$$\operatorname{essup}_{\omega \in \Omega} \|\tilde{u}_t\|_{k+\alpha} \leqslant \Lambda_k \exp(\lambda_k (T-t)), \quad t \in [0,T].$$

Proof. Consider the source term in the backward equation in (4.10):

$$\varphi_t := \vartheta \tilde{H}_t(\cdot, D\tilde{u}_t) - \varpi \tilde{F}_t(\cdot, \tilde{m}_t) + \sum_{i=1}^N [a^i \tilde{H}_t(\cdot, D\tilde{u}_t^i) - b^i \tilde{F}_t(\cdot, \tilde{m}_t^i)] + f_t^0.$$

Then, for any $k \in \{1, \dots, n\}$, we can find a constant C_k and a continuous nondecreasing function Φ_k, independent of $(\tilde{m}^i, \tilde{u}^i)$, $i = 1, \dots, N$, and of $(\tilde{m}, \tilde{u})$ (but depending on the inputs $(b_t^0)_{t \in [0,T]}$, $(f_t^0)_{t \in [0,T]}$ and g_T), such that

$$\begin{aligned}
\|\varphi_t\|_{k+\alpha-1} &\leqslant C_k \left[1 + \Phi_k \left(\|\tilde{u}_t\|_{k+\alpha-1} + \max_{i=1,\dots,N} \|\tilde{u}_t^i\|_{k+\alpha-1} \right) \right] \\
&\quad \times \left[1 + \|\tilde{u}_t\|_{k+\alpha} + \max_{i=1,\dots,N} \|\tilde{u}_t^i\|_{k+\alpha} \right].
\end{aligned} \tag{4.21}$$

When $k = 1$, the above bound holds true with $\Phi_1 \equiv 0$: it then follows from **(HF1(n-1))** and from the fact that H (or equivalently $\tilde{H}_t$) is globally Lipschitz in (x, p) (uniformly in t if dealing with $\tilde{H}_t$ instead of H). When $k \in \{2, \dots, n\}$, it follows from the standard Faà di Bruno formula for the higher-order derivatives of the composition of two functions (together with the fact that D_pH is globally bounded and that the higher-order derivatives of H are locally bounded). Faà di Bruno's formula says that each Φ_k may be chosen as a polynomial function.

Therefore, by (4.21) and by (4.19) in the proof of Lemma 4.3.5 (and modifying the constant C_k in such a way that $\|g_T^0\|_{k+\alpha} + \sup_{m \in \mathcal{P}(\mathbb{T}^d)} \|G(\cdot, m)\|_{k+\alpha} \leqslant C_k$), we deduce that

$$\begin{aligned}
&\operatorname{essup}_{\omega \in \Omega} \|\tilde{u}_t\|_{k+\alpha} \\
&\leqslant \tilde{C}_k \left[1 + \int_t^T \frac{1}{\sqrt{s-t}} \left[\operatorname{essup}_{\omega \in \Omega} \|\tilde{u}_s\|_{k+\alpha} + \operatorname{essup}_{\omega \in \Omega} \max_{i=1,\dots,N} \|\tilde{u}_s^i\|_{k+\alpha} \right] ds \right],
\end{aligned}$$

with

$$\tilde{C}_k := C_k \left[1 + \operatorname*{essup}_{\omega \in \Omega} \sup_{s \in [0,T]} \Phi_k \left(\|\tilde{u}_s\|_{k+\alpha-1} + \max_{i=1,\dots,N} \|\tilde{u}_s^i\|_{k+\alpha-1} \right) \right]. \quad (4.22)$$

Now (independently of the above bound), by (4.21) and Lemma 4.3.5, we can modify C_k in such a way that

$$\int_t^T \frac{\operatorname{essup}_{\omega \in \Omega} \|\tilde{u}_s\|_{k+\alpha}}{\sqrt{s-t}} \, ds$$

$$\leqslant \tilde{C}_k \left[1 + \int_t^T \left(\operatorname{essup}_{\omega \in \Omega} \|\tilde{u}_s\|_{k+\alpha} + \operatorname{essup}_{\omega \in \Omega} \max_{i=1,\dots,N} \|\tilde{u}_s^i\|_{k+\alpha} \right) ds \right],$$

so that, collecting the two last inequalities (and allowing the constant C_k to increase from line to line as long as the definition of $\tilde{C}_k$ in terms of C_k [see (4.22)] remains valid),

$$\operatorname{essup}_{\omega \in \Omega} \|\tilde{u}_t\|_{k+\alpha}$$

$$\leqslant \tilde{C}_k \left[1 + \int_t^T \left(\operatorname{essup}_{\omega \in \Omega} \|\tilde{u}_s\|_{k+\alpha} + \frac{\operatorname{essup}_{\omega \in \Omega} \max_{i=1,\dots,N} \|\tilde{u}_s^i\|_{k+\alpha}}{\sqrt{s-t}} \right) ds \right]$$

$$\leqslant \tilde{C}_k \left[1 + \int_t^T \left(\operatorname{essup}_{\omega \in \Omega} \|\tilde{u}_s\|_{k+\alpha} \right. \right. \qquad (4.23)$$

$$\left. \left. + \frac{\operatorname{essup}_{\omega \in \Omega} \max_{i=1,\dots,N} \sup_{r \in [s,T]} \|\tilde{u}_r^i\|_{k+\alpha}}{\sqrt{s-t}} \right) ds \right].$$

Now, notice that the last term in the above right-hand side may be rewritten

$$\int_t^T \frac{\operatorname{essup}_{\omega \in \Omega} \max_{i=1,\dots,N} \sup_{r \in [s,T]} \|\tilde{u}_r^i\|_{k+\alpha}}{\sqrt{s-t}} \, ds$$

$$= \int_0^{T-t} \frac{\operatorname{essup}_{\omega \in \Omega} \max_{i=1,\dots,N} \sup_{r \in [t+s,T]} \|\tilde{u}_r^i\|_{k+\alpha}}{\sqrt{s}} \, ds,$$

which is clearly nonincreasing in t. Returning to (4.23), this permits application of Gronwall's lemma, from which we get

$$\operatorname{essup}_{\omega \in \Omega} \|\tilde{u}_t\|_{k+\alpha}$$

$$\leqslant \tilde{C}_k \left[1 + \int_t^T \frac{\operatorname{essup}_{\omega \in \Omega} \max_{i=1,\dots,N} \sup_{r \in [s,T]} \|\tilde{u}_r^i\|_{k+\alpha}}{\sqrt{s-t}} \, ds \right]. \qquad (4.24)$$

In particular, if, for any $s \in [0,T]$ and any $i \in \{1, \ldots, N\}$, $\mathrm{essup}_{\omega \in \Omega} \|\tilde{u}_s^i\|_{k+\alpha} \leqslant \Lambda_k \exp(\lambda_k(T-s))$, then, for all $t \in [0,T]$,

$$\int_t^T \frac{\mathrm{essup}_{\omega \in \Omega} \max_{i=1,\ldots,N} \sup_{r \in [s,T]} \|\tilde{u}_r^i\|_{k+\alpha}}{\sqrt{s-t}} \, ds$$

$$\leqslant \Lambda_k \int_t^T \frac{\exp(\lambda_k(T-s))}{\sqrt{s-t}} \, ds \qquad (4.25)$$

$$\leqslant \Lambda_k \exp(\lambda_k(T-t)) \int_0^{T-t} \frac{\exp(-\lambda_k s)}{\sqrt{s}} \, ds,$$

the passage from the first to the second line following from a change of variable. Write now

$$\Lambda_k \exp(\lambda_k(T-t)) \int_0^{T-t} \frac{\exp(-\lambda_k s)}{\sqrt{s}} \, ds$$

$$= \Lambda_k \exp(\lambda_k(T-t)) \int_0^\infty \frac{\exp(-\lambda_k s)}{\sqrt{s}} \, ds - \Lambda_k \int_{T-t}^{+\infty} \frac{\exp(-\lambda_k(s-(T-t)))}{\sqrt{s}} \, ds$$

$$= \Lambda_k \exp(\lambda_k(T-t)) \int_0^\infty \frac{\exp(-\lambda_k s)}{\sqrt{s}} \, ds - \Lambda_k \int_0^{+\infty} \frac{\exp(-\lambda_k s)}{\sqrt{T-t+s}} \, ds$$

$$\leqslant \Lambda_k \exp(\lambda_k(T-t)) \int_0^\infty \frac{\exp(-\lambda_k s)}{\sqrt{s}} \, ds - \Lambda_k \int_0^{+\infty} \frac{\exp(-\lambda_k s)}{\sqrt{T+s}} \, ds,$$

and deduce, from (4.24) and (4.25), that we can find two constants $\gamma_1(\lambda_k)$ and $\gamma_2(\lambda_k)$ that tend to 0 as λ_k tend to $+\infty$ such that

$$\mathrm{essup}_{\omega \in \Omega} \|\tilde{u}_t\|_{k+\alpha} \leqslant \tilde{C}_k \left[1 - \Lambda_k \gamma_1(\lambda_k) + \gamma_2(\lambda_k) \Lambda_k \exp(\lambda_k(T-t))\right].$$

Choosing λ_k first such that $\gamma_2(\lambda_k)\tilde{C}_k \leqslant 1$ and then Λ_k such that

$$1 \leqslant \gamma_1(\lambda_k)\Lambda_k,$$

we finally get that

$$\mathrm{essup}_{\omega \in \Omega} \|\tilde{u}_t\|_{k+\alpha} \leqslant \Lambda_k \exp(\lambda_k(T-t)).$$

The proof is easily completed by induction (on k). $\square$

4.3.3   Case $(\vartheta, \varpi) = (1, 0)$

Using a standard contraction argument, we are going to prove:

Proposition 4.3.8. *Given some adapted inputs* $(b_t)_{t\in[0,T]}$, $(f_t)_{t\in[0,T]}$ *and* g_T
satisfying

$$\mathrm{essup}_{\omega\in\Omega}\ \sup_{t\in[0,T]}\ \|b_t\|_1,\ \mathrm{essup}_{\omega\in\Omega}\ \sup_{t\in[0,T]}\ \|f_t\|_{n+\alpha-1},\ \mathrm{essup}_{\omega\in\Omega}\|g_T\|_{n+\alpha} < \infty,$$

the system (4.10), *with* $\vartheta = 1$ *and* $\varpi = 0$, *admits a unique adapted solution*
$(\tilde{m}_t, \tilde{u}_t)_{t\in[0,T]}$, *with paths in* $\mathcal{C}^0([0,T], \mathcal{P}(\mathbb{T}^d)) \times \mathcal{C}^0([0,T], \mathcal{C}^n(\mathbb{T}^d))$. *As required,*
it satisfies

$$\mathrm{essup}_{\omega\in\Omega}\ \sup_{t\in[0,T]}\ \|\tilde{u}_t\|_{n+\alpha} < \infty.$$

Proof.   Actually, the only difficulty is to solve the backward equation. Once
the backward equation has been solved, the forward equation may be solved by
means of Lemma 4.3.2.

To solve the backward equation, we make use of the Picard fixed-point the-
orem. Given an $(\mathcal{F}_t)_{t\in[0,T]}$ adapted process $(\tilde{u}_t)_{t\in[0,T]}$, with paths in $\mathcal{C}^0([0,T],$
$\mathcal{C}^n(\mathbb{T}^d))$ and satisfying $\mathrm{essup}_{\omega\in\Omega} \sup_{t\in[0,T]} \|\tilde{u}_t\|_{n+\alpha} < \infty$, we denote by $(\tilde{u}'_t)_{t\in[0,T]}$
the solution to the backward equation in (4.10), with $\vartheta = \varpi = 0$ and with
$(f_t)_{t\in[0,T]}$ replaced by $(f_t + H_t(\cdot, D\tilde{u}_t))_{t\in[0,T]}$. By Lemma 4.3.2, the process
$(\tilde{u}'_t)_{t\in[0,T]}$ belongs to $\mathcal{C}^0([0,T], \mathcal{C}^n(\mathbb{T}^d))$ and satisfies $\mathrm{essup}_{\omega\in\Omega} \sup_{t\in[0,T]}$
$\|\tilde{u}'_t\|_{n+\alpha} < \infty$. This defines a mapping (with obvious domain and codomain)

$$\Psi : (\tilde{u}_t)_{t\in[0,T]} \mapsto (\tilde{u}'_t)_{t\in[0,T]}.$$

The point is to exhibit a norm for which it is a contraction.

Given two adapted inputs $(\tilde{u}^i_t)_{t\in[0,T]}$, $i = 1, 2$, with paths in $\mathcal{C}^0([0,T], \mathcal{C}^n(\mathbb{T}^d))$
and with $\mathrm{essup}_{\omega\in\Omega} \sup_{t\in[0,T]} \|\tilde{u}^i_t\|_{n+\alpha} < \infty$, $i = 1, 2$, we call $(\tilde{u}'^{,i}_t)_{t\in[0,T]}$, $i = 1, 2$,
the images by Ψ. By Lemma 4.3.7 (with $N = 1$, $\vartheta = \varpi = 0$, $a^1 = 1$, and $b^1 = 0$),
we can find constants $(\lambda_k, \Lambda_k)_{k=1,\dots,n}$ such that the set

$$\mathcal{B} = \Big\{ w \in \mathcal{C}^0([0,T], \mathcal{C}^n(\mathbb{T}^d)) :$$

$$\forall k \in \{0, \dots, n\},\ \forall t \in [0,T],\ \|w_t\|_{k+\alpha} \leqslant \Lambda_k \exp\big(\lambda_k(T - t)\big) \Big\}$$

is stable by Ψ. We shall prove that Ψ is a contraction on $\mathcal{B}$.

We let $\tilde{w}_t = \tilde{u}^1_t - \tilde{u}^2_t$ and $\tilde{w}'_t = \tilde{u}'^{,1}_t - \tilde{u}'^{,2}_t$, for $t \in [0,T]$. We notice that

$$-d\tilde{w}'_t = \big[\Delta\tilde{w}'_t - \langle \tilde{V}_t, D\tilde{w}_t \rangle\big]\,dt - d\tilde{N}_t,$$

with the terminal boundary condition $\tilde{w}'_T = 0$. Above, $(\tilde{N}_t)_{t\in[0,T]}$ is a process
with paths in $\mathcal{C}^0([0,T], \mathcal{C}^{n-2}(\mathbb{T}^d))$ and, for any $x \in \mathbb{T}^d$, $(\tilde{N}_t(x))_{t\in[0,T]}$ is a martingale. Moreover, $(\tilde{V}_t)_{t\in[0,T]}$ is given by

$$\tilde{V}_t(x) = \int_0^1 D_p \tilde{H}_t\big(x, r D\tilde{u}_t^1(x) + (1-r)D\tilde{u}_t^2(x)\big)dr.$$

We can find a constant C such that, for any $\tilde{u}^1, \tilde{u}^2 \in \mathcal{B}$,

$$\sup_{t\in[0,T]} \|\tilde{V}_t\|_{n+\alpha-1} \leqslant C.$$

Therefore, for any $\tilde{u}^1, \tilde{u}^2 \in \mathcal{B}$, for any $k \in \{0,\dots,n-1\}$

$$\forall t \in [0,T], \quad \|\langle \tilde{V}_t, D\tilde{w}_t \rangle\|_{k+\alpha} \leqslant C\|\tilde{w}_t\|_{k+1+\alpha}, \quad \tilde{w} := \tilde{u}^1 - \tilde{u}^2.$$

Now, following (4.19), we deduce that, for any $k \in \{1,\dots,n\}$,

$$\operatorname{essup}_{\omega\in\Omega}\|\tilde{w}_t'\|_{k+\alpha} \leqslant C \int_t^T \frac{\operatorname{essup}_{\omega\in\Omega}\|\tilde{w}_s\|_{k+\alpha}}{\sqrt{s-t}}\,ds, \qquad (4.26)$$

so that, for any $\mu > 0$,

$$\int_0^T \operatorname{essup}_{\omega\in\Omega}\|\tilde{w}_t'\|_{k+\alpha}\exp(\mu t)\,dt$$

$$\leqslant C \int_0^T \operatorname{essup}_{\omega\in\Omega}\|\tilde{w}_s\|_{k+\alpha}\left(\int_0^s \frac{\exp(\mu t)}{\sqrt{s-t}}\,dt\right)ds$$

$$\leqslant \left(C \int_0^{+\infty} \frac{\exp(-\mu s)}{\sqrt{s}}\,ds\right)\int_0^T \operatorname{essup}_{\omega\in\Omega}\|\tilde{w}_s\|_{k+\alpha}\exp(\mu s)\,ds.$$

Choosing μ large enough, we easily deduce that Ψ has at most one fixed point in $\mathcal{B}$. Moreover, letting $\tilde{u}^0 \equiv 0$ and defining by induction $\tilde{u}^{i+1} = \Psi(\tilde{u}^i)$, $i \in \mathbb{N}$, we easily deduce that, for μ large enough, for any $i,j \in \mathbb{N}$,

$$\int_0^T \operatorname{essup}_{\omega\in\Omega}\|\tilde{u}_t^{i+j} - \tilde{u}_t^i\|_{n+\alpha}\exp(\mu t)\,dt \leqslant \frac{C}{2^i},$$

so that (modifying the value of C)

$$\int_0^T \operatorname{essup}_{\omega\in\Omega}\|\tilde{u}_t^{i+j} - \tilde{u}_t^i\|_{n+\alpha}dt \leqslant \frac{C}{2^i}.$$

Therefore, by definition of $\mathcal{B}$ and by (4.26), we deduce that, for any $\varepsilon > 0$,

$$\forall i \in \mathbb{N}, \quad \sup_{j\in\mathbb{N}} \operatorname{essup}_{\omega\in\Omega} \sup_{t\in[0,T]} \|\tilde{u}_t^{i+j} - \tilde{u}_t^i\|_{n+\alpha} \leqslant C\sqrt{\varepsilon} + \frac{C}{2^i\sqrt{\varepsilon}},$$

from which we deduce that $(\tilde{u}^i)_{i\in\mathbb{N}}$ converges in $L^\infty(\Omega, \mathcal{C}^0([0,T], \mathcal{C}^n(\mathbb{T}^d)))$. The limit is in $\mathcal{B}$ and is a fixed point of Ψ.

Actually, by Lemma 4.3.7 (with $N = 1$, $\vartheta = 1$, $\varpi = 0$ and $a^1 = b^1 = 0$), any fixed point must be in $\mathcal{B}$, so that Ψ has a unique fixed point in the whole space. $\square$

4.3.4 Stability Estimates

Lemma 4.3.9. *Consider two sets of inputs (b, f, g) and (b', f', g') to (4.10), when driven by two parameters $\vartheta, \varpi \in [0, 1]$. Assume that $(\tilde{m}, \tilde{u})$ and $(\tilde{m}', \tilde{u}')$ are associated solutions (with adapted paths that take values in $\mathcal{C}^0([0, T], \mathcal{P}(\mathbb{T}^d)) \times \mathcal{C}^0([0, T], \mathcal{C}^n(\mathbb{T}^d)))$ that satisfy the conclusions of Lemma 4.3.7 with respect to some vectors of constants $\mathbf{\Lambda} = (\Lambda_1, \ldots, \Lambda_n)$ and $\mathbf{\lambda} = (\lambda_1, \ldots, \lambda_n)$. Then, we can find a constant $C \geqslant 1$, depending on the inputs and the outputs through $\mathbf{\Lambda}$ and $\mathbf{\lambda}$ only, such that, provided that*

$$\mathrm{essup}_{\omega \in \Omega} \sup_{t \in [0, T]} \|b_t\|_1 \leqslant \frac{1}{C}$$

it holds that

$$\mathbb{E}\Big[\sup_{t \in [0, T]} \|\tilde{u}_t - \tilde{u}'_t\|_{n+\alpha}^2 + \mathbf{d}_1^2(\tilde{m}_t, \tilde{m}'_t)\Big]$$
$$\leqslant C\Big\{ \mathbf{d}_1^2(m_0, m'_0) + \mathbb{E}\Big[\sup_{t \in [0, T]} \|b_t - b'_t\|_0^2$$
$$+ \sup_{t \in [0, T]} \|f_t - f'_t\|_{n+\alpha-1}^2 + \|g_T - g'_T\|_{n+\alpha}^2\Big]\Big\}.$$

Remark 4.3.10. *The precise knowledge of $\mathbf{\Lambda}$ and $\mathbf{\lambda}$ is crucial in order to make use of the convexity assumption of the Hamiltonian.*

The proof relies on the following stochastic integration by parts formula:

Lemma 4.3.11. *Let $(m_t)_{t \in [0, T]}$ be an adapted process with paths in the space $\mathcal{C}^0([0, T], \mathcal{P}(\mathbb{T}^d))$ such that, with n as in the statement of Theorem 4.3.1, for any smooth test function $\varphi \in \mathcal{C}^n(\mathbb{T}^d)$, $\mathbb{P}$ almost surely,*

$$d_t\Big[\int_{\mathbb{T}^d} \varphi(x) \, dm_t(x)\Big] = \Big\{ \int_{\mathbb{T}^d} [\Delta\varphi(x) - \langle \beta_t(x), D\varphi(x)\rangle] \, dm_t(x)\Big\} \, dt, \quad t \in [0, T],$$

for some adapted process $(\beta_t)_{0 \leqslant t \leqslant T}$ with paths in $\mathcal{C}^0([0, T], [\mathcal{C}^0(\mathbb{T}^d)]^d)$. (Notice, by separability of $\mathcal{C}^n(\mathbb{T}^d)$, that the above holds true, $\mathbb{P}$ almost surely, for any smooth test function $\varphi \in \mathcal{C}^n(\mathbb{T}^d)$.)
    Let $(u_t)_{t \in [0, T]}$ be an adapted process with paths in $\mathcal{C}^0([0, T], \mathcal{C}^n(\mathbb{T}^d))$ such that, for any $x \in \mathbb{T}^d$,

$$d_t u_t(x) = \gamma_t(x) \, dt + dM_t(x), \quad t \in [0, T],$$

where $(\gamma_t)_{t\in[0,T]}$ and $(M_t)_{t\in[0,T]}$ are adapted processes with paths in the space $\mathcal{C}^0([0,T],\mathcal{C}^0(\mathbb{T}^d))$ and, for any $x\in\mathbb{T}^d$, $(M_t(x))_{t\in[0,T]}$ is a martingale. Assume that

$$\operatorname{essup}_{\omega\in\Omega}\sup_{0\leqslant t\leqslant T}\left(\|u_t\|_n+\|\beta_t\|_0+\|\gamma_t\|_0+\|M_t\|_0\right)<\infty. \qquad (4.27)$$

Then, the process

$$\left(\int_{\mathbb{T}^d}u_t(x)\,dm_t(x)\right.$$

$$\left.-\int_0^t\left\{\int_{\mathbb{T}^d}[\gamma_s(x)+\Delta u_s(x)-\langle\beta_s(x),Du_s(x)\rangle]\,dm_s(x)\right\}ds\right)_{t\in[0,T]}$$

is a continuous martingale.

Proof. Although slightly technical, the proof is quite standard. Given two reals $s<t$ in $[0,T]$, we consider a mesh $s=r_0<r_1<\cdots<r_N=t$ of the interval $[s,t]$. Then,

$$\int_{\mathbb{T}^d}u_t(x)\,dm_t(x)-\int_{\mathbb{T}^d}u_s(x)\,dm_s(x)$$

$$=\sum_{i=0}^{N-1}\left[\int_{\mathbb{T}^d}u_{r_{i+1}}(x)\,dm_{r_{i+1}}(x)-\int_{\mathbb{T}^d}u_{r_i}(x)\,dm_{r_i}(x)\right]$$

$$=\sum_{i=0}^{N-1}\left[\int_{\mathbb{T}^d}u_{r_{i+1}}(x)\,dm_{r_{i+1}}(x)-\int_{\mathbb{T}^d}u_{r_{i+1}}(x)\,dm_{r_i}(x)\right]$$

$$+\sum_{i=0}^{N-1}\left[\int_{\mathbb{T}^d}u_{r_{i+1}}(x)\,dm_{r_i}(x)-\int_{\mathbb{T}^d}u_{r_i}(x)\,dm_{r_i}(x)\right] \qquad (4.28)$$

$$=\sum_{i=0}^{N-1}\int_{r_i}^{r_{i+1}}\left\{\int_{\mathbb{T}^d}[\Delta u_{r_{i+1}}(x)-\langle\beta_r(x),Du_{r_{i+1}}(x)\rangle]\,dm_r(x)\right\}dr$$

$$+\sum_{i=0}^{N-1}\int_{\mathbb{T}^d}\left\{\int_{r_i}^{r_{i+1}}\gamma_r(x)dr+M_{r_{i+1}}(x)-M_{r_i}(x)\right\}dm_{r_i}(x).$$

By conditional Fubini's theorem and by (4.27),

$$\mathbb{E}\left[\sum_{i=0}^{N-1}\int_{\mathbb{T}^d}\{M_{r_{i+1}}(x)-M_{r_i}(x)\}\,dm_{r_i}(x)\,|\,\mathcal{F}_s\right]$$

$$=\sum_{i=0}^{N-1}\mathbb{E}\left[\left(\int_{\mathbb{T}^d}\{\mathbb{E}[M_{r_{i+1}}(x)-M_{r_i}(x)|\mathcal{F}_{r_i}]\}\,dm_{r_i}(x)\right)|\,\mathcal{F}_s\right]=0,$$

so that

$$\mathbb{E}[S^N|\mathcal{F}_s]=0,$$

where we have let

$$S^N := \int_{\mathbb{T}^d} u_t(x)\,dm_t(x) - \int_{\mathbb{T}^d} u_s(x)\,dm_s(x)$$

$$- \sum_{i=0}^{N-1} \int_{r_i}^{r_{i+1}} \left\{ \int_{\mathbb{T}^d} \left[\Delta u_{r_{i+1}}(x) - \langle \beta_r(x), Du_{r_{i+1}}(x) \rangle \right] dm_r(x) \right\} dr$$

$$- \sum_{i=0}^{N-1} \int_{\mathbb{T}^d} \left\{ \int_{r_i}^{r_{i+1}} \gamma_r(x)dr \right\} dm_{r_i}(x).$$

Now, we notice that the sequence $(S^N)_{N \geqslant 1}$ converges pointwise to

$$S^\infty := \int_{\mathbb{T}^d} u_t(x)\,dm_t(x) - \int_{\mathbb{T}^d} u_s(x)\,dm_s(x)$$

$$- \int_s^t \left\{ \int_{\mathbb{T}^d} \left[\Delta u_r(x) - \langle \beta_r(x), Du_r(x) \rangle + \gamma_r(x) \right] dm_r(x) \right\} dr.$$

As the sequence $(S^N)_{N \geqslant 1}$ is bounded in $L^\infty(\Omega, \mathcal{A}, \mathbb{P})$, it is straightforward to deduce that, $\mathbb{P}$ almost surely,

$$\mathbb{E}\left[S^\infty | \mathcal{F}_s \right] = \lim_{N \to \infty} \mathbb{E}\left[S^N | \mathcal{F}_s \right] = 0. \qquad \square$$

We now switch to

Proof of Lemma 4.3.9. Following the deterministic case, the idea is to use the monotonicity condition. Using the same duality argument as in the deterministic case, we thus compute by means of Lemma 4.3.11:

$$d_t \int_{\mathbb{T}^d} \left(\tilde{u}_t' - \tilde{u}_t \right) d\left(\tilde{m}_t' - \tilde{m}_t \right)$$

$$= \left\{ -\vartheta \int_{\mathbb{T}^d} \left\langle D\tilde{u}_t' - D\tilde{u}_t, D_p \tilde{H}_t(\cdot, D\tilde{u}_t') d\tilde{m}_t' - D_p \tilde{H}_t(\cdot, D\tilde{u}_t) d\tilde{m}_t \right\rangle \right.$$

$$- \int_{\mathbb{T}^d} \left\langle D\tilde{u}_t' - D\tilde{u}_t, b_t' d\tilde{m}_t' - b_t d\tilde{m}_t \right\rangle$$

$$+ \vartheta \int_{\mathbb{T}^d} \left(\tilde{H}_t(\cdot, D\tilde{u}_t') - \tilde{H}_t(\cdot, D\tilde{u}_t) \right) d(\tilde{m}_t' - \tilde{m}_t)$$

$$- \varpi \int_{\mathbb{T}^d} \left(\tilde{F}_t(\cdot, m_t') - \tilde{F}_t(\cdot, m_t) \right) d(\tilde{m}_t' - \tilde{m}_t)$$

$$+ \int_{\mathbb{T}^d} \left(f_t' - f_t \right) d(\tilde{m}_t' - \tilde{m}_t) \right\} dt + dM_t,$$

where $(M_t)_{t \in [0,T]}$ is a martingale, with the terminal boundary condition

$$\int_{\mathbb{T}^d} (\tilde{u}'_T - \tilde{u}_T) d(\tilde{m}'_T - \tilde{m}_T) = \varpi \int_{\mathbb{T}^d} (\tilde{G}(\cdot, m'_T) - \tilde{G}(\cdot, m_T)) d(\tilde{m}'_T - \tilde{m}_T)$$
$$+ \int_{\mathbb{T}^d} (g'_T - g_T) d(\tilde{m}'_T - \tilde{m}_T).$$

Making use of the convexity and monotonicity assumptions and taking the expectation, we can find a constant $c > 0$, depending on the inputs and the outputs through Λ and λ only, such that

$$\vartheta c \mathbb{E} \int_0^T \left[\int_{\mathbb{T}^d} |D\tilde{u}'_t - D\tilde{u}_t|^2 d(\tilde{m}_t + \tilde{m}'_t) \right] dt$$
$$\leqslant \|u'_0 - u_0\|_1 \mathbf{d}_1(\tilde{m}_0, \tilde{m}'_0) + \mathbb{E} \left[\|g'_T - g_T\|_1 \mathbf{d}_1(\tilde{m}_T, \tilde{m}'_T) \right]$$
$$+ \mathbb{E} \int_0^T \|b'_t - b_t\|_0 \|\tilde{u}'_t - \tilde{u}_t\|_1 \, dt \qquad (4.29)$$
$$+ \mathbb{E} \int_0^T \left(\|\langle b_t, D\tilde{u}'_t - D\tilde{u}_t \rangle\|_1 + \|f'_t - f_t\|_1 \right) \mathbf{d}_1(\tilde{m}_t, \tilde{m}'_t) \, dt.$$

(As for the first term in the right-hand side, recall that u_0 and u'_0 are $\mathcal{F}_0$-measurable and thus almost-surely deterministic since $\mathcal{F}_0$ is almost surely trivial—which is Blumenthal's zero-one law–.) We now implement the same strategy as in the proof of Proposition 3.2.1 in the deterministic case. Following (3.7), we get that there exists a constant C, depending on T, the Lipschitz constant of $D_p H$, and the parameters Λ and λ, such that

$$\sup_{t \in [0,T]} \mathbf{d}_1(\tilde{m}'_t, \tilde{m}_t) \leqslant C \left(\mathbf{d}_1(\tilde{m}'_0, \tilde{m}_0) + \sup_{t \in [0,T]} \|b'_t - b_t\|_0 \right.$$
$$\left. + \vartheta \int_0^T \left[\int_{\mathbb{T}^d} |D\tilde{u}'_s - D\tilde{u}_s| d(\tilde{m}_s + \tilde{m}'_s) \right] ds \right), \qquad (4.30)$$

which holds pathwise.

Taking the square and the expectation and then plugging (4.29) into (4.30), we deduce that, for any small $\eta > 0$ and for a possibly new value of C,

$$\mathbb{E} \left[\sup_t \mathbf{d}_1^2(\tilde{m}_t, \tilde{m}'_t) \right]$$
$$\leqslant C \left\{ \eta^{-1} \mathbf{d}_1^2(m_0, m'_0) + \eta \mathbb{E} \left[\sup_{t \in [0,T]} \|\tilde{u}_t - \tilde{u}'_t\|_1^2 \right] \right.$$
$$+ \eta^{-1} \mathrm{essup}_{\omega \in \Omega} \sup_{t \in [0,T]} \|b_t\|_1 \mathbb{E} \left[\sup_{t \in [0,T]} \|\tilde{u}_t - \tilde{u}'_t\|_2^2 \right] \qquad (4.31)$$
$$\left. + \eta^{-1} \mathbb{E} \left[\sup_{t \in [0,T]} \|b_t - b'_t\|_0^2 + \sup_{t \in [0,T]} \|f_t - f'_t\|_1^2 + \|g_T - g'_T\|_1^2 \right] \right\}.$$

Following the deterministic case, we let $\tilde{w}_t = \tilde{u}_t - \tilde{u}'_t$, for $t \in [0, T]$, so that

$$- d\tilde{w}_t = \left[\Delta\tilde{w}_t - \vartheta\langle\tilde{V}_t, D\tilde{w}_t\rangle + \varpi\tilde{R}_t^1 - (f_t - f'_t) \right] dt - d\tilde{N}_t, \qquad (4.32)$$

with the terminal boundary condition $\tilde{w}_T = \varpi\tilde{R}^T + g'_T - g_T$. Above, $(\tilde{N}_t)_{t\in[0,T]}$ is a process with paths in $\mathcal{C}^0([0,T], \mathcal{C}^0(\mathbb{T}^d))$, with $\mathrm{essup}_{\omega\in\Omega} \sup_{t\in[0,T]} \|\tilde{N}_t\|_0 < \infty$, and, for any $x \in \mathbb{T}^d$, $(\tilde{N}_t(x))_{t\in[0,T]}$ is a martingale. Moreover, the coefficients $(\tilde{V}_t)_{t\in[0,T]}$, $(\tilde{R}_t^1)_{t\in[0,T]}$ and $\tilde{R}^T$ are given by

$$\tilde{V}_t(x) = \int_0^1 D_p\tilde{H}_t\left(x, rD\tilde{u}(x) + (1-r)D\tilde{u}'(x)\right) dr,$$

$$\tilde{R}_t^1(x) = \int_0^1 \frac{\delta\tilde{F}_t}{\delta m}\left(x, r\tilde{m}_t + (1-r)\tilde{m}'_t\right)\left(\tilde{m}_t - \tilde{m}'_t\right) dr,$$

$$\tilde{R}^T(x) = \int_0^1 \frac{\delta G}{\delta m}\left(x, r\tilde{m}_T + (1-r)\tilde{m}'_T\right)\left(\tilde{m}_T - \tilde{m}'_T\right) dr.$$

Following the deterministic case, we have

$$\sup_{t\in[0,T]} \|\tilde{R}_t^1\|_{n+\alpha-1} + \|\tilde{R}^T\|_{n+\alpha} \leqslant C \sup_{t\in[0,T]} \mathbf{d}_1(\tilde{m}_t, \tilde{m}'_t). \qquad (4.33)$$

Moreover, recalling that the outputs $\tilde{u}$ and $\tilde{u}'$ are assumed to satisfy the conclusion of Lemma 4.3.7, we deduce that

$$\sup_{t\in[0,T]} \|\tilde{V}_t\|_{n+\alpha-1} \leqslant C.$$

In particular, for any $k \in \{0, \dots, n-1\}$

$$\forall t \in [0, T], \quad \|\langle\tilde{V}_t, D\tilde{w}_t\rangle\|_{k+\alpha} \leqslant C\|\tilde{w}_t\|_{k+1+\alpha}.$$

Now, following (4.20) (with $p = 1$) in the proof of Lemma 4.3.7 and implementing (4.33), we get, for any $t \in [0, T]$,

$$\|\tilde{w}_t\|_{k+\alpha} \leqslant C\mathbb{E}\left[\|g_T - g'_T\|_{k+\alpha} + \int_t^T \frac{\|\tilde{w}_s\|_{k+\alpha}}{\sqrt{s-t}} ds + \sup_{s\in[0,T]} \|f_s - f'_s\|_{k+\alpha-1} \right.$$

$$\left. + \sup_{s\in[0,T]} \mathbf{d}_1(\tilde{m}_s, \tilde{m}'_s) \,\Big|\, \mathcal{F}_t \right]$$

$$\leqslant C\mathbb{E}\left[\|g_T - g'_T\|_{k+\alpha} + \int_t^T \|\tilde{w}_s\|_{k+\alpha} ds + \sup_{s\in[0,T]} \|f_s - f'_s\|_{k+\alpha-1} \right.$$

$$\left. + \sup_{s\in[0,T]} \mathbf{d}_1(\tilde{m}_s, \tilde{m}'_s) \,\Big|\, \mathcal{F}_t \right],$$

the second line following from Lemma 4.3.6 (with $p = 1$). By Doob's inequality, we deduce that

$$\mathbb{E}\big[\sup_{s\in[t,T]}\|\tilde{w}_s\|^2_{k+\alpha}\big] \leqslant \mathbb{E}\bigg[\|g_T - g'_T\|^2_{k+\alpha} + \int_t^T \|\tilde{w}_s\|^2_{k+\alpha}ds$$

$$+ \sup_{s\in[0,T]}\|f_s - f'_s\|^2_{k+\alpha-1} + \sup_{s\in[0,T]}\mathbf{d}_1^2(\tilde{m}_s, \tilde{m}'_s)\bigg].$$

By Gronwall's lemma, we deduce that, for any $k \in \{1, \ldots, n\}$,

$$\mathbb{E}\big[\sup_{t\in[0,T]}\|\tilde{w}_t\|^2_{k+\alpha}\big] \leqslant C\mathbb{E}\bigg[\|g_T - g'_T\|^2_{k+\alpha}$$

$$+ \sup_{t\in[0,T]}\|f_t - f'_t\|^2_{k+\alpha-1} + \sup_{t\in[0,T]}\mathbf{d}_1^2(\tilde{m}_t, \tilde{m}'_t)\bigg]. \qquad (4.34)$$

We finally go back to (4.31). Choosing η small enough and assuming that $\mathrm{essup}_{\omega\in\Omega}\|b_t\|_1$ is also small enough, we finally obtain (modifying the constant C):

$$\mathbb{E}\big[\sup_{t\in[0,T]}\big(\|\tilde{u}_t - \tilde{u}'_t\|^2_{n+\alpha} + \mathbf{d}_1^2(\tilde{m}_t, \tilde{m}'_t)\big)\big] \leqslant C\Big\{\mathbf{d}_1^2(m_0, m'_0)$$

$$+ \mathbb{E}\big[\sup_{t\in[0,T]}\|b_t - b'_t\|^2_0 + \sup_{t\in[0,T]}\|f_t - f'_t\|^2_{n+\alpha-1} + \|g_T - g'_T\|^2_{n+\alpha}\big]\Big\},$$

which completes the proof. $\qquad\qquad\qquad\qquad\qquad\qquad\qquad\qquad\qquad\quad$ $\square$

4.3.5   Proof of Theorem 4.3.1

We now finish the proof of Theorem 4.3.1.

First step. We first notice that the L^2 stability estimate in the statement is a direct consequence of Lemma 4.3.7 (in order to bound the solutions) and of Lemma 4.3.9 (in order to get the stability estimate itself), provided that existence and uniqueness hold true.

Second step (a). We now prove that, given an initial condition $m_0 \in \mathcal{P}(\mathbb{T}^d)$, the system (4.7) is uniquely solvable.

The strategy consists in increasing inductively the value of ϖ, step by step, from $\varpi = 0$ to $\varpi = 1$, and to prove, at each step, that existence and uniqueness hold true. At each step of the induction, the strategy relies on a fixed-point argument. It works as follows. Given some $\varpi \in [0, 1)$, we assume that, for any input (f, g) in a certain class, we can (uniquely) solve (in the same sense as in the statement of Theorem 4.3.1)

$$d_t \tilde{m}_t = \left\{ \Delta \tilde{m}_t + \mathrm{div}\big[\tilde{m}_t D_p \tilde{H}_t(\cdot, D\tilde{u}_t)\big] \right\} dt,$$
$$d_t \tilde{u}_t = \left\{ -\Delta \tilde{u}_t + \tilde{H}_t(\cdot, D\tilde{u}_t) - \varpi \tilde{F}_t(\cdot, m_t) + f_t \right\} dt + d\tilde{M}_t, \tag{4.35}$$

with $\tilde{m}_0 = m_0$ as the initial condition and $\tilde{u}_T = \varpi \tilde{G}(\cdot, m_T) + g_T$ as the boundary condition. Then, the objective is to prove that the same holds true for ϖ replaced by $\varpi + \epsilon$, for $\epsilon > 0$ small enough (independent of ϖ). Freezing an input $(\bar{f}, \bar{g})$ in the admissible class, the point is to show that the mapping

$$\Phi : (\tilde{m}_t)_{t \in [0,T]} \mapsto \left\{ \begin{array}{l} \big(f_t = -\epsilon \tilde{F}_t(\cdot, m_t) + \bar{f}_t\big)_{t \in [0,T]} \\ g_T = \epsilon \tilde{G}(\cdot, m_T) + \bar{g}_T \end{array} \right\} \mapsto (\tilde{m}'_t)_{t \in [0,T]},$$

is a contraction on the space of adapted processes $(\tilde{m}_t)_{t \in [0,T]}$ with paths in $\mathcal{C}^0([0,T], \mathcal{P}(\mathbb{T}^d))$, where the last output is given as the forward component of the solution of the system (4.35).

The value of ϖ being given, we assume that the input $(\bar{f}, \bar{g})$ is of the form

$$\bar{f}_t = -\sum_{i=1}^{N} b^i \tilde{F}_t(\cdot, m_t^i), \quad \bar{g}_T = \sum_{i=1}^{N} b^i \tilde{G}(\cdot, m_T^i), \tag{4.36}$$

where $N \geqslant 1$, $b^1, \ldots, b^N \geqslant 0$, with $\epsilon + b^1 + \cdots + b^N \leqslant 2$, and $(\tilde{m}^i)_{i=1,\ldots,N}$ (or equivalently $(m^i)_{i=1,\ldots,N}$) is a family of N adapted processes with paths in $\mathcal{C}^0([0,T], \mathcal{P}(\mathbb{T}^d))$.

Such an input $((\bar{f}_t)_{t \in [0,T]}, \bar{g}_T)$ being given, we consider two adapted processes $(\tilde{m}_t^{(1)})_{t \in [0,T]}$ and $(\tilde{m}_t^{(2)})_{t \in [0,T]}$ with paths in $\mathcal{C}^0([0,T], \mathcal{P}(\mathbb{T}^d))$ (or equivalently $(m_t^{(1)})_{t \in [0,T]}$ and $(m_t^{(2)})_{t \in [0,T]}$ without the push-forwards by each of the mappings $(\mathbb{T}^d \ni x \mapsto x - \sqrt{2}W_t \in \mathbb{T}^d)_{t \in [0,T]}$, cf. Remark 4.1.2), and we let

$$f_t^{(l)} := -\epsilon \tilde{F}_t(\cdot, m_t^{(l)}) + \bar{f}_t, \ t \in [0,T]; \quad g_T^{(l)} := -\epsilon \tilde{G}(\cdot, m_T^{(l)}) + \bar{g}_T; \quad l = 1, 2.$$

and

$$\tilde{m}^{(l\prime)} := \Phi(\tilde{m}^{(l)}), \quad l = 1, 2.$$

Second step (b). By Lemma 4.3.7, we can find positive constants $(\lambda_k)_{k=1,\ldots,n}$ and $(\Lambda_k)_{k=1,\ldots,n}$ such that, whenever $(\tilde{m}_t, \tilde{u}_t)_{t \in [0,T]}$ solves (4.35) with respect to an input $((\bar{f}_t)_{t \in [0,T]}, \bar{g}_T)$ of the same type as in (4.36), then

$$\forall k \in \{1, \ldots, n\}, \ \forall t \in [0,T], \quad \mathrm{essup}_{\omega \in \Omega} \|\tilde{u}_t\|_{k+\alpha} \leqslant \Lambda_k \exp\big(\lambda_k (T - t)\big).$$

It is worth mentioning that the values of $(\lambda_k)_{k=0,\ldots,n}$ and $(\Lambda_k)_{k=0,\ldots,n}$ are universal in the sense that they depend neither on ϖ nor on the precise shape of the inputs $(\bar{f}, \bar{g})$ when taken in the class (4.36). In particular, any output $(\tilde{m}'_t, u'_t)_{t \in [0,T]}$ obtained by solving (4.35) with the same input (f, g) as in the definition of the mapping Φ must satisfy the same bound.

Second step (c). We apply Lemma 4.3.9 with $b = b' = 0$, $(f_t, f'_t)_{0 \leqslant t \leqslant T} = (\bar{f}_t^{(1)}, \bar{f}_t^{(2)})_{0 \leqslant t \leqslant T}$, and $(g_T, g'_T) = (\bar{g}_T^{(1)}, \bar{g}_T^{(2)})$. We deduce that

$$\mathbb{E}\Big[\sup_{t \in [0,T]} \mathbf{d}_1^2(\tilde{m}_t^{(1')}, \tilde{m}_t^{(2')}) \Big] \leqslant \epsilon^2 C \Big\{ \mathbb{E}\Big[\sup_{t \in [0,T]} \|\tilde{F}_t(\cdot, m_t^{(1)}) - \tilde{F}_t(\cdot, m_t^{(2)})\|_{n+\alpha-1}^2 \Big]$$
$$+ \|\tilde{G}_T(\cdot, m_T^{(1)}) - \tilde{G}_T(\cdot, m_T^{(2)})\|_{n+\alpha}^2 \Big] \Big\},$$

the constant C being independent of ϖ and of the precise shape of the input $(\bar{f}, \bar{g})$ in the class (4.36). Up to a modification of C, we deduce that

$$\mathbb{E}\Big[\sup_{t \in [0,T]} \mathbf{d}_1^2(\tilde{m}_t^{(1')}, \tilde{m}_t^{(2')}) \Big] \leqslant \epsilon^2 C \mathbb{E}\Big[\sup_{t \in [0,T]} \mathbf{d}_1^2(\tilde{m}_t^{(1)}, \tilde{m}_t^{(2)}) \Big],$$

which shows that Φ is a contraction on the space $L^2(\Omega, \mathcal{A}, \mathbb{P}; \mathcal{C}^0([0,T], \mathcal{P}(\mathbb{T}^d)))$, when ϵ is small enough (independently of ϖ and of $(\bar{f}, \bar{g})$ in the class (4.36)). By the Picard fixed-point theorem, we deduce that the system (4.35) is solvable when ϖ is replaced by $\varpi + \varepsilon$ (and for the same input $(\bar{f}, \bar{g})$ in the class (4.36)). By Lemmata 4.3.7 and 4.3.9, the solution must be unique.

Third step. We finally establish the L^∞ version of the stability estimates. The trick is to derive the L^∞ estimate from the L^2 version of the stability estimates, which seems rather surprising at first sight but that is quite standard in the theory of backward stochastic differential equations (SDEs).

The starting point is to notice that the expectation in the proof of the L^2 version permits getting rid of the martingale part when applying Itô's formula in the proof of Lemma 4.3.9 (see, for instance, (4.29)). Actually, it would suffice to use the conditional expectation given $\mathcal{F}_0$ in order to get rid of it, which means that the L^2 estimate may be written as

$$\mathbb{E}\Big[\sup_{t \in [0,T]} \big(\mathbf{d}_1^2(\tilde{m}_t, \tilde{m}'_t) + \|\tilde{u}_t - \tilde{u}'_t\|_{n+\alpha}^2 \big) \big| \mathcal{F}_0 \Big] \leqslant C \mathbf{d}_1^2(m_0, m'_0),$$

which holds $\mathbb{P}$ almost surely. Of course, when m_0 and m'_0 are deterministic the foregoing conditional bound does not say anything more in comparison with the original one: when m_0 and m'_0 are deterministic, the σ-field $\mathcal{F}_0$ contains no information and is almost surely trivial. Actually, the inequality is especially meaningful when the initial time 0 is replaced by another time $t \in (0, T]$, in which case the initial conditions become $\tilde{m}_t$ and $\tilde{m}'_t$ and are thus random. The trick is thus to say that the same inequality as earlier holds with any time $t \in [0, T]$ as initial condition instead of 0. This proves that

$$\mathbb{E}\Big[\sup_{s \in [t,T]} \big(\mathbf{d}_1^2(\tilde{m}_s, \tilde{m}'_s) + \|\tilde{u}_s - \tilde{u}'_s\|_{n+\alpha}^2 \big) \big| \mathcal{F}_t \Big] \leqslant C \mathbf{d}_1^2(m_t, m'_t).$$

Since $\|\tilde{u}_t - \tilde{u}'_t\|_{n+\alpha}$ is $\mathcal{F}_t$-measurable, we deduce that

$$\|\tilde{u}_t - \tilde{u}'_t\|_{n+\alpha} \leqslant C\mathbf{d}_1(m_t, m'_t).$$

Plugging the above bound in (4.30), we deduce that (modifying C if necessary)

$$\sup_{t\in[0,T]} \mathbf{d}_1(m_t, m'_t) \leqslant C\mathbf{d}_1(m_0, m'_0).$$

Collecting the two last bounds, the proof is easily completed.   □

4.4   LINEARIZATION

Assumption. Throughout the section, α stands for a Hölder exponent in $(0, 1)$.

The purpose here is to follow Section 3.3 and to discuss the following linearized version of the system (4.7):

$$d_t\tilde{z}_t = \left\{-\Delta\tilde{z}_t + \langle\tilde{V}_t(\cdot), D\tilde{z}_t\rangle - \frac{\delta\tilde{F}_t}{\delta m}(\cdot, m_t)(\rho_t) + \tilde{f}^0_t\right\} dt + d\tilde{M}_t, \qquad (4.37)$$

$$\partial_t\tilde{\rho}_t - \Delta\tilde{\rho}_t - \operatorname{div}(\tilde{\rho}_t\tilde{V}_t) - \operatorname{div}(\tilde{m}_t\Gamma_t D\tilde{z}_t + \tilde{b}^0_t) = 0,$$

with a boundary condition of the form

$$\tilde{z}_T = \frac{\delta\tilde{G}}{\delta m}(\cdot, m_T)(\rho_t) + \tilde{g}^0_T,$$

where $(\tilde{M}_t)_{t\in[0,T]}$ is the so-called martingale part of the backward equation, that is, $(\tilde{M}_t)_{t\in[0,T]}$ is an $(\mathcal{F}_t)_{t\in[0,T]}$-adapted process with paths in the space $\mathcal{C}^0([0,T], \mathcal{C}^0(\mathbb{T}^d))$, such that, for any $x \in \mathbb{T}^d$, $(\tilde{M}_t(x))_{t\in[0,T]}$ is an $(\mathcal{F}_t)_{t\in[0,T]}$ martingale.

Remark 4.4.1. *Above, we used the same convention as in Remark 4.1.2. For $(\tilde{\rho}_t)_{t\in[0,T]}$ with paths in $\mathcal{C}^0([0,T], (\mathcal{C}^k(\mathbb{T}^d))')$ for some $k \geqslant 0$, we let $(\rho_t)_{t\in[0,T]}$ be the Schwartz distributional-valued random function with paths in the space $\mathcal{C}^0([0,T], (\mathcal{C}^k(\mathbb{T}^d))')$ defined by*

$$\langle\varphi, \rho_t\rangle_{\mathcal{C}^k(\mathbb{T}^d),(\mathcal{C}^k(\mathbb{T}^d))'} = \langle\varphi(\cdot + \sqrt{2}W_t), \tilde{\rho}_t\rangle_{\mathcal{C}^k(\mathbb{T}^d),(\mathcal{C}^k(\mathbb{T}^d))'}.$$

Generally speaking, the framework is the same as that used in Section 3.3, namely we can find a constant $C \geqslant 1$ such that

1. The initial condition $\tilde{\rho}_0 = \rho_0$ takes values in $(\mathcal{C}^{n+\alpha'}(\mathbb{T}^d))'$, for some $\alpha' \in (0, \alpha)$, and, unless it is explicitly stated, it is deterministic.

2. $(\tilde{V}_t)_{t \in [0,T]}$ is an adapted process with paths in $\mathcal{C}^0([0,T], \mathcal{C}^n(\mathbb{T}^d, \mathbb{R}^d))$, with

$$\operatorname{essup}_{\omega \in \Omega} \sup_{t \in [0,T]} \|\tilde{V}_t\|_{n+\alpha} \leqslant C.$$

3. $(\tilde{m}_t)_{t \in [0,T]}$ is an adapted process with paths in $\mathcal{C}^0([0,T], \mathcal{P}(\mathbb{T}^d))$.
4. $(\Gamma_t)_{t \in [0,T]}$ is an adapted process with paths in $\mathcal{C}^0([0,T], [\mathcal{C}^1(\mathbb{T}^d)]^{d \times d})$ such that, with probability 1,

$$\sup_{t \in [0,T]} \|\Gamma_t\|_1 \leqslant C,$$

$$\forall (t,x) \in [0,T] \times \mathbb{T}^d, \quad C^{-1} I_d \leqslant \Gamma_t(x) \leqslant C I_d.$$

5. $(\tilde{b}_t^0)_{t \in [0,T]}$ is an adapted process with paths in $\mathcal{C}^0([0,T], [(\mathcal{C}^{n+\alpha-1}(\mathbb{T}^d))']^d)$, and $(\tilde{f}_t^0)_{t \in [0,T]}$ is an adapted process with paths in $\mathcal{C}^0([0,T], \mathcal{C}^n(\mathbb{T}^d))$, with

$$\operatorname{essup}_{\omega \in \Omega} \sup_{t \in [0,T]} \left(\|\tilde{b}_t^0\|_{-(n+\alpha'-1)} + \|\tilde{f}_t^0\|_{n+\alpha} \right) < \infty.$$

6. $\tilde{g}_T^0$ is an $\mathcal{F}_T$-measurable random variable with values in $\mathcal{C}^{n+1}(\mathbb{T}^d)$, with

$$\operatorname{essup}_{\omega \in \Omega} \|\tilde{g}_T^0\|_{n+1+\alpha} < \infty.$$

Here is the analogue of Lemma 3.3.1:

Theorem 4.4.2. *Under the assumptions (1–6) right above and* **(HF1(n))** *and* **(HG1(n+1))**, *for* $n \geqslant 2$ *and* $\beta \in (\alpha', \alpha)$, *the system (4.37) admits a unique solution* $(\tilde{\rho}, \tilde{z}, \tilde{M})$, *adapted with respect to the filtration* $(\mathcal{F}_t)_{t \in [0,T]}$, *with paths in the space* $\mathcal{C}^0([0,T], (\mathcal{C}^{n+\beta}(\mathbb{T}^d))' \times \mathcal{C}^{n+1+\beta}(\mathbb{T}^d) \times \mathcal{C}^{n+\beta}(\mathbb{T}^d))$ *and with* $\operatorname{essup}_\omega \sup_{t \in [0,T]} (\|\tilde{\rho}_t\|_{-(n+\beta)} + \|\tilde{z}_t\|_{n+1+\beta} + \|\tilde{M}_t\|_{n-1+\beta}) < \infty$. *It satisfies*

$$\operatorname{essup}_{\omega \in \Omega} \sup_{t \in [0,T]} \left(\|\tilde{\rho}_t\|_{-(n+\alpha')} + \|\tilde{z}_t\|_{n+1+\alpha} + \|\tilde{M}_t\|_{n+\alpha-1} \right) < \infty.$$

The proof imitates that of Theorem 4.3.1 and relies on a continuation argument. For a parameter $\vartheta \in [0,1]$, we consider the system

$$d_t \tilde{z}_t = \left\{ -\Delta \tilde{z}_t + \langle \tilde{V}_t(\cdot), D\tilde{z}_t \rangle - \vartheta \frac{\delta \tilde{F}_t}{\delta m}(\cdot, m_t)(\rho_t) + \tilde{f}_t^0 \right\} dt + d\tilde{M}_t,$$

$$\partial_t \tilde{\rho}_t - \Delta \tilde{\rho}_t - \operatorname{div}(\tilde{\rho}_t \tilde{V}_t) - \operatorname{div}(\vartheta \tilde{m}_t \Gamma_t D\tilde{z}_t + \tilde{b}_t^0) = 0, \tag{4.38}$$

with the boundary conditions

$$\tilde{\rho}_0 = \rho_0, \quad \tilde{z}_T = \vartheta \frac{\delta \tilde{G}}{\delta m}(\cdot, m_T)(\rho_T) + \tilde{g}_T^0. \tag{4.39}$$

As above, the goal is to prove, by increasing step by step the value of ϑ, that the system (4.38), with the boundary condition (4.39) has a unique solution for any $\vartheta \in [0, 1]$.

Following the discussion after Theorem 4.3.1, notice that, whenever $(b_t)_{t\in[0,T]}$ is a process with paths in $\mathcal{C}^0([0, T], \mathcal{C}^{-(n+\beta)}(\mathbb{T}^d))$, for some $\beta \in (\alpha', \alpha)$, the quantity $\sup_{t\in[0,T]} \|b_t\|_{-(n+\alpha')}$ is a random variable, equal to $\sup_{t\in[0,T]\cap\mathbb{Q}} \|b_t\|_{-(n+\alpha')}$. Moreover,

$$\text{essup}_{\omega\in\Omega} \sup_{t\in[0,T]} \|b_t\|_{-(n+\alpha')} = \sup_{t\in[0,T]} \text{essup}_{\omega\in\Omega} \|b_t\|_{-(n+\alpha')}.$$

Below, we often omit the process $(\tilde{M}_t)_{t\in[0,T]}$ when denoting a solution; namely, we often write $(\tilde{\rho}_t, \tilde{z}_t)_{t\in[0,T]}$ instead of $(\tilde{\rho}_t, \tilde{z}_t, \tilde{M}_t)_{t\in[0,T]}$ so that the backward component is understood implicitly. We feel that the rule is quite clear now: in a systematic way, the martingale component has two degrees of regularity less than $(\tilde{z}_t)_{t\in[0,T]}$.

Throughout the subsection, we assume that the assumption of Theorem 4.4.2 is in force.

4.4.1   Case $\vartheta = 0$

We start with the case $\vartheta = 0$:

Lemma 4.4.3.   *Assume that $\vartheta = 0$ in the system (4.38) with the boundary condition (4.39). Then, for any $\beta \in (\alpha', \alpha)$, there is a unique solution $(\tilde{\rho}, \tilde{z})$, adapted with respect to $(\mathcal{F}_t)_{t\in[0,T]}$, with paths in $\mathcal{C}^0([0, T], (\mathcal{C}^{n+\beta}(\mathbb{T}^d))' \times \mathcal{C}^{n+1+\beta}(\mathbb{T}^d)))$ and with $\text{essup}_\omega \sup_{t\in[0,T]}(\|\tilde{\rho}_t\|_{-(n+\beta)} + \|\tilde{z}_t\|_{n+1+\beta}) < \infty$. Moreover, we can find a constant C', depending only C, T, d, and the bounds in **(HF1(n))** and **(HG1(n+1))**, such that*

$$\text{essup}_{\omega\in\Omega} \sup_{t\in[0,T]} \|\tilde{\rho}_t\|_{-(n+\alpha')} \leqslant C'\Big(\|\rho_0\|_{-(n+\alpha')} + \text{essup}_{\omega\in\Omega} \sup_{t\in[0,T]} \|\tilde{b}_t^0\|_{-(n+\alpha'-1)}\Big),$$

$$\text{essup}_{\omega\in\Omega} \sup_{t\in[0,T]} \|\tilde{z}_t\|_{n+1+\alpha} \leqslant C'\Big(\text{essup}_{\omega\in\Omega}\|\tilde{g}_T^0\|_{n+1+\alpha} + \text{essup}_{\omega\in\Omega} \sup_{t\in[0,T]} \|\tilde{f}_t^0\|_{n+\alpha}\Big).$$

Proof.   When $\vartheta = 0$, there is no coupling in the equation and it simply reads

$$\begin{aligned}
&(i) \ d_t\tilde{z}_t = \big\{-\Delta\tilde{z}_t + \langle \tilde{V}_t(\cdot), D\tilde{z}_t\rangle + \tilde{f}_t^0\big\} \, dt + d\tilde{M}_t, \\
&(ii) \ \partial_t\tilde{\rho}_t - \Delta\tilde{\rho}_t - \text{div}\big(\tilde{\rho}_t\tilde{V}_t\big) - \text{div}\big(\tilde{b}_t^0\big) = 0,
\end{aligned} \tag{4.40}$$

with the boundary condition $\tilde{\rho}_0 = \rho_0$ and $\tilde{z}_T = \tilde{g}_T^0$.

First step. Let us first consider the forward equation (4.40-(ii)). We notice that, whenever ρ_0 and $(\tilde{b}_t^0)_{t\in[0,T]}$ are smooth in the space variable, the forward equation may be solved pathwise in the classical sense. Then, by the same duality technique as in Lemma 3.3.1 (with the restriction that the role played by n in the statement of Lemma 3.3.1 is now played by $n-1$ and that the coefficients c and b in the statement of Lemma 3.3.1 are now respectively denoted by $\tilde{b}^0$ and $\tilde{f}^0$), for any $\beta \in [\alpha', \alpha]$, it holds, $\mathbb{P}$ almost surely, that

$$\sup_{t \in [0,T]} \|\tilde{\rho}_t\|_{-(n+\beta)} \leqslant C' \left(\|\tilde{\rho}_0\|_{-(n+\beta)} + \sup_{t \in [0,T]} \|\tilde{b}_t^0\|_{-(n-1+\beta)} \right). \qquad (4.41)$$

Whenever ρ_0 and $(\tilde{b}_t^0)_{t \in [0,T]}$ are not smooth but take values in $(\mathcal{C}^{n+\alpha'}(\mathbb{T}^d))'$ and $(\mathcal{C}^{n+\alpha'-1}(\mathbb{T}^d))'$ only, we can mollify them by a standard convolution argument. Denoting the mollified sequences by $(\rho_0^N)_{N \geqslant 1}$ and $((\tilde{b}_t^{0,N})_{t \in [0,T]})_{N \geqslant 1}$, it is standard to check that, for any $\beta \in (\alpha', \alpha)$, $\mathbb{P}$ almost surely,

$$\lim_{N \to +\infty} \left(\|\rho_0^N - \rho_0\|_{-(n+\beta)} + \sup_{t \in [0,T]} \|\tilde{b}_t^{0,N} - \tilde{b}_t\|_{-(n-1+\beta)} \right) = 0, \qquad (4.42)$$

from which, together with (4.41), we deduce that, $\mathbb{P}$ almost surely, the sequence $((\tilde{\rho}_t^N)_{t \in [0,T]})_{N \geqslant 1}$ is Cauchy in the space $\mathcal{C}([0,T], (\mathcal{C}^{n+\beta}(\mathbb{T}^d))')$, where each $(\tilde{\rho}_t^N)_{t \in [0,T]}$ denotes the solution of the forward equation (4.40-(ii)) with inputs $(\rho_0^N, (\tilde{b}_t^{0,N})_{t \in [0,T]})$. With probability 1 under $\mathbb{P}$, the limit of the Cauchy sequence belongs to $\mathcal{C}([0,T], (\mathcal{C}^{n+\beta}(\mathbb{T}^d))')$ and satisfies (4.41). Pathwise, it solves the forward equation.

Note that the duality techniques of Lemma 3.3.1 are valid for any solution $(\tilde{\rho}_t)_{t \in [0,T]}$ of the forward equation in (4.40-(ii)), with paths in the space $\mathcal{C}^0([0,T], (\mathcal{C}^{n+\beta}(\mathbb{T}^d))')$. This proves uniqueness to the forward equation.

Finally, it is plain to see that the solution is adapted with respect to the filtration $(\mathcal{F}_t)_{t \in [0,T]}$. The reason is that the solutions are constructed as limits of Cauchy sequences, which may be shown to be adapted by means of a Duhamel type formula.

Second step. For the backward component of (4.40), we can adapt Proposition 4.3.8: the solution is adapted, has paths in $\mathcal{C}^0([0,T], \mathcal{C}^{n+1+\beta}(\mathbb{T}^d))$, for any $\beta \in (\alpha', \alpha)$, and, following (4.11), it satisfies

$$\operatorname{essup}_{\omega \in \Omega} \sup_{0 \leqslant t \leqslant T} \|\tilde{z}_t\|_{n+1+\alpha}$$

$$\leqslant C' \left(\operatorname{essup}_{\omega \in \Omega} \|\tilde{g}_T^0\|_{n+1+\alpha} + \operatorname{essup}_{\omega \in \Omega} \sup_{t \in [0,T]} \|\tilde{f}_t^0\|_{n+\alpha} \right),$$

which completes the proof. $\square$

4.4.2   Stability Argument

The purpose is now to increase ϑ step by step in order to prove that (4.38)–(4.39) has a unique solution.

We start with the following consequence of Lemma 4.4.3:

Lemma 4.4.4. *Given some* $\vartheta \in [0,1]$; *an initial condition* $\tilde{\rho}_0$ *in* $(\mathcal{C}^{n+\alpha'}(\mathbb{T}^d))'$; *a set of coefficients* $(\tilde{V}_t, \tilde{m}_t, \Gamma_t)_{t\in[0,T]}$ *as in points 2, 3, and 4 of the introduction of Section 4.4; and a set of inputs* $((\tilde{b}_t^0, \tilde{f}_t^0)_{t\in[0,T]}, \tilde{g}_T^0)$ *as in points 5 and 6 of the introduction to Section 4.4, consider a solution* $(\tilde{\rho}_t, \tilde{z}_t)_{t\in[0,T]}$ *of the system (4.38) with the boundary condition (4.39), the solution being adapted with respect to the filtration* $(\mathcal{F}_t)_{t\in[0,T]}$, *having paths in the space* $\mathcal{C}^0([0,T],(\mathcal{C}^{n+\beta}(\mathbb{T}^d))') \times \mathcal{C}^0([0,T],\mathcal{C}^{n+1+\beta}(\mathbb{T}^d))$, *for some* $\beta \in (\alpha',\alpha)$, *and satisfying*

$$\operatorname{essup}_{\omega\in\Omega} \sup_{t\in[0,T]} (\|\tilde{\rho}_t\|_{-(n+\beta)} + \|\tilde{z}_t\|_{n+1+\beta}) < \infty.$$

Then,

$$\operatorname{essup}_{\omega\in\Omega} \sup_{t\in[0,T]} \left[\|\tilde{\rho}_t\|_{-(n+\alpha')} + \|\tilde{z}_t\|_{n+1+\alpha}\right] < \infty.$$

Proof. Given a solution $(\tilde{\rho}_t, \tilde{z}_t)_{t\in[0,T]}$ as in the statement, we let

$$\hat{b}_t^0 = \tilde{b}_t^0 + \vartheta \tilde{m}_t \Gamma_t D\tilde{z}_t, \quad \hat{f}_t^0 = \tilde{f}_t^0 - \vartheta \frac{\delta \tilde{F}_t}{\delta m}(\cdot, m_t)(\rho_t), \quad t \in [0,T] \,;$$

$$\hat{g}_T^0 = \tilde{g}_T^0 + \vartheta \frac{\delta \tilde{G}}{\delta m}(\cdot, m_T)(\rho_T).$$

Taking advantage of the assumption **(HF1(n))**, we can check that $(\hat{b}_t^0)_{t\in[0,T]}$, $(\hat{f}_t^0)_{t\in[0,T]}$ and $\hat{g}_T^0$ satisfy the same assumptions as $(\tilde{b}_t^0)_{t\in[0,T]}$, $(\tilde{f}_t^0)_{t\in[0,T]}$ and $\tilde{g}_T^0$ in the introduction to Section 4.4. The result then follows from Lemma 4.4.3. □

The strategy now relies on a new stability argument, which is the analogue of Lemma 4.3.9:

Proposition 4.4.5. *Given some* $\vartheta \in [0,1]$; *two initial conditions* $\tilde{\rho}_0$ *and* $\tilde{\rho}_0'$ *in* $(\mathcal{C}^{n+\alpha'}(\mathbb{T}^d))'$; *two sets of coefficients* $(\tilde{V}_t, \tilde{m}_t, \Gamma_t)_{t\in[0,T]}$ *and* $(\tilde{V}_t', \tilde{m}_t', \Gamma_t')_{t\in[0,T]}$ *as in points 2, 3, and 4 of the introduction of Section 4.4; and two sets of inputs* $((\tilde{b}_t^0, \tilde{f}_t^0)_{t\in[0,T]}, \tilde{g}_T^0)$ *and* $((\tilde{b}_t^{0\prime}, \tilde{f}_t^{0\prime})_{t\in[0,T]}, \tilde{g}_T^{0\prime})$ *as in points 5 and 6 of the introduction to Section 4.4, consider two solutions* $(\tilde{\rho}_t, \tilde{z}_t)_{t\in[0,T]}$ *and* $(\tilde{\rho}_t', \tilde{z}_t')_{t\in[0,T]}$ *of the system (4.38) with the boundary condition (4.39), both being adapted with respect to the filtration* $(\mathcal{F}_t)_{t\in[0,T]}$, *having paths in the space* $\mathcal{C}^0([0,T],(\mathcal{C}^{n+\beta}(\mathbb{T}^d))') \times \mathcal{C}^0([0,T],\mathcal{C}^{n+1+\beta}(\mathbb{T}^d))$, *for some* $\beta \in (\alpha',\alpha)$, *and satisfying*

$$\operatorname{essup}_{\omega\in\Omega} \sup_{t\in[0,T]} \left(\|\tilde{\rho}_t\|_{-(n+\beta)} + \|\tilde{z}_t\|_{n+1+\beta} + \|\tilde{\rho}_t'\|_{-(n+\beta)} + \|\tilde{z}_t'\|_{n+1+\beta}\right) < \infty.$$

Then, it holds that

$$\mathbb{E}\left[\sup_{t\in[0,T]}\|\tilde{z}_t - \tilde{z}'_t\|^2_{n+1+\alpha} + \sup_{t\in[0,T]}\|\tilde{\rho}_t - \tilde{\rho}'_t\|^2_{-(n+\alpha')}\right]$$

$$\leqslant C'\left\{\|\tilde{\rho}_0 - \tilde{\rho}'_0\|^2_{-(n+\alpha')} + \mathbb{E}\left[\sup_{t\in[0,T]}\|\tilde{b}^0_t - \tilde{b}^{0'}_t\|^2_{-(n+\alpha'-1)}\right.\right.$$

$$+ \sup_{t\in[0,T]}\|\tilde{f}^0_t - \tilde{f}^{0'}_t\|^2_{n+\alpha} + \|\tilde{g}^0_T - \tilde{g}^{0'}_T\|^2_{n+1+\alpha}$$

$$+ \sup_{t\in[0,T]}\left\{\left(\|\tilde{z}'_t\|^2_{n+1+\alpha} + \|\tilde{\rho}'_t\|^2_{-(n+\alpha')}\right) \times \left(\|\tilde{V}_t - \tilde{V}'_t\|^2_{n+\alpha}\right.\right.$$

$$\left.\left.\left.\left. + [\mathbf{d}_1(m_t, m'_t)]^2 + \|\Gamma_t - \Gamma'_t\|^2_0\right)\right\}\right]\right\},$$

the constant C' depending only upon C in the introduction to Section 4.4, T, d, α, and α'.

Proof. *First step.* The first step is to make use of a duality argument.

We start with the case when $\tilde{\rho}_0$, $\tilde{\rho}'_0$, $\tilde{b}^0$ and $\tilde{b}^{0'}$ are smooth. Letting $\hat{b}^0_t = \vartheta\tilde{m}_t\Gamma_t D\tilde{z}_t + \tilde{b}^0_t$ and $\hat{b}^{0'}_t = \vartheta\tilde{m}'_t\Gamma'_t D\tilde{z}'_t + \tilde{b}^{0'}_t$, for $t \in [0,T]$, $(\tilde{\rho}_t)_{t\in[0,T]}$ and $(\tilde{\rho}'_t)_{t\in[0,T]}$ solve the linear equation (ii) in (4.40) with $(\tilde{b}^0_t)_{t\in[0,T]}$ and $(\tilde{b}^{0'}_t)_{t\in[0,T]}$ replaced by $(\hat{b}^0_t)_{t\in[0,T]}$ and $(\hat{b}^{0'}_t)_{t\in[0,T]}$ respectively. By Lemma 4.4.3 with $(\tilde{b}^0_t)_{t\in[0,T]}$ in (4.40) equal to $(\hat{b}^0_t)_{t\in[0,T]}$ and with n in the statement of Lemma 4.4.3 replaced by $n - 1$, we deduce that $(\tilde{\rho}_t)_{t\in[0,T]}$ and $(\tilde{\rho}'_t)_{t\in[0,T]}$ have bounded paths in $\mathcal{C}^0([0,T],(\mathcal{C}^{n-1+\beta}(\mathbb{T}^d))')$, for the same $\beta \in (\alpha',\alpha)$ as in the statement of Proposition 4.4.5.

With a suitable adaptation of Lemma 4.3.11 and with the same kind of notations as in Section 3.3, this allows us to expand the infinitesimal variation of the duality bracket $\langle\tilde{z}_t - \tilde{z}'_t, \tilde{\rho}_t - \tilde{\rho}'_t\rangle_{n+\beta}$, where $\lg\cdot,\cdot\rangle_{n+\beta}$ stands for $\lg\cdot,\cdot\rangle_{\mathcal{C}^{n+\beta}(\mathbb{T}^d),(\mathcal{C}^{n+\beta}(\mathbb{T}^d))}$. We compute

$$d_t\langle\tilde{z}_t - \tilde{z}'_t, \tilde{\rho}_t - \tilde{\rho}'_t\rangle_{n+\beta}$$

$$= \left\{-\left\langle\tilde{D}(\tilde{z}_t - \tilde{z}'_t), \tilde{\rho}'_t(\tilde{V}_t - \tilde{V}'_t)\right\rangle_{n+\beta} + \left\langle D\tilde{z}'_t, (\tilde{V}_t - \tilde{V}'_t)(\tilde{\rho}_t - \tilde{\rho}'_t)\right\rangle_{n+\beta}\right\}dt$$

$$+ \left\{\left\langle\tilde{f}^0_t - \tilde{f}^{0'}_t, \tilde{\rho}_t - \tilde{\rho}'_t\right\rangle_{n+\beta}dt - \left\langle D(\tilde{z}_t - \tilde{z}'_t), \tilde{b}^0_t - \tilde{b}^{0'}_t\right\rangle_{n-1+\beta}\right\}dt$$

$$- \vartheta\left\{\left\langle\frac{\delta\tilde{F}_t}{\delta m}(\cdot, m_t)(\rho_t - \rho'_t), \tilde{\rho}_t - \tilde{\rho}'_t\right\rangle_{n+\beta}\right.$$

$$\left. + \left\langle\left(\frac{\delta\tilde{F}_t}{\delta m}(\cdot, m_t) - \frac{\delta\tilde{F}_t}{\delta m}(\cdot, m'_t)\right)(\rho'_t), \tilde{\rho}_t - \tilde{\rho}'_t\right\rangle_{n+\beta}\right\}dt$$

$$- \vartheta\left\{\left\langle D(\tilde{z}_t - \tilde{z}'_t), \tilde{m}_t\Gamma_t D(\tilde{z}_t - \tilde{z}'_t)\right\rangle_{n+\beta}\right.$$

$$\left. + \left\langle D(\tilde{z}_t - \tilde{z}'_t), (\tilde{m}_t\Gamma_t - \tilde{m}'_t\Gamma'_t)D\tilde{z}'_t\right\rangle_{n+\beta}\right\}dt + d_tM_t,$$

where $(M_t)_{0 \leqslant t \leqslant T}$ is a martingale and where we applied Remark 4.4.1 to define $(\rho_t)_{t \in [0,T]}$ and $(\rho'_t)_{t \in [0,T]}$. An important fact in the proof is that the martingale part in (4.40) has continuous paths in $\mathcal{C}^0([0,T], \mathcal{C}^{n-1+\beta}(\mathbb{T}^d))$, which allows us to give a sense to the duality bracket (in x) with $(\tilde{\rho}_t - \tilde{\rho}'_t)_{t \in [0,T]}$, since $(\tilde{\rho}_t - \tilde{\rho}'_t)_{t \in [0,T]}$ is here assumed to have continuous paths in $\mathcal{C}^0([0,T], (\mathcal{C}^{n-1+\beta}(\mathbb{T}^d))')$. Similarly, the duality bracket of $(\tilde{z}_t - \tilde{z}'_t)_{t \in [0,T]}$ with the Laplacian of $(\tilde{\rho}_t - \tilde{\rho}'_t)_{t \in [0,T]}$ makes sense and, conversely, the duality bracket of $(\tilde{\rho}_t)_{t \in [0,T]}$ with the Laplacian of $(\tilde{z}_t - \tilde{z}'_t)_{t \in [0,T]}$ makes sense as well, the two of them canceling each other.

Of course, the goal is to relax the smoothness assumption made on $\tilde{\rho}_0$, $\tilde{\rho}'_0$, $\tilde{b}^0$, and $\tilde{b}^{0'}$. Although it was pretty straightforward to do in the deterministic case, it is more difficult here because of the additional martingale term. As already mentioned, the martingale term is defined as a duality bracket between a path with values in $\mathcal{C}^0([0,T], \mathcal{C}^{n-1+\beta}(\mathbb{T}^d))$ and a path with values in $\mathcal{C}^0([0,T], \mathcal{C}^{-(n-1+\beta)}(\mathbb{T}^d))$. Of course, the problem is that this is no longer true in the general case that $(\tilde{\rho}_t - \tilde{\rho}'_t)_{t \in [0,T]}$ has paths in $\mathcal{C}^0([0,T], \mathcal{C}^{-(n-1+\beta)}(\mathbb{T}^d))$. To circumvent the difficulty, one way is to take first the expectation in order to cancel the martingale part and then to relax the smoothness conditions. Taking the expectation in the above formula, we get (in the mollified setting):

$$
\begin{aligned}
\frac{d}{dt} &\mathbb{E}\big[\langle \tilde{z}_t - \tilde{z}'_t, \tilde{\rho}_t - \tilde{\rho}'_t \rangle_{n+\beta}\big] \\
&= \Big\{ -\mathbb{E}\Big[\big\langle \tilde{D}(\tilde{z}_t - \tilde{z}'_t), \tilde{\rho}'_t(\tilde{V}_t - \tilde{V}'_t)\big\rangle_{n+\beta}\Big] \\
&\quad + \mathbb{E}\Big[\big\langle D\tilde{z}'_t, (\tilde{V}_t - \tilde{V}'_t)(\tilde{\rho}_t - \tilde{\rho}'_t)\big\rangle_{n+\beta}\Big] \Big\} \\
&\quad + \Big\{ \mathbb{E}\Big[\big\langle \tilde{f}^0_t - \tilde{f}^{0'}_t, \tilde{\rho}_t - \tilde{\rho}'_t \big\rangle_{n+\beta}\Big] - \mathbb{E}\Big[\big\langle D(\tilde{z}_t - \tilde{z}'_t), \tilde{b}^0_t - \tilde{b}^{0'}_t \big\rangle_{n-1+\beta}\Big] \Big\} \\
&\quad - \vartheta \Big\{ \mathbb{E}\Big[\big\langle \frac{\delta \tilde{F}_t}{\delta m}(\cdot, m_t)(\rho_t - \rho'_t), \tilde{\rho}_t - \tilde{\rho}'_t \big\rangle_{n+\beta}\Big] \quad\quad\quad (4.43) \\
&\quad + \mathbb{E}\Big[\big\langle \big(\frac{\delta \tilde{F}_t}{\delta m}(\cdot, m_t) - \frac{\delta \tilde{F}_t}{\delta m}(\cdot, m'_t)\big)(\rho'_t), \tilde{\rho}_t - \tilde{\rho}'_t \big\rangle_{n+\beta}\Big] \Big\} \\
&\quad - \vartheta \Big\{ \mathbb{E}\Big[\big\langle D(\tilde{z}_t - \tilde{z}'_t), \tilde{m}_t \Gamma_t D(\tilde{z}_t - \tilde{z}'_t)\big\rangle_{n+\beta}\Big] \\
&\quad + \mathbb{E}\Big[\big\langle D(\tilde{z}_t - \tilde{z}'_t), (\tilde{m}_t \Gamma_t - \tilde{m}'_t \Gamma'_t) D\tilde{z}'_t \big\rangle_{n+\beta}\Big] \Big\}.
\end{aligned}
$$

Whenever $\tilde{\rho}_0$, $\tilde{\rho}'_0$, $\tilde{b}^0$, and $\tilde{b}^{0'}$ are not smooth (and thus just satisfy the assumption in the statement of Proposition 4.4.5), we can mollify them in the same way as in the first step of Lemma 4.4.3. We respectively call $(\tilde{\rho}^N_0)_{N \geqslant 1}$, $(\tilde{\rho}'^{,N}_0)_{N \geqslant 1}$, $(\tilde{b}^{0,N}_t)_{N \geqslant 1}$, and $(\tilde{b}^{0',N}_t)_{N \geqslant 1}$ the mollifying sequences. For any $\beta' \in (\alpha', \alpha)$ and $\mathbb{P}$ almost surely, the first two sequences respectively converge to $\tilde{\rho}_0$ and $\tilde{\rho}'_0$ in norm $\|\cdot\|_{-(n+\beta')}$ and the last two ones respectively converge to $\tilde{b}^0_t$ and $\tilde{b}^{0'}_t$ in norm $\|\cdot\|_{-(n-1+\beta')}$, uniformly in $t \in [0,T]$. With $(\tilde{\rho}_t, \tilde{z}_t)_{t \in [0,T]}$ and $(\tilde{\rho}'_t, \tilde{z}'_t)_{t \in [0,T]}$ the original solutions given by the statement of Proposition 4.4.5, we denote, for each $N \geqslant 1$, by $(\tilde{\rho}_t^N, \tilde{z}_t^N)_{t \in [0,T]}$ and $(\tilde{\rho}_t^{\prime,N}, \tilde{z}_t^{\prime,N})_{t \in [0,T]}$ the respective solutions to (4.40), but with $(\tilde{b}_t^0, \tilde{f}_t^0, \tilde{g}_T^0)_{t \in [0,T]}$ respectively replaced by

$$\begin{pmatrix} \hat{b}_t^{0,N} := \tilde{b}_t^{0,N} + \vartheta \tilde{m}_t \Gamma_t D \tilde{z}_t \\ \hat{f}_t^0 := \tilde{f}_t^0 - \vartheta \dfrac{\delta F}{\delta m}(\cdot, m_t)(\rho_t) \\ \hat{g}_T^0 := \tilde{g}_T^0 + \vartheta \dfrac{\delta G}{\delta m}(\cdot, m_T)(\rho_T) \end{pmatrix}_{t \in [0,T]}$$

and

$$\begin{pmatrix} \hat{b}_t^{0\prime,N} := \tilde{b}_t^{0\prime,N} + \vartheta \tilde{m}_t' \Gamma_t' D \tilde{z}_t' \\ \hat{f}_t^{0\prime} := \tilde{f}_t^{0\prime} - \vartheta \dfrac{\delta F}{\delta m}(\cdot, m_t')(\rho_t') \\ \hat{g}_T^{0\prime} := \tilde{g}_T^{0\prime} + \vartheta \dfrac{\delta G}{\delta m}(\cdot, m_T')(\rho_T') \end{pmatrix}_{t \in [0,T]}.$$

By linearity of (4.40) and by Lemma 4.4.3, we have that $(\tilde{\rho}_t^N)_{N \geqslant 1}$ and $(\tilde{\rho}_t^{\prime,N})_{N \geqslant 1}$ converge to $\tilde{\rho}_t$ and $\tilde{\rho}_t'$ in norm $\| \cdot \|_{-(n+\beta)}$, uniformly in $t \in [0,T]$, and that $(\tilde{z}_t^N)_{N \geqslant 1}$ and $(\tilde{z}_t^{\prime,N})_{N \geqslant 1}$ converge to $\tilde{z}_t$ and $\tilde{z}_t'$ in norm $\| \cdot \|_{n+1+\beta}$, uniformly in $t \in [0,T]$.

Then, we may write down the analogue of (4.43) for any mollified solution $(\tilde{\rho}_t^N, \tilde{z}_t^N)_{N \geqslant 1}$ (take note that the formulation of (4.43) for the mollified solutions is slightly different since the mollified solutions satisfy only an approximate version of (4.38)). Following (4.42), we can pass to the limit under the symbol $\mathbb{E}$. By Lemma 4.4.3, we can easily exchange the almost sure convergence and the symbol $\mathbb{E}$, proving that the identity (4.43) holds true under the standing assumption on $\tilde{\rho}_0$, $\tilde{\rho}_{0\prime}$, $(\tilde{b}_t^0)_{t \in [0,T]}$, and $(\tilde{b}_t^{0\prime})_{t \in [0,T]}$.

Using the positive definiteness of Γ and the monotonicity of $\tilde{F}$, we deduce that

$$\mathbb{E}\Big[\langle \tilde{z}_T - \tilde{z}_T', \tilde{\rho}_T - \tilde{\rho}_T' \rangle_{n+\beta} \Big] + C^{-1}\vartheta \mathbb{E}\Big[\int_t^T \Big(\int_{\mathbb{T}^d} |D(\tilde{z}_s - \tilde{z}_s')|^2 d\tilde{m}_s \Big) ds \Big]$$
$$\leqslant \mathbb{E}\big[\langle \tilde{z}_0 - \tilde{z}_0', \tilde{\rho}_0 - \tilde{\rho}_0' \rangle_{n+\beta} \big]$$
$$+ C' \mathbb{E}\Big[\int_t^T \Theta \Big(\|\tilde{\rho}_s - \tilde{\rho}_s'\|_{-(n+\alpha')} + \|\tilde{z}_s - \tilde{z}_s'\|_{n+1+\alpha} \Big) ds \Big],$$

where

$$\Theta := \|\tilde{\rho}_0 - \tilde{\rho}_0'\|_{-(n+\alpha')} + \|\tilde{g}_T^0 - \tilde{g}_T^{0\prime}\|_{n+1+\alpha}$$
$$+ \sup_{s \in [0,T]} \Big[\|\tilde{f}_s^0 - \tilde{f}_s^{0\prime}\|_{n+\alpha} + \|\tilde{b}_s^0 - \tilde{b}_s^{0\prime}\|_{-(n+\alpha'-1)} \Big]$$
$$+ \big(\|\tilde{z}_s'\|_{n+1+\alpha} + \|\tilde{\rho}_s'\|_{-(n+\alpha')} \big)$$
$$\times \big(\|\tilde{V}_s - \tilde{V}_s'\|_{n+\alpha} + \mathbf{d}_1(m_s, m_s') + \|\Gamma_s - \Gamma_s'\|_0 \big) \Big].$$

Recalling that

$$
\langle \tilde{z}_T - \tilde{z}_T', \tilde{\rho}_T - \tilde{\rho}_T' \rangle_{n+\beta}
$$

$$
= \vartheta \Big\langle \frac{\delta \tilde{G}}{\delta m}(\cdot, m_T)(\rho_T - \rho_T'), \tilde{\rho}_T - \tilde{\rho}_T' \Big\rangle_{n+\beta}
$$

$$
+ \vartheta \Big\langle \big(\frac{\delta \tilde{G}}{\delta m}(\cdot, m_T) - \frac{\delta \tilde{G}}{\delta m}(\cdot, m_T') \big)(\rho_T'), \tilde{\rho}_T - \tilde{\rho}_T' \Big\rangle_{n+\beta}
$$

$$
+ \langle \tilde{g}_T^0 - \tilde{g}_T^{0'}, \tilde{\rho}_T - \tilde{\rho}_T' \rangle_{n+\beta}
$$

$$
\geqslant -C'\Theta \|\tilde{\rho}_T - \tilde{\rho}_T'\|_{-(n+\alpha')},
$$

where we have used the monotonicity of G to deduce the second line, we thus get

$$
\vartheta \mathbb{E}\Big[\int_0^T \Big(\int_{\mathbb{T}^d} |D(\tilde{z}_s - \tilde{z}_s')|^2 d\tilde{m}_s\Big)\, ds\Big]
$$

$$
\leqslant C'\mathbb{E}\Big[\Theta\Big(\|\tilde{z}_0 - \tilde{z}_0'\|_{n+1+\alpha} + \|\tilde{\rho}_T - \tilde{\rho}_T'\|_{-(n+\alpha')} \tag{4.44}
$$

$$
+ \int_0^T \big(\|\tilde{\rho}_s - \tilde{\rho}_s'\|_{-(n+\alpha')} + \|\tilde{z}_s - \tilde{z}_s'\|_{n+1+\alpha}\big)\, ds\Big)\Big].
$$

Second step. As a second step, we follow the strategy used in the deterministic case in order to estimate $(\|\tilde{\rho}_t - \tilde{\rho}_t'\|_{-(n+\alpha')})_{t\in[0,T]}$ in terms of $\int_0^T (\int_{\mathbb{T}^d} |D(\tilde{z}_s - \tilde{z}_s')|^2 d\tilde{m}_s)\, ds$ on the left-hand side of (4.44).

We use again a duality argument. Given $\xi \in \mathcal{C}^{n+\alpha}(\mathbb{T}^d)$ and $\tau \in [0,T]$, we consider the solution $(\tilde{w}_t)_{t\in[0,\tau]}$, with paths in $\mathcal{C}^0([0,\tau], \mathcal{C}^{n+\beta}(\mathbb{T}^d))$, to the backward PDE:

$$
\partial_t \tilde{w}_t = \{ -\Delta \tilde{w}_t + \langle \tilde{V}_t(\cdot), D\tilde{w}_t \rangle \}, \tag{4.45}
$$

with the terminal boundary condition $\tilde{w}_\tau = \xi$. Take note that the solution is not adapted. It satisfies (see the proof in the last step below), with probability 1,

$$
\forall t \in [0,\tau], \quad \|\tilde{w}_t\|_{n+\alpha'} \leqslant C' \|\xi\|_{n+\alpha'},
$$

$$
\forall t \in [0,\tau), \quad \|\tilde{w}_t\|_{n+1+\alpha'} \leqslant \frac{C'}{\sqrt{\tau - t}} \|\xi\|_{n+\alpha'}. \tag{4.46}
$$

Then following the end of the proof of Lemma 3.3.1, we have

$$
d_t \langle \tilde{w}_t, \tilde{\rho}_t - \tilde{\rho}_t' \rangle_{n+\alpha'}
$$

$$
= -\Big\langle D\tilde{w}_t, \tilde{b}_t^0 - \tilde{b}_t^{0'} \Big\rangle_{n-1+\alpha'} dt + \Big\langle D\tilde{w}_t, (\tilde{V}_t' - \tilde{V}_t)\tilde{\rho}_t' \Big\rangle_{n+\alpha'} dt
$$

$$
- \vartheta \Big\langle D\tilde{w}_t, \tilde{m}_t \Gamma_t D(\tilde{z}_t - \tilde{z}_t') \Big\rangle_{n+\alpha'} dt
$$

$$
- \vartheta \Big\langle D\tilde{w}_t, (\tilde{m}_t \Gamma_t - \tilde{m}_t' \Gamma_t') D\tilde{z}_t' \Big\rangle_{n+\alpha'} dt,
$$

so that

$$\langle \xi, \tilde{\rho}_\tau - \tilde{\rho}'_\tau \rangle_{n+\alpha'} \leqslant C' \|\xi\|_{n+\alpha'} \left[\Theta + \vartheta \int_0^\tau \left(\int_{\mathbb{T}^d} |D(\tilde{z}_s - \tilde{z}'_s)|^2 d\tilde{m}_s \right)^{1/2} ds \right].$$

Therefore,

$$\|\tilde{\rho}_\tau - \tilde{\rho}'_\tau\|_{-(n+\alpha')} \leqslant C' \left[\Theta + \vartheta \left(\int_0^T \int_{\mathbb{R}^d} |D(\tilde{z}_s - \tilde{z}'_s)|^2 d\tilde{m}_s \right)^{1/2} ds \right]. \qquad (4.47)$$

Plugging (4.47) into (4.44), we obtain

$$\vartheta \mathbb{E} \left[\int_0^T \left(\int_{\mathbb{T}^d} |D(\tilde{z}_s - \tilde{z}'_s)|^2 d\tilde{m}_s \right) ds \right]$$

$$\leqslant C' \mathbb{E} \left[\Theta \left(\Theta + \sup_{t \in [0,T]} \|\tilde{z}_t - \tilde{z}'_t\|_{n+1+\alpha} \right) \right]. \qquad (4.48)$$

Therefore,

$$\mathbb{E} \left[\sup_{t \in [0,T]} \|\tilde{\rho}_t - \tilde{\rho}'_t\|^2_{-(n+\alpha')} \right] \leqslant C' \mathbb{E} \left[\Theta \left(\Theta + \sup_{t \in [0,T]} \|\tilde{z}_t - \tilde{z}'_t\|_{n+1+\alpha} \right) \right]. \qquad (4.49)$$

Third step. We now combine the two first steps to get an estimate of $(\|\tilde{z}_t - \tilde{z}'_t\|_{n+1+\alpha})_{t \in [0,T]}$. Following the proof of (4.34) on the linear equation (4.32) and using the assumptions **(HF1(n))** and **(HG1(n+1))**, we get that

$$\mathbb{E} \left[\sup_{t \in [0,T]} \|\tilde{z}_t - \tilde{z}'_t\|^2_{n+1+\alpha} \right]$$
$$\leqslant \mathbb{E} \left[\Theta^2 + \|\tilde{\rho}_T - \tilde{\rho}'_T\|^2_{-(n+\alpha')} + \int_0^T \|\tilde{\rho}_s - \tilde{\rho}'_s\|^2_{-(n+\alpha')} ds \right]. \qquad (4.50)$$

By (4.49), we easily complete the proof.

It just remains to prove (4.46). The first line follows from Lemma 3.2.2. The second line may be proved as follows. Following (4.26), we have, with probability 1,

$$\forall t \in [0,\tau), \quad \|\tilde{w}_t\|_{n+1+\alpha'} \leqslant C' \left(\frac{\|\xi\|_{n+\alpha'}}{\sqrt{\tau - t}} + \int_t^\tau \frac{\|\tilde{w}_s\|_{n+1+\alpha'}}{\sqrt{s - t}} ds \right). \qquad (4.51)$$

Integrating and allowing the constant C' to increase from line to line, we have, for all $t \in [0,\tau)$,

$$\int_t^\tau \frac{\|\tilde{w}_s\|_{n+1+\alpha'}}{\sqrt{s-t}} \, ds \leqslant C' \left[\|\xi\|_{n+\alpha'} \int_t^\tau \frac{1}{\sqrt{\tau-s}\sqrt{s-t}} \, ds \right.$$
$$\left. + \int_t^\tau \|\tilde{w}_r\|_{n+1+\alpha'} \left(\int_t^r \frac{1}{\sqrt{r-s}\sqrt{s-t}} ds \right) dr \right]$$
$$\leqslant C' \left[\|\xi\|_{n+\alpha'} + \int_t^\tau \|\tilde{w}_r\|_{n+1+\alpha'} dr \right].$$

Plugging the above estimate into (4.51), we get that

$$\forall t \in [0,\tau), \quad \sqrt{\tau-t}\|\tilde{w}_t\|_{n+1+\alpha'} \leqslant C' \left(\|\xi\|_{n+\alpha'} + \int_t^\tau \sqrt{\tau-r}\|\tilde{w}_r\|_{n+1+\alpha'} dr \right),$$

which yields, by Gronwall's lemma,

$$\forall t \in [0,\tau), \quad \|\tilde{w}_t\|_{n+1+\alpha'} \leqslant \frac{C'}{\sqrt{\tau-t}}\|\xi\|_{n+\alpha'},$$

which is the required bound. $\qquad\qquad\qquad\qquad\qquad\qquad\qquad\qquad\qquad\qquad\quad\Box$

4.4.3   A Priori Estimate

A typical example of application of Proposition 4.4.5 is to choose $\tilde{\rho}'_0 = 0$, $(\tilde{b}^{0\prime}, \tilde{f}^{0\prime}, \tilde{g}^{0\prime}) \equiv (0,0,0)$, $\tilde{V} \equiv \tilde{V}'$, $\Gamma \equiv \Gamma'$, in which case

$$\left(\tilde{\rho}', \tilde{z}'\right) \equiv (0,0).$$

Then, Proposition 4.4.5 provides an a priori L^2 estimate of the solutions to (4.37). (Take note that the constant C in the statement depends on the smoothness assumptions satisfied by $\tilde{V}$.) The following corollary shows that the L^2 bound can be turned into an L^∞ bound. It reads as an extension of Lemma 4.4.3 to the case in which ϑ may be nonzero:

Corollary 4.4.6. *Given $\vartheta \in [0,1]$, an initial condition $\tilde{\rho}_0$ in $(\mathcal{C}^{n+\alpha'}(\mathbb{T}^d))'$ and a set of inputs $((\tilde{b}^0_t, \tilde{f}^0_t)_{t\in[0,T]}, \tilde{g}^0_T)$ as in points 1–6 in the introduction of Section 4.4, consider an adapted solution $(\tilde{\rho}_t, \tilde{z}_t)_{t\in[0,T]}$ of the system (4.38)– (4.39), with paths in the space $\mathcal{C}^0([0,T], (\mathcal{C}^{n+\beta}(\mathbb{T}^d))') \times \mathcal{C}^0([0,T], \mathcal{C}^{n+1+\beta}(\mathbb{T}^d))$ for some $\beta \in (\alpha', \alpha)$, such that*

$$\operatorname{essup}_{\omega\in\Omega} \sup_{t\in[0,T]} \left(\|\tilde{z}_t\|_{n+1+\beta} + \|\tilde{\rho}_t\|_{-(n+\beta)} \right) < \infty.$$

Then, we can find a constant C', depending only on C, T, d, α, and α', such that

$$\text{essup}_{\omega\in\Omega}\sup_{t\in[0,T]}\left(\|\tilde{z}_t\|_{n+1+\alpha} + \|\tilde{\rho}_t\|_{-(n+\alpha')}\right)$$

$$\leqslant C'\Bigg(\|\tilde{\rho}_0\|_{-(n+\alpha')} \tag{4.52}$$

$$+ \text{essup}_{\omega\in\Omega}\left[\|\tilde{g}_T^0\|_{n+1+\alpha} + \sup_{t\in[0,T]}\left(\|\tilde{f}_t^0\|_{n+\alpha} + \|\tilde{b}_t^0\|_{-(n+\alpha'-1)}\right)\right]\Bigg).$$

For another initial condition in the space $\tilde{\rho}_0'$ in $(\mathcal{C}^{n+\alpha'}(\mathbb{T}^d))'$, another set of coefficients $(\tilde{V}_t', \tilde{m}_t', \Gamma_t')_{t\in[0,T]}$, and another set of inputs $((\tilde{b}_t^{0\prime}, \tilde{f}_t^{0\prime})_{t\in[0,T]}, \tilde{g}_T^{0,\prime})$ as in points 1–6 in the introduction to Section 4.4, consider an adapted solution $(\tilde{\rho}_t', \tilde{z}_t')_{t\in[0,T]}$ of the system (4.38)–(4.39), with paths in $\mathcal{C}^0([0,T], (\mathcal{C}^{n+\beta}(\mathbb{T}^d))') \times \mathcal{C}^0([0,T], \mathcal{C}^{n+1+\beta}(\mathbb{T}^d))$ for the same $\beta \in (\alpha', \alpha)$ as above, such that

$$\text{essup}_{\omega\in\Omega}\sup_{t\in[0,T]}\left(\|\tilde{z}_t'\|_{n+1+\beta} + \|\tilde{\rho}_t'\|_{-(n+\beta)}\right) < \infty.$$

Then, we can find a constant C', depending only on C, T, d, α, and α' and on

$$\|\tilde{\rho}_0\|_{-(n+\alpha')} + \|\tilde{\rho}_0'\|_{-(n+\alpha')} + \text{essup}_{\omega\in\Omega}\left[\|\tilde{g}_T^0\|_{n+1+\alpha} + \|\tilde{g}_T^{0\prime}\|_{n+1+\alpha}\right]$$

$$+ \text{essup}_{\omega\in\Omega}\sup_{t\in[0,T]}\left[\|\tilde{f}_t^0\|_{n+\alpha} + \|\tilde{f}_t^{0\prime}\|_{n+\alpha} + \|\tilde{b}_t^0\|_{-(n+\alpha'-1)} + \|\tilde{b}_t^{0\prime}\|_{-(n+\alpha'-1)}\right],$$

such that

$$\text{essup}_{\omega\in\Omega}\sup_{t\in[0,T]}\left[\|\tilde{z}_t - \tilde{z}_t'\|_{n+1+\alpha}^2 + \|\tilde{\rho}_t - \tilde{\rho}_t'\|_{-(n+\alpha')}^2\right]$$

$$\leqslant C'\Bigg\{\|\tilde{\rho}_0 - \tilde{\rho}_0'\|_{-(n+\alpha')}^2 + \text{essup}_{\omega\in\Omega}\Bigg(\|\tilde{g}_T^0 - \tilde{g}_T^{0\prime}\|_{n+1+\alpha}^2$$

$$+ \sup_{t\in[0,T]}\left[\|\tilde{b}_t^0 - \tilde{b}_t^{0\prime}\|_{-(n+\alpha'-1)}^2 + \|\tilde{f}_t^0 - \tilde{f}_t^{0\prime}\|_{n+\alpha}^2\right] \tag{4.53}$$

$$+ \text{essup}_{\omega\in\Omega}\sup_{t\in[0,T]}\left[\|\tilde{V}_t - \tilde{V}_t'\|_{n+\alpha}^2 + \mathbf{d}_1^2(m_t, m_t') + \|\Gamma_t - \Gamma_t'\|_0^2\right]\Bigg)\Bigg\}.$$

Proof. We start with the proof of (4.52).

First step. The proof relies on the same trick as that used in the third step of the proof of Theorem 4.3.1. In the statement of Proposition 4.4.5, the initial conditions $\tilde{\rho}_0$ and $\tilde{\rho}_0'$ are assumed to be deterministic. It can be checked that the same argument holds when both are random and the expectation is replaced by a conditional expectation given the initial condition. More generally, given some time $t \in [0, T]$, we may see the pair $(\tilde{\rho}_s, \tilde{z}_s)_{s \in [t,T]}$ as the solution of the system (4.38) with the boundary condition (4.39), but on the interval $[t, T]$ instead of $[0, T]$. In particular, when $\tilde{\rho}_0' = 0$, $(\tilde{b}^{0\prime}, \tilde{f}^{0\prime}, \tilde{g}^{0\prime}) \equiv (0, 0, 0)$, $\tilde{V} \equiv \tilde{V}'$, $\Gamma \equiv \Gamma'$ (in which case $(\tilde{\rho}', \tilde{z}') \equiv (0, 0)$), we get

$$\mathbb{E}\left[\sup_{s \in [t,T]} \left(\|\tilde{z}_s\|_{n+1+\alpha}^2 + \|\tilde{\rho}_s\|_{-(n+\alpha')}^2 \right) \Big| \mathcal{F}_t \right] \leqslant C' \left[\|\tilde{\rho}_t\|_{-(n+\alpha')}^2 + \mathbb{E}\left[\Theta^2 | \mathcal{F}_t \right] \right],$$

where we have let

$$\Theta = \sup_{s \in [t,T]} \|\tilde{b}_s^0\|_{-(n+\alpha'-1)} + \sup_{s \in [t,T]} \|\tilde{f}_s^0\|_{n+\alpha} + \|\tilde{g}_T^0\|_{n+1+\alpha}.$$

Second step. We now prove the estimate on $\tilde{\rho}$. From the first step, we deduce that

$$\begin{aligned} \|\tilde{z}_t\|_{n+1+\alpha}^2 &\leqslant C' \left[\|\tilde{\rho}_t\|_{-(n+\alpha')}^2 + \mathbb{E}\left[\Theta^2 | \mathcal{F}_t \right] \right] \\ &\leqslant C' \left[\|\tilde{\rho}_t\|_{-(n+\alpha')}^2 + \operatorname{essup}_{\omega \in \Omega} \Theta^2 \right]. \end{aligned} \tag{4.54}$$

The above inequality holds true for any $t \in [0, T]$, $\mathbb{P}$ almost surely. By continuity of both sides, we can exchange the "$\mathbb{P}$ almost sure" and the "for all $t \in [0, T]$." Now we can use the same duality trick as in the proof of Proposition 4.4.5. With the same notations as in (4.45) and (4.46), we have

$$\forall t \in [0, \tau], \quad \|\tilde{w}_t\|_{n+\alpha'} \leqslant C' \|\xi\|_{n+\alpha'}.$$

Then, we have

$$\begin{aligned} \langle \tilde{w}_\tau, \tilde{\rho}_\tau \rangle_{n+\alpha'} &\leqslant \langle \tilde{w}_0, \tilde{\rho}_0 \rangle_{n+\alpha'} \\ &\quad + \int_0^\tau \|D\tilde{w}_s\|_{n+\alpha'-1} \left(\|\tilde{b}_s^0\|_{-(n+\alpha'-1)} + \|\tilde{z}_s\|_{n+\alpha} \right) ds \\ &\leqslant C' \|\xi\|_{n+\alpha'} \left(\|\tilde{\rho}_0\|_{-(n+\alpha')} + \int_0^\tau \left[\|\tilde{\rho}_s\|_{-(n+\alpha')} + \operatorname{essup}_{\omega \in \Omega} \Theta \right] ds \right), \end{aligned}$$

from which we deduce, by Gronwall's lemma, that

$$\|\tilde{\rho}_\tau\|_{-(n+\alpha')} \leqslant C' \left(\|\tilde{\rho}_0\|_{-(n+\alpha')} + \sup_{t \in [0,T]} \operatorname{essup}_{\omega \in \Omega} \Theta \right),$$

and thus

$$\operatorname{essup}_{\omega \in \Omega} \sup_{t \in [0,T]} \|\tilde{\rho}_t\|_{-(n+\alpha')} \leqslant C' \left(\|\tilde{\rho}_0\|_{-(n+\alpha')} + \operatorname{essup}_{\omega \in \Omega} \Theta \right). \tag{4.55}$$

By (4.54) and (4.55), we easily get a bound for $\tilde{z}$.

Last step. It then remains to prove (4.53). By means of the first step, we have bounds for

$$\text{essup}_{\omega \in \Omega} \sup_{t \in [0,T]} \left(\|\tilde{z}_t\|_{n+1+\alpha} + \|\tilde{z}_t'\|_{n+1+\alpha} + \|\tilde{\rho}_t\|_{-(n+\alpha')} + \|\tilde{\rho}_t'\|_{-(n+\alpha')} \right).$$

Plugging the bound into the stability estimate in Proposition 4.4.5, we may proceed in the same way as in the first two steps in order to complete the proof. □

4.4.4   Proof of Theorem 4.4.2

We now complete the proof of Theorem 4.4.2. It suffices to prove

Proposition 4.4.7. *There is an $\varepsilon_0 > 0$ such that if, for some $\vartheta \in [0,1)$ and $\beta \in (\alpha', \alpha)$, for any initial condition $\tilde{\rho}_0$ in $(\mathcal{C}^{n+\alpha'}(\mathbb{T}^d))'$, any set of coefficients $(\tilde{V}_t, \tilde{m}_t, \Gamma_t)_{t \in [0,T]}$ and any input $((\tilde{b}_t^0, \tilde{f}_t^0)_{t \in [0,T]}, \tilde{g}_T^0)$ as in the introduction to Section 4.4, the system (4.38)–(4.39) has a unique solution $(\tilde{\rho}_t, \tilde{z}_t)_{t \in [0,T]}$ with paths in the space $\mathcal{C}^0([0,T], (\mathcal{C}^{n+\beta}(\mathbb{T}^d))') \times \mathcal{C}^0([0,T], \mathcal{C}^{n+1+\beta}(\mathbb{T}^d))$ such that $\text{essup}_{\omega} \sup_{t \in [0,T]} \|\tilde{\rho}_t\|_{-(n+\beta)} + \|\tilde{z}_t\|_{n+1+\beta}) < \infty$, $(\tilde{\rho}_t, \tilde{z}_t)_{t \in [0,T]}$ also satisfying $\text{essup}_{\omega} \sup_{t \in [0,T]} (\|\tilde{\rho}_t\|_{-(n+\alpha')} + \|\tilde{z}_t\|_{n+1+\alpha}) < \infty$, then, for any $\varepsilon \in (0, \varepsilon_0]$, unique solvability also holds with ϑ replaced by $\vartheta + \varepsilon$, for the same class of initial conditions and of inputs and in the same space; moreover, solutions also lie (almost surely) in a bounded subset of the space $L^{\infty}([0,T], (\mathcal{C}^{(n+\alpha')}(\mathbb{T}^d))') \times L^{\infty}([0,T], \mathcal{C}^{n+1+\alpha}(\mathbb{T}^d))$.*

Proof. Given $\vartheta \in [0,1)$ in the statement, $\varepsilon > 0$, an initial condition $\tilde{\rho}_0 \in (\mathcal{C}^{n+\alpha'}(\mathbb{T}^d))'$, an input $((\tilde{b}_t^0)_{t \in [0,T]}, (\tilde{f}_t^0)_{t \in [0,T]}, \tilde{g}_T^0)$ satisfying the prescription described in the introduction to Section 4.4, and an adapted process $(\tilde{\rho}_t, \tilde{z}_t)_{t \in [0,T]}$ ($\tilde{\rho}$ having $\tilde{\rho}_0$ as initial condition) with paths in the space $\mathcal{C}^0([0,T], (\mathcal{C}^{n+\beta}(\mathbb{T}^d))') \times \mathcal{C}^0([0,T], \mathcal{C}^{n+1+\beta}(\mathbb{T}^d))$ such that

$$\text{essup}_{\omega \in \Omega} \sup_{t \in [0,T]} \left(\|\tilde{\rho}_t\|_{-(n+\alpha')} + \|\tilde{z}_t\|_{n+1+\alpha} \right) < \infty, \tag{4.56}$$

we call $\Phi_{\varepsilon}(\tilde{\rho}, \tilde{z})$ the pair $(\tilde{\rho}_t', \tilde{z}_t')_{0 \leqslant t \leqslant T}$ solving the system (4.38) with respect to the initial condition $\tilde{\rho}_0$ and to the input:

$$\tilde{b}_t^{0\prime} = \varepsilon \tilde{m}_t \Gamma_t D\tilde{z}_t + \tilde{b}_t^0,$$

$$\tilde{f}_t^{0\prime} = -\varepsilon \frac{\delta \tilde{F}_t}{\delta m}(\cdot, m_t)(\rho_t) + \tilde{f}_t^0,$$

$$\tilde{g}_T^{0\prime} = \varepsilon \frac{\delta \tilde{G}}{\delta m}(\cdot, m_T)(\rho_T) + \tilde{g}_T^0.$$

By assumption, it satisfies

$$\operatorname{essup}_{\omega\in\Omega} \sup_{t\in[0,T]} \left(\|\tilde{\rho}'_t\|_{-(n+\alpha')} + \|\tilde{z}'_t\|_{n+1+\alpha} \right) < \infty,$$

By Corollary 4.4.6,

$$\operatorname{essup}_{\omega\in\Omega} \sup_{t\in[0,T]} \left(\|\tilde{z}'_t\|_{n+1+\alpha} + \|\tilde{\rho}'_t\|_{-(n+\alpha')} \right)$$

$$\leqslant C'\left[\|\tilde{\rho}_0\|_{-(n+\alpha')} + c\varepsilon \operatorname{essup}_{\omega\in\Omega} \sup_{t\in[0,T]} \left(\|\tilde{\rho}_t\|_{-(n+\alpha')} + \|\tilde{z}_t\|_{n+1+\alpha} \right) \right.$$

$$\left. + \operatorname{essup}_{\omega\in\Omega} \left[\sup_{t\in[0,T]} \left(\|\tilde{b}^0_t\|_{-(n+\alpha'-1)} + \|\tilde{f}^0_t\|_{n+\alpha} \right) + \|\tilde{g}^0_T\|_{n+1+\alpha} \right] \right],$$

where c is a constant, which depends only on the constant C appearing in points 1–6 in the introduction to Section 4.4 and on the bounds appearing in $(\mathbf{HF1}(n))$ and $(\mathbf{HG1}(n+1))$.

In particular, if

$$\operatorname{essup}_{\omega\in\Omega} \sup_{t\in[0,T]} \left(\|\tilde{z}_t\|_{n+1+\alpha} + \|\tilde{\rho}_t\|_{-(n+\alpha')} \right)$$

$$\leqslant 2C'\left(\|\tilde{\rho}_0\|_{-(n+\alpha')} + \operatorname{essup}_{\omega\in\Omega} \left[\|\tilde{g}^0_T\|_{n+1+\alpha} \right. \right. \tag{4.57}$$

$$\left. \left. + \sup_{t\in[0,T]} \left(\|\tilde{b}^0_t\|_{-(n+\alpha'-1)} + \|\tilde{f}^0_t\|_{n+\alpha} \right) \right] \right),$$

and $2C'c\varepsilon \leqslant 1$, then

$$\operatorname{essup}_{\omega\in\Omega} \sup_{t\in[0,T]} \left(\|\tilde{z}'_t\|_{n+1+\alpha} + \|\tilde{\rho}'_t\|_{-(n+\alpha')} \right)$$

$$\leqslant 2C'\left(\|\tilde{\rho}_0\|_{-(n+\alpha')} + \operatorname{essup}_{\omega\in\Omega} \left[\|\tilde{g}^0_T\|_{n+1+\alpha} \right. \right.$$

$$\left. \left. + \sup_{t\in[0,T]} \left(\|\tilde{b}^0_t\|_{-(n+\alpha'-1)} + \|\tilde{f}^0_t\|_{n+\alpha} \right) \right] \right),$$

so that the set of pairs $(\tilde{\rho}, \tilde{z})$ that satisfy (4.56) and (4.57) is stable by Φ_ε for ε small enough.

Now, given two pairs $(\tilde{\rho}^1_t, \tilde{z}^1_t)_{t\in[0,T]}$ and $(\tilde{\rho}^2_t, \tilde{z}^2_t)_{t\in[0,T]}$ satisfying (4.57), we let $(\tilde{\rho}^{1\prime}_t, \tilde{z}^{1\prime}_t)_{t\in[0,T]}$ and $(\tilde{\rho}^{2\prime}_t, \tilde{z}^{2\prime}_t)_{t\in[0,T]}$ be their respective images by Φ_ε. We deduce from Proposition 4.4.5 that

$$\mathbb{E}\left[\sup_{t\in[0,T]}\|\tilde{z}_t^{1\prime}-\tilde{z}_t^{2\prime}\|_{n+1+\alpha}^2+\sup_{t\in[0,T]}\|\tilde{\rho}_t^{1\prime}-\tilde{\rho}_t^{2\prime}\|_{-(n+\alpha')}^2\right]$$

$$\leqslant C'\varepsilon^2\mathbb{E}\left[\sup_{t\in[0,T]}\|\tilde{z}_t^1-\tilde{z}_t^2\|_{n+1+\alpha}^2+\sup_{t\in[0,T]}\|\tilde{\rho}_t^1-\tilde{\rho}_t^2\|_{-(n+\alpha')}^2\right],$$

for a possibly new value of the constant C', but still independent of ϑ and ε. Therefore, for $C'\varepsilon^2 < 1$ and $2C''c\varepsilon \leqslant 1$, Φ_ε is a contraction on the set of adapted processes $(\tilde{\rho}_t, \tilde{z}_t)_{t\in[0,T]}$ having paths in $\mathcal{C}^0([0,T],(\mathcal{C}^{n+\beta}(\mathbb{T}^d))') \times \mathcal{C}^0([0,T], \mathcal{C}^{n+1+\beta}(\mathbb{T}^d))$ and satisfying (4.57) (and thus (4.56) as well), which is a closed set of the Banach space $\mathcal{C}^0([0,T],(\mathcal{C}^{n+\beta}(\mathbb{T}^d))')\times\mathcal{C}^0([0,T],\mathcal{C}^{n+\beta}(\mathbb{T}^d))$. By the Picard fixed-point theorem, we deduce that Φ_ε has a unique fixed point satisfying (4.57). The fixed point solves (4.38)–(4.39), with ϑ replaced by $\vartheta + \varepsilon$.

Consider now another solution to (4.38)–(4.39) with ϑ replaced by $\vartheta+\varepsilon$, with paths in a bounded subset of $\mathcal{C}^0([0,T],(\mathcal{C}^{n+\beta}(\mathbb{T}^d))')\times\mathcal{C}^0([0,T],\mathcal{C}^{n+1+\beta}(\mathbb{T}^d))$. By Proposition 4.4.5, it must coincide with the solution we just constructed. $\qquad\square$

Chapter Five

The Second-Order Master Equation

TAKING ADVANTAGE of the analysis performed in the previous chapter on the unique solvability of the mean field game (MFG) system, we are now ready to define and investigate the solution of the master equation. The principle is the same as in the first-order case: the forward component of the MFG system has to be seen as the characteristics of the master equation. The regularity of the solution of the master equation is then investigated through the *tangent process* that solves the linearized MFG system.

As in the previous chapter, the level of common noise β is set to 1 throughout this chapter. This is without loss of generality and this makes the notation a little bit simpler.

5.1 CONSTRUCTION OF THE SOLUTION

Assumption. Throughout the section, we assume that the assumption of Theorem 4.3.1 is in force, with $\alpha \in (0,1)$. Namely, we assume that F, G, and H satisfy (2.4) and (2.5) in Section 2.3, and that, for some integer $n \geqslant 2$ and some $\alpha \in (0,1)$, **(HF1(n-1))** and **(HG1(n))** hold true.

For any initial distribution $m_0 \in \mathcal{P}(\mathbb{T}^d)$, the system (4.7) admits a unique solution so that, following the analysis performed in the deterministic setting, we may let
$$U(0, x, m_0) = \tilde{u}_0(x), \quad x \in \mathbb{T}^d.$$

The initialization is here performed at time 0, but, of course, there is no difficulty in replacing 0 by any arbitrary time $t_0 \in [0, T]$, in which case we rewrite the system (4.7) as

$$
\begin{aligned}
d_t \tilde{m}_t &= \big\{ \Delta \tilde{m}_t + \operatorname{div}\big(\tilde{m}_t D_p \tilde{H}_{t_0,t}(\cdot, D\tilde{u}_t)\big) \big\} dt, \\
d_t \tilde{u}_t &= \big\{ -\Delta \tilde{u}_t + \tilde{H}_{t_0,t}(\cdot, D\tilde{u}_t) - \tilde{F}_{t_0,t}(\cdot, m_{t_0,t}) \big\} dt + d\tilde{M}_t,
\end{aligned}
\tag{5.1}
$$

with the initial condition $\tilde{m}_{t_0} = m_0$ and the terminal boundary condition $\tilde{u}_T =$

$\tilde{G}_{t_0}(\cdot, m_{t_0,T})$, under the prescription that

$$
\begin{aligned}
m_{t_0,t} &= \big(id + \sqrt{2}(W_t - W_{t_0})\big)\sharp\tilde{m}_t, \\
\tilde{F}_{t_0,t}(x,\mu) &= F\big(x + \sqrt{2}(W_t - W_{t_0}),\mu\big), \\
\tilde{G}_{t_0}(x,\mu) &= G\big(x + \sqrt{2}(W_T - W_{t_0}),\mu\big), \\
\tilde{H}_{t_0,t}(x,p) &= H\big(x + \sqrt{2}(W_t - W_{t_0}),p\big), \quad x \in \mathbb{T}^d, \ p \in \mathbb{R}^d, \ \mu \in \mathcal{P}(\mathbb{T}^d).
\end{aligned}
\tag{5.2}
$$

It is then possible to let

$$
U(t_0, x, m_0) = \tilde{u}_{t_0}(x), \quad x \in \mathbb{T}^d.
$$

Note that, following Theorem 4.3.1, $U : (t_0, x, m_0) \mapsto U(t_0, x, m_0)$ is Lipschitz continuous in the last two variables.

We shall often use the following important fact:

Lemma 5.1.1. *Given an initial condition $(t_0, m_0) \in [0, T] \times \mathcal{P}(\mathbb{T}^d)$, denote by $(\tilde{m}_t, \tilde{u}_t)_{t \in [t_0, T]}$ the solution of (5.1) with the prescription (5.2) and with $\tilde{m}_{t_0} = m_0$ as initial condition. Call $m_{t_0,t}$ the image of $\tilde{m}_t$ by the random mapping $\mathbb{T}^d \ni x \mapsto x + \sqrt{2}(W_t - W_{t_0})$, that is, $m_{t_0,t} = [id + \sqrt{2}(W_t - W_{t_0})]\sharp\tilde{m}_t$. Then, for any $t_0 + h \in [t_0, T]$, $\mathbb{P}$ almost surely,*

$$
\tilde{u}_{t_0+h}(x) = U\big(t_0 + h, x + \sqrt{2}(W_{t_0+h} - W_{t_0}), m_{t_0,t_0+h}\big), \quad x \in \mathbb{T}^d.
$$

Proof. Given t_0 and h as above, we let

$$
\begin{aligned}
\bar{m}_t &= \big[id + \sqrt{2}\big(W_{t_0+h} - W_{t_0}\big)\big]\sharp\tilde{m}_t, \\
\bar{u}_t(x) &= \tilde{u}_t\big[x - \sqrt{2}\big(W_{t_0+h} - W_{t_0}\big)\big], \ t \in [t_0 + h, T], \ x \in \mathbb{T}^d.
\end{aligned}
$$

We claim that $(\bar{m}_t, \bar{u}_t)_{t \in [t_0+h,T]}$ is a solution of (5.1)–(5.2), with t_0 replaced by $t_0 + h$ and with m_{t_0,t_0+h} as initial condition.

The proof is as follows. We start with a preliminary remark. For $t \in [t_0+h, T]$,

$$
\big[id + \sqrt{2}\big(W_t - W_{t_0+h}\big)\big]\sharp\bar{m}_t = \big[id + \sqrt{2}\big(W_t - W_{t_0}\big)\big]\sharp\tilde{m}_t = m_{t_0,t}. \tag{5.3}
$$

We now prove that the pair $(\bar{m}_t, \bar{u}_t)_{t_0+h \leqslant t \leqslant T}$ solves the forward equation in (5.1). To this end, denote by $(X_{t_0,t})_{t \in [t_0,T]}$ the solution of the stochastic differential equation (SDE)

$$
dX_{t_0,t} = -D_p\tilde{H}_{t_0,t}\big(X_{t_0,t}, D\tilde{u}_t(X_{t_0,t})\big)dt + \sqrt{2}dB_t, \quad t \in [t_0, T],
$$

the initial condition X_{t_0,t_0} having m_0 as distribution. (Notice that the equation is well-posed, as $D\tilde{u}$ is known to be Lipschitz in space.) Then, the process $(\tilde{X}_t = X_{t_0,t} + \sqrt{2}(W_{t_0+h} - W_{t_0}))_{t \in [t_0+h,T]}$ has $(\bar{m}_t = (id + \sqrt{2}(W_{t_0+h} -$

$W_{t_0}))\sharp \tilde{m}_t)_{t \in [t_0+h,T]}$ as marginal conditional distributions (given $(W_t)_{t \in [0,T]}$). The process satisfies the SDE

$$d\tilde{X}_t = -D_p \tilde{H}_{t_0,t}\Big(\tilde{X}_t - \sqrt{2}(W_{t_0+h} - W_{t_0}),$$

$$D\tilde{u}_t\big(\tilde{X}_t - \sqrt{2}(W_{t_0+h} - W_{t_0})\big)\Big) dt + \sqrt{2}dB_t$$

$$= -D_p \tilde{H}_{t_0+h,t}\Big(\tilde{X}_t, D\bar{u}_t(\tilde{X}_t)\Big) dt + \sqrt{2}dB_t,$$

which is enough to check that the forward equation holds true, with $\bar{m}_{t_0+h} = m_{t_0,t_0+h}$ as the initial condition; see (5.3).

We now have

$$d_t \bar{u}_t = \big[-\Delta \bar{u}_t + \{\tilde{H}_{t_0,t}(\cdot, D\tilde{u}_t) - \tilde{F}_{t_0,t}(\cdot, m_{t_0,t})\}\big(\cdot - \sqrt{2}(W_{t_0+h} - W_{t_0})\big)\big] dt$$

$$+ d\tilde{M}_t\big(\cdot - \sqrt{2}(W_{t_0+h} - W_{t_0})\big)$$

$$= \big[-\Delta \bar{u}_t + \{\tilde{H}_{t_0+h,t}(\cdot, D\bar{u}_t) - \tilde{F}_{t_0+h,t}(\cdot, m_{t_0,t})\}\big] dt$$

$$+ d\tilde{M}_t\big(\cdot - \sqrt{2}(W_{t_0+h} - W_{t_0})\big).$$

Now, (5.3) says that $m_{t_0,t}$ reads $[id + \sqrt{2}(W_t - W_{t_0+h})]\sharp \bar{m}_t$, where $(\bar{m}_t)_{t_0+h \leqslant t \leqslant T}$ is the current forward component. This matches exactly the prescription on the backward equation in (5.1) and (5.2).

If m_{t_0,t_0+h} was deterministic, we would have, by definition of U, $U(t_0 + h, x, m_{t_0,t_0+h}) = \bar{u}_{t_0+h}(x)$, $x \in \mathbb{T}^d$, and thus, by definition of $\bar{u}_{t_0+h}$,

$$\tilde{u}_{t_0+h}(x) = U\big(t_0 + h, x + \sqrt{2}(W_{t_0+h} - W_{t_0}), m_{t_0,t_0+h}\big), \quad x \in \mathbb{T}^d. \qquad (5.4)$$

Although the result is indeed correct, the argument is false, as m_{t_0,t_0+h} is random.

To prove (5.4), we proceed as follows. By compactness of $\mathcal{P}(\mathbb{T}^d)$, we can find, for any ε, a family of N disjoint Borel subsets $A^1, \ldots, A^N \subset \mathcal{P}(\mathbb{T}^d)$, each of them being of diameter less than ε, that covers $\mathcal{P}(\mathbb{T}^d)$.

For each $i \in \{1, \ldots, N\}$, we may find $\mu_i \in A^i$. We then call $(\hat{m}_t^i, \hat{u}_t^i)_{t \in [t_0+h,T]}$ the solution of (5.1)–(5.2), with t_0 replaced by $t_0 + h$ and with μ_i as initial condition. We let

$$\hat{m}_t := \sum_{i=1}^{N} \hat{m}_t^i \mathbf{1}_{A^i}\big(m_{t_0,t_0+h}\big),$$

$$\hat{u}_t := \sum_{i=1}^{N} \hat{u}_t^i \mathbf{1}_{A^i}\big(m_{t_0,t_0+h}\big).$$

Since the events $\{m_{t_0,t_0+h} \in A^i\}$, for each $i = 1, \ldots, N$, are independent of the Brownian motion $(W_t - W_{t_0+h})_{t \in [t_0+h,T]}$, the process $(\hat{m}_t, \hat{u}_t)_{t \in [t_0+h,T]}$ is a solution of (5.1)–(5.2), with t_0 replaced by $t_0 + h$ and with $\hat{m}_{t_0, t_0+h}$ as the initial condition. With an obvious generalization of Theorem 4.3.1 to cases when the initial conditions are random, we deduce that

$$
\mathbb{E}\big[\|\bar{u}_{t_0+h} - \hat{u}_{t_0+h}\|_{n+\alpha}^2\big] \leqslant C\mathbb{E}\big[\mathbf{d}_1^2(\bar{m}_{t_0+h}, \hat{m}_{t_0+h})\big]
$$
$$
= C\sum_{i-1}^{N} \mathbb{E}\big[\mathbf{1}_{A^i}\big(m_{t_0, t_0+h}\big)\mathbf{d}_1^2(m_{t_0, t_0+h}, \mu^i)\big].
$$

Obviously, the right-hand side is less than $C\varepsilon^2$. The trick is then to say that $\hat{u}_{t_0+h}^i$ reads $U(t_0 + h, \cdot, \mu_i)$. Therefore,

$$
\sum_{i=1}^{N} \mathbb{E}\big[\mathbf{1}_{A^i}\big(m_{t_0, t_0+h}\big)\|\bar{u}_{t_0+h} - U(t_0 + h, \cdot, \mu^i)\|_{n+\alpha}^2\big] \leqslant C\varepsilon^2.
$$

Using the Lipschitz property of $U(t_0 + h, \cdot, \cdot)$ in the measure argument (see Theorem 4.3.1), we deduce that

$$
\mathbb{E}\Big[\big\|\bar{u}_{t_0+h} - U\big(t_0 + h, \cdot, m_{t_0, t_0+h}\big)\big\|_{n+\alpha}^2\Big] \leqslant C\varepsilon^2.
$$

Letting ε tend to 0, we complete the proof. $\square$

Corollary 5.1.2. *For any $\alpha' \in (0, \alpha)$, we can find a constant C such that, for any $t_0 \in [0, T]$, $h \in [0, T - t_0]$, and $m_0 \in \mathcal{P}(\mathbb{T}^d)$,*

$$
\big\|U(t_0 + h, \cdot, m_0) - U(t_0, \cdot, m_0)\big\|_{n+\alpha'} \leqslant Ch^{(\alpha-\alpha')/2}.
$$

Proof. Using the backward equation in (5.1), we have that

$$
\tilde{u}_{t_0}(\cdot) = \mathbb{E}\bigg[P_h \tilde{u}_{t_0+h}(\cdot) - \int_{t_0}^{t_0+h} P_{s-t_0}\big(\tilde{H}_{t_0, s}(\cdot, D\tilde{u}_s) - \tilde{F}_{t_0, s}(\cdot, m_{t_0, s})\big)ds\bigg].
$$

Therefore,

$$
\tilde{u}_{t_0}(\cdot) - \mathbb{E}\big(\tilde{u}_{t_0+h}(\cdot)\big) = \mathbb{E}\bigg[(P_h - id)\tilde{u}_{t_0+h}(\cdot)
$$
$$
- \int_{t_0}^{t_0+h} P_{s-t_0}\big(\tilde{H}_{t_0, s}(\cdot, D\tilde{u}_s) - \tilde{F}_{t_0, s}(\cdot, m_{t_0, s})\big)ds\bigg].
$$

So that

$$
\|\tilde{u}_{t_0} - \mathbb{E}(\tilde{u}_{t_0+h})\|_{n+\alpha'} \leqslant \mathbb{E}\Big[\big\|(P_h - id)\tilde{u}_{t_0+h}\big\|_{n+\alpha'}\Big]
$$
$$
+ C \int_{t_0}^{t_0+h} (s - t_0)^{-1/2}\big\|\tilde{H}_{t_0,s}(\cdot, D\tilde{u}_s) - \tilde{F}_{t_0,s}(\cdot, m_{t_0,s})\big\|_{n+\alpha'-1} ds.
$$

It is well checked that

$$
\mathbb{E}\Big[\big\|(P_h - id)\tilde{u}_{t_0+h}\big\|_{n+\alpha'}\Big] \leqslant C h^{(\alpha-\alpha')/2} \mathbb{E}\Big[\big\|\tilde{u}_{t_0+h}\big\|_{n+\alpha}\Big]
$$
$$
\leqslant C h^{(\alpha-\alpha')/2},
$$

the last line following from Lemma 4.3.7.

Now, by Lemma 5.1.1,

$$
\mathbb{E}\big[\tilde{u}_{t_0+h}\big]
$$
$$
= \mathbb{E}\big[U\big(t_0 + h, \cdot + \sqrt{2}(W_{t_0+h} - W_{t_0}), m_{t_0,t_0+h}\big)\big]
$$
$$
= \mathbb{E}\big[U\big(t_0 + h, \cdot + \sqrt{2}(W_{t_0+h} - W_{t_0}), m_{t_0,t_0+h}\big)
$$
$$
- U\big(t_0 + h, \cdot, m_0\big)\big] + U\big(t_0 + h, \cdot, m_0\big),
$$

where, by Theorem 4.3.1, it holds that

$$
\Big\|\mathbb{E}\big[U\big(t_0 + h, \cdot + \sqrt{2}(W_{t_0+h} - W_{t_0}), m_{t_0,t_0+h}\big) - U\big(t_0 + h, \cdot, m_0\big)\big]\Big\|_{n+\alpha'}
$$
$$
\leqslant C\mathbb{E}\big[|\mathbf{d}_1(m_{t_0,t_0+h}, m_0)|\big] + \mathbb{E}\Big[\big\|U\big(t_0 + h, \cdot
$$
$$
+ \sqrt{2}(W_{t_0+h} - W_{t_0}), m_0\big) - U\big(t_0 + h, \cdot, m_0\big)\big\|_{n+\alpha'}\Big],
$$

which is less than $C h^{(\alpha-\alpha')/2}$. $\square$

5.2   FIRST-ORDER DIFFERENTIABILITY

Assumption. Throughout the section, we assume that F, G, and H satisfy (2.4) and (2.5) in Subsection 2.3 and that, for some integer $n \geqslant 2$ and some $\alpha \in (0,1)$, **(HF1(n))** and **(HG1($n+1$))** hold true.

The purpose is here to follow Section 3.4 in order to establish the differentiability of U with respect to the argument m_0. The analysis is performed at t_0 fixed, so that, without any loss of generality, t_0 can be chosen as $t_0 = 0$.

The initial distribution $m_0 \in \mathcal{P}(\mathbb{T}^d)$ being given, we call $(\tilde{m}, \tilde{u})$ the solution of the system (4.7) with m_0 as initial distribution. Following (3.29), the strategy is to investigate the linearized system (of the same type as (4.37)):

$$d_t \tilde{z}_t = \left\{ -\Delta \tilde{z}_t + \langle D_p \tilde{H}_t(\cdot, D\tilde{u}_t), D\tilde{z}_t \rangle - \frac{\delta \tilde{F}_t}{\delta m}(\cdot, m_t)(\rho_t) \right\} dt + d\tilde{M}_t,$$

$$\partial_t \tilde{\rho}_t - \Delta \tilde{\rho}_t - \mathrm{div}\big(\tilde{\rho}_t D_p \tilde{H}_t(\cdot, D\tilde{u}_t) \big) - \mathrm{div}\big(\tilde{m}_t D_{pp}^2 \tilde{H}_t(\cdot, D\tilde{u}_t) D\tilde{z}_t \big) = 0,$$

(5.5)

with a boundary condition of the form

$$\tilde{z}_T = \frac{\delta \tilde{G}}{\delta m}(\cdot, m_T)(\rho_T).$$

As explained later on, the initial condition of the forward equation will be chosen in an appropriate way. In that framework, we shall repeatedly apply the results from Section 4.4 with

$$\tilde{V}_t(\cdot) = D_p \tilde{H}_t(\cdot, D\tilde{u}_t), \quad \Gamma_t = D_{pp}^2 \tilde{H}_t(\cdot, D\tilde{u}_t), \quad t \in [0, T],$$

(5.6)

which motivates the following lemma:

Lemma 5.2.1. *There exists a constant C such that, for any initial condition $m_0 \in \mathcal{P}(\mathbb{T}^d)$, the processes $(\tilde{V}_t)_{t \in [0,T]}$ and $(\Gamma_t)_{t \in [0,T]}$ in (5.6) satisfy points 2 and 4 in the introduction to Section 4.4.*

Proof. By Theorem 4.3.1 and Lemma 4.3.7, we can find a constant C such that any solution $(\tilde{m}_t, \tilde{u}_t)_{t \in [0,T]}$ to (4.7) satisfies, independently of the initial condition m_0,

$$\operatorname*{essup}_{\omega \in \Omega} \sup_{t \in [0,T]} \|\tilde{u}_t\|_{n+1+\alpha} \leqslant C.$$

In particular, allowing the constant C to increase from line to line, it must hold that

$$\operatorname*{essup}_{\omega \in \Omega} \sup_{t \in [0,T]} \big\| D_p \tilde{H}_t(\cdot, D\tilde{u}_t) \big\|_{n+\alpha} \leqslant C.$$

Moreover, implementing the local coercivity condition (2.4), we deduce that (assuming $C \geqslant 1$), with probability 1, for all $t \in [0, T]$,

$$\|\Gamma_t\|_1 \leqslant C \ ; \quad \forall x \in \mathbb{T}^d, \quad C^{-1} I_d \leqslant \Gamma_t(x) \leqslant C I_d,$$

which completes the proof.   □

Given $y \in \mathbb{T}^d$ and a d-tuple $\ell \in \{0, \ldots, n\}^d$ such that $|\ell| = \sum_{i=1}^n \ell_i \leqslant n$, we call $\mathbb{T}^d \ni x \mapsto v^{(\ell)}(x, m_0, y) \in \mathbb{R}$ the value at time 0 of the backward component of the solution to (5.5) when the forward component is initialized with the Schwartz distribution $(-1)^{|\ell|} D^\ell \delta_y$. Clearly, $D^\ell \delta_y \in (\mathcal{C}^{n+\alpha'}(\mathbb{T}^d))'$ for any $\alpha' \in (0, 1)$, so that, by Theorem 4.4.2, $v^{(\ell)}(\cdot, m_0, y)$ belongs to $\mathcal{C}^{n+\alpha}(\mathbb{T}^d)$. (Recall that, for a test function $\varphi \in \mathcal{C}^n(\mathbb{T}^d)$, $(D^\ell \delta_y)\varphi = (-1)^{|\ell|} D^\ell_{y_1^{\ell_1} \ldots y_d^{\ell_d}} \varphi(y)$.) Similarly, we may denote by $(\tilde{\rho}_t^{\ell,y}, \tilde{z}_t^{\ell,y})_{t\in[0,T]}$ the solution of (5.5) with $\tilde{\rho}_0^{\ell,y} = (-1)^{|\ell|} D^\ell \delta_y$ as the initial condition. For simplicity, we omit m_0 in the notation. We then have

$$\tilde{z}_0^{\ell,y} = v^{(\ell)}(\cdot, m_0, y). \tag{5.7}$$

We then claim

Lemma 5.2.2. *Let $m_0 \in \mathcal{P}(\mathbb{T}^d)$. Then, with the same notation as above, we have, for any $\alpha' \in (0, \alpha)$ and any d-tuple $\ell \in \{0, \dots, n\}^d$ such that $|\ell| \leqslant n$,*

$$\lim_{\mathbb{T}^d \ni h \to 0} \operatorname{essup}_{\omega \in \Omega} \sup_{t\in[0,T]} \left(\left\| \tilde{\rho}_t^{\ell,y+h} - \tilde{\rho}_t^{\ell,y} \right\|_{-(n+\alpha')} \right.$$
$$+ \left\| \tilde{z}_t^{\ell,y+h} - \tilde{z}_t^{\ell,y} \right\|_{n+1+\alpha} \right) = 0. \tag{5.8}$$

Moreover, for any $\ell \in \{0, \dots, n-1\}^d$ with $|\ell| \leqslant n-1$ and any $i \in \{1, \dots, d\}$,

$$\lim_{\mathbb{R}\backslash\{0\} \ni h \to 0} \operatorname{essup}_{\omega \in \Omega} \sup_{t\in[0,T]} \left(\left\| \frac{1}{h}\left(\tilde{\rho}_t^{\ell,y+he_i} - \tilde{\rho}_t^{\ell,y} \right) - \tilde{\rho}_t^{\ell+e_i,y} \right\|_{-(n+\alpha')} \right.$$
$$+ \left\| \frac{1}{h}\left(\tilde{z}_t^{\ell,y+he_i} - \tilde{z}_t^{\ell,y} \right) - \tilde{z}_t^{\ell+e_i,y} \right\|_{n+1+\alpha} \right) = 0,$$

where e_i denotes the i-th vector of the canonical basis and $\ell + e_i$ is understood as $(\ell + e_i)_j = \ell_j + \delta_i^j$, for $j \in \{1, \dots, d\}$, δ_i^j denoting the Kronecker symbol.

In particular, the function $[\mathbb{T}^d]^2 \ni (x, y) \mapsto v^{(0)}(x, m_0, y)$ is n-times differentiable with respect to y and, for any $\ell \in \{0, \dots, n\}^d$ with $|\ell| \leqslant n$, the derivative $D_y^\ell v^{(0)}(\cdot, m_0, y) : \mathbb{T}^d \ni x \mapsto D_y^\ell v^{(0)}(x, m_0, y)$ belongs to $\mathcal{C}^{n+1+\alpha}(\mathbb{T}^d)$ and we write

$$D_y^\ell v^{(0)}(x, m_0, y) = v^{(\ell)}(x, m_0, y), \quad (x, y) \in \mathbb{T}^d.$$

Moreover,

$$\sup_{m_0 \in \mathcal{P}(\mathbb{T}^d)} \sup_{y \in \mathbb{T}^d} \| D_y^\ell v^{(0)}(\cdot, m_0, y) \|_{n+1+\alpha} < \infty.$$

Proof. By Corollary 4.4.6, we can find a constant C such that, for all $y \in \mathbb{T}^d$, for all $m_0 \in \mathcal{P}(\mathbb{T}^d)$, and all $\ell \in \{0, \dots, n\}^d$ with $|\ell| \leqslant n$,

$$\operatorname{essup}_{\omega \in \Omega} \sup_{t\in[0,T]} \left(\| \tilde{z}_t^{\ell,y} \|_{n+1+\alpha} + \| \tilde{\rho}_t^{\ell,y} \|_{-(n+\alpha')} \right) \leqslant C.$$

In particular,

$$\| v^{(\ell)}(\cdot, m_0, y) \|_{n+1+\alpha} \leqslant C.$$

Now, we make use of Proposition 4.4.5. We know that, for any $\alpha' \in (0, 1)$,

$$\lim_{h \to 0} \left\| D^\ell \delta_{y+h} - D^\ell \delta_y \right\|_{-(n+\alpha')} = 0.$$

Therefore, for $\alpha' < \alpha$, Corollary 4.4.6 gives (5.8). This yields

$$\lim_{h \to 0} \left\| v^{(\ell)}(\cdot, m_0, y + h) - v^{(\ell)}(\cdot, m_0, y) \right\|_{n+1+\alpha} = 0,$$

proving that the mapping $\mathbb{T}^d \ni y \mapsto v^{(\ell)}(\cdot, m_0, y) \in \mathcal{C}^{n+1+\alpha}(\mathbb{T}^d)$ is continuous. Similarly, for $|\ell| \leqslant n - 1$ and $i \in \{1, \dots, d\}$,

$$\lim_{\mathbb{R} \backslash \{0\} \ni h \to 0} \left\| \frac{1}{h} \left(D^\ell \delta_{y+he_i} - D^\ell \delta_y \right) + D^{\ell+e_i} \delta_y \right\|_{-(n+\alpha')} = 0,$$

or equivalently,

$$\lim_{\mathbb{R} \backslash \{0\} \ni h \to 0} \left\| \frac{1}{h} \left((-1)^{|\ell|} D^\ell \delta_{y+he_i} - (-1)^{|\ell|} D^\ell \delta_y \right) - (-1)^{|\ell+e_i|} D^{\ell+e_i} \delta_y \right\|_{-(n+\alpha')} = 0.$$

As a byproduct, we get

$$\lim_{\mathbb{R} \backslash \{0\} \ni h \to 0} \left\| \frac{1}{h} \left[v^{(\ell)}(\cdot, m_0, y + he_i) - v^{(\ell)}(\cdot, m_0, y) \right] - v^{(\ell+e_i)}(\cdot, m_0, y) \right\|_{n+1+\alpha} = 0,$$

which proves, by induction, that

$$D_y^\ell v^{(0)}(x, m_0, y) = v^{(\ell)}(x, m_0, y), \quad x, y \in \mathbb{T}^d.$$

This completes the proof. $\square$

Now, we prove

Lemma 5.2.3.   *Given a finite signed measure μ on $\mathbb{T}^d$, the solution $\tilde{z}$ to (5.5) with μ as initial condition reads, when taken at time 0,*

$$\tilde{z}_0 : \mathbb{R}^d \ni x \mapsto \tilde{z}_0(x) = \int_{\mathbb{T}^d} v^{(0)}(x, m_0, y) d\mu(y).$$

Proof.   By compactness of the torus, we can find, for a given $\varepsilon > 0$, a covering $(U_i)_{1 \leqslant i \leqslant N}$ of $\mathbb{T}^d$, made of disjoint Borel subsets, such that each U_i, $i = 1, \dots, N$, has a diameter less than ε. Choosing, for each $i \in \{1, \dots, N\}$, $y_i \in U_i$, we then let

$$\mu^\varepsilon := \sum_{i=1}^N \mu(U_i) \delta_{y_i}.$$

Then, for any $\varphi \in \mathcal{C}^1(\mathbb{T}^d)$, with $\|\varphi\|_1 \leqslant 1$, we have

$$\left| \int_{\mathbb{T}^d} \varphi(y) d(\mu - \mu^\varepsilon)(y) \right| = \left| \sum_{i=1}^N \int_{U_i} (\varphi(y) - \varphi(y_i)) d\mu(y) \right| \leqslant C \|\mu\| \varepsilon,$$

where we have denoted by $\|\mu\|$ the total mass of μ.

Therefore, by Proposition 4.4.5,

$$\left\| \tilde{z}_0 - \sum_{i=1}^{N} \int_{U_i} v^{(0)}(\cdot, m_0, y_i) d\mu(y) \right\|_{n+1+\alpha} \leqslant C\|\mu\|\varepsilon,$$

where we have used the fact that, by linearity, the value at time 0 of the backward component of the solution to (5.5), when the forward component is initialized with μ^ε, reads

$$\sum_{i=1}^{N} \mu(U_i) v^{(0)}(\cdot, m_0, y_i) = \sum_{i=1}^{N} \int_{U_i} v^{(0)}(\cdot, m_0, y_i) d\mu(y).$$

By smoothness of $v^{(0)}$ in y, we easily deduce that

$$\left\| \tilde{z}_0 - \int_{\mathbb{T}^d} v^{(0)}(\cdot, m_0, y) d\mu(y) \right\|_{n+1+\alpha} \leqslant C\|\mu\|\varepsilon.$$

The result follows by letting ε tend to 0. $\square$

On the model of Corollary 3.4.4, we now claim

Proposition 5.2.4. *Given two initial conditions* $m_0, m_0' \in \mathcal{P}(\mathbb{T}^d)$, *we denote by* $(\tilde{m}_t, \tilde{u}_t)_{t \in [0,T]}$ *and* $(\tilde{m}_t', \tilde{u}_t')_{t \in [0,T]}$ *the respective solutions of (4.7) with* m_0 *and* m_0' *as initial conditions and by* $(\tilde{\rho}_t, \tilde{z}_t)_{t \in [0,T]}$ *the solution of (5.5) with* $m_0' - m_0$ *as the initial condition, so that we can let*

$$\delta\tilde{\rho}_t = \tilde{m}_t' - \tilde{m}_t - \tilde{\rho}_t, \quad \delta\tilde{z}_t = \tilde{u}_t' - \tilde{u}_t - \tilde{z}_t, \quad t \in [0,T].$$

Then, for any $\alpha' \in (0, \alpha)$, *we can find a constant* C, *independent of* m_0 *and* m_0', *such that*

$$\operatorname{essup}_{\omega \in \Omega} \sup_{0 \leqslant t \leqslant T} \left(\|\delta\tilde{\rho}_t\|_{-(n+\alpha')} + \|\delta\tilde{z}_t\|_{n+1+\alpha} \right) \leqslant C\mathbf{d}_1^2(m_0, m_0').$$

In particular,

$$\left\| U(0, \cdot, m_0') - U(0, \cdot, m_0) - \int_{\mathbb{T}^d} v^{(0)}(x, m_0, y) d(m_0' - m_0)(y) \right\|_{n+1+\alpha}$$
$$\leqslant C\mathbf{d}_1^2(m_0, m_0'),$$

and, thus, for any $x \in \mathbb{T}^d$, *the mapping* $\mathcal{P}(\mathbb{T}^d) \ni m \mapsto U(0, x, m)$ *is differentiable with respect to* m *and the derivative reads, for any* $m \in \mathcal{P}(\mathbb{T}^d)$,

$$\frac{\delta U}{\delta m}(0, x, m, y) = v^{(0)}(x, m, y), \quad y \in \mathbb{T}^d.$$

The normalization condition holds:

$$\int_{\mathbb{T}^d} v^{(0)}(x, m, y) dm(y) = 0.$$

The proof is the same as in the deterministic case (see Remark 3.4.5).

Proof. We have

$$d_t(\delta \tilde{z}_t) = \{-\Delta(\delta \tilde{z}_t) + \langle D_p \tilde{H}_t(\cdot, D\tilde{u}_t), D(\delta \tilde{z}_t) \rangle$$
$$- \frac{\delta \tilde{F}_t}{\delta m}(\cdot, m_t)(\delta \rho_t) + \tilde{f}_t\} dt + d\tilde{M}_t,$$

together with

$$\partial_t(\delta \tilde{\rho}_t) - \Delta(\delta \tilde{\rho}_t) - \text{div}\left[(\delta \tilde{\rho}_t) D_p \tilde{H}_t(\cdot, D\tilde{u}_t)\right]$$
$$- \text{div}\left[\tilde{m}_t D_{pp}^2 \tilde{H}_t(\cdot, D\tilde{u}_t)(D\delta \tilde{z}_t) + \tilde{b}_t\right] = 0,$$

with a boundary condition of the form

$$\delta \tilde{z}_T = \frac{\delta \tilde{G}}{\delta m}(\cdot, m_T)(\delta \rho_T) + \tilde{g}_T,$$

where

$$\tilde{b}_t = \tilde{m}_t'\big(D_p \tilde{H}_t(\cdot, D\tilde{u}_t') - D_p \tilde{H}_t(\cdot, D\tilde{u}_t)\big) - \tilde{m}_t D_{pp}^2 \tilde{H}_t(\cdot, D\tilde{u}_t)(D\tilde{u}_t' - D\tilde{u}_t)$$
$$\tilde{f}_t = \tilde{H}_t(\cdot, D\tilde{u}_t') - \tilde{H}_t(\cdot, D\tilde{u}_t) - \langle D_p \tilde{H}_t(\cdot, D\tilde{u}_t), D\tilde{u}_t' - D\tilde{u}_t \rangle$$
$$- \Big(\tilde{F}_t(\cdot, m_t') - \tilde{F}_t(\cdot, m_t) - \frac{\delta \tilde{F}_t}{\delta m}(\cdot, m_t)(m_t' - m_t)\Big),$$
$$\tilde{g}_T = \tilde{G}(\cdot, m_T') - \tilde{G}(\cdot, m_T) - \frac{\delta \tilde{G}}{\delta m}(\cdot, m_T)(m_T' - m_T).$$

Now,

$$\tilde{b}_t = (\tilde{m}_t' - \tilde{m}_t)\big(D_p \tilde{H}_t(\cdot, D\tilde{u}_t') - D_p \tilde{H}_t(\cdot, D\tilde{u}_t)\big)$$
$$+ \tilde{m}_t \int_0^1 \left[D_{pp}^2 \tilde{H}_t\big(\cdot, \lambda D\tilde{u}_t' + (1-\lambda)D\tilde{u}_t\big) - D_{pp}^2 \tilde{H}_t(\cdot, D\tilde{u}_t)\right](D\tilde{u}_t' - D\tilde{u}_t)d\lambda$$
$$= (\tilde{m}_t' - \tilde{m}_t) \int_0^1 D_{pp}^2 \tilde{H}_t\big(\cdot, \lambda D\tilde{u}_t' + (1-\lambda)D\tilde{u}_t\big)(D\tilde{u}_t' - D\tilde{u}_t)d\lambda$$
$$+ \tilde{m}_t \int_0^1 \int_0^1 \lambda D_{ppp}^3 \tilde{H}_t\big(\cdot, \lambda s D\tilde{u}_t' + (1-\lambda+\lambda(1-s))D\tilde{u}_t\big)(D\tilde{u}_t' - D\tilde{u}_t)^{\otimes 2} d\lambda \, ds.$$

Also,

$$\tilde{f}_t = \int_0^1 \langle D_p \tilde{H}_t(\cdot, \lambda D\tilde{u}_t' + (1-\lambda)D\tilde{u}_t) - D_p \tilde{H}_t(\cdot, D\tilde{u}_t), D\tilde{u}_t' - D\tilde{u}_t \rangle d\lambda$$

$$- \int_0^1 \Big(\frac{\delta \tilde{F}_t}{\delta m}(\cdot, \lambda m_t' + (1-\lambda)m_t) - \frac{\delta \tilde{F}_t}{\delta m}(\cdot, m_t) \Big)(m_t' - m_t)d\lambda$$

$$= \int_0^1 \int_0^1 \lambda \langle D_{pp}^2 \tilde{H}_t(\cdot, \lambda s D\tilde{u}_t' + (1 - \lambda + \lambda(1-s))D\tilde{u}_t)(D\tilde{u}_t' - D\tilde{u}_t),$$

$$D\tilde{u}_t' - D\tilde{u}_t \rangle d\lambda\, ds$$

$$- \int_0^1 \Big(\frac{\delta \tilde{F}_t}{\delta m}(\cdot, \lambda m_t' + (1-\lambda)m_t) - \frac{\delta \tilde{F}_t}{\delta m}(\cdot, m_t) \Big)(m_t' - m_t)d\lambda.$$

And,

$$\tilde{g}_T = \int_0^1 \Big(\frac{\delta \tilde{G}}{\delta m}(\cdot, \lambda m_T' + (1-\lambda)m_T) - \frac{\delta \tilde{G}}{\delta m}(\cdot, m_T) \Big)(m_T' - m_T)d\lambda.$$

By Lemma 4.3.7, we have a universal bound for

$$\operatorname{essup}_{\omega \in \Omega} \sup_{t \in [0,T]} \big(\|\tilde{u}_t\|_{n+1+\alpha} + \|\tilde{u}_t'\|_{n+1+\alpha} \big).$$

We deduce that

$$\|\tilde{b}_t\|_{-1} \leqslant C \Big(\mathbf{d}_1(\tilde{m}_t', \tilde{m}_t) \|\tilde{u}_t' - \tilde{u}_t\|_2 + \|\tilde{u}_t' - \tilde{u}_t\|_1^2 \Big),$$

$$\|\tilde{f}_t\|_{n+\alpha} \leqslant C \Big(\|\tilde{u}_t' - \tilde{u}_t\|_{n+1+\alpha}^2 + \mathbf{d}_1^2(\tilde{m}_t', \tilde{m}_t) \Big),$$

$$\|\tilde{g}_T\|_{n+1+\alpha} \leqslant C \mathbf{d}_1^2(\tilde{m}_T', \tilde{m}_T).$$

Therefore, by Theorem 4.3.1, we deduce that

$$\operatorname{essup}_{\omega \in \Omega} \sup_{0 \leqslant t \leqslant T} \|\tilde{b}_t\|_{-1} + \operatorname{essup}_{\omega \in \Omega} \sup_{0 \leqslant t \leqslant T} \|\tilde{f}_t\|_{n+\alpha} + \operatorname{essup}_{\omega \in \Omega} \|\tilde{g}_T\|_{n+1+\alpha}$$

$$\leqslant C \mathbf{d}_1^2(m_0', m_0).$$

By Corollary 4.4.6, we get the first of the two inequalities in the statement. We deduce that

$$\big\| U(0, \cdot, m_0') - U(0, \cdot, m_0) - \tilde{z}_0 \big\|_{n+1+\alpha} \leqslant C \mathbf{d}_1^2(m_0, m_0').$$

By Lemma 5.2.3, we complete the proof. $\square$

Proposition 5.2.5. *For any $\alpha' \in (0, \alpha)$, we can find a constant C such that, for any $m_0, m_0' \in \mathcal{P}(\mathbb{T}^d)$, any $y, y' \in \mathbb{T}^d$ and any index $\ell \in \{0, \ldots, n\}^d$ with*

$|\ell| \leqslant n$, *denoting by* $(\tilde{m}_t, \tilde{u}_t)_{t \in [0,T]}$ *and* $(\tilde{m}'_t, \tilde{u}'_t)_{t \in [0,T]}$ *the respective solutions of* (4.7), *and then* $(\tilde{\rho}_t, \tilde{z}_t)_{t \in [0,T]}$ *and* $(\tilde{\rho}'_t, \tilde{z}'_t)_{t \in [0,T]}$ *the corresponding solutions of* (5.5) *when driven by two initial conditions* $(-1)^{|\ell|} D^\ell \delta_y$ *and* $(-1)^{|\ell|} D^\ell \delta_{y'}$, *it holds that*

$$\text{essup}_{\omega \in \Omega} \left[\sup_{t \in [0,T]} \|\tilde{z}_t - \tilde{z}'_t\|_{n+1+\alpha} + \sup_{t \in [0,T]} \|\tilde{\rho}_t - \tilde{\rho}'_t\|_{-(n+\alpha')} \right]$$
$$\leqslant C \Big(\mathbf{d}_1(m_0, m'_0) + |y - y'|^{\alpha'} \Big).$$

In particular,

$$\forall y, y' \in \mathbb{T}^d, \quad \left\| D_y^\ell \frac{\delta U}{\delta m}(0, \cdot, m_0, y) - D_y^\ell \frac{\delta U}{\delta m}(0, \cdot, m'_0, y') \right\|_{n+1+\alpha}$$
$$\leqslant C \Big(\mathbf{d}_1(m_0, m'_0) + |y - y'|^{\alpha'} \Big).$$

Proof. Given two initial conditions m_0 and m'_0, we call $(\tilde{m}_t, \tilde{u}_t)_{t \in [0,T]}$ and $(\tilde{m}'_t, \tilde{u}'_t)_{t \in [0,T]}$ the respective solutions of (4.7). With $(\tilde{m}_t, \tilde{u}_t)_{t \in [0,T]}$ and $(\tilde{m}'_t, \tilde{u}'_t)_{t \in [0,T]}$, we associate the solutions $(\tilde{\rho}_t, \tilde{z}_t)_{t \in [0,T]}$ and $(\tilde{\rho}'_t, \tilde{z}'_t)_{t \in [0,T]}$ of (5.5) when driven by two initial conditions $(-1)^{|\ell|} D^\ell \delta_y$ and $(-1)^{|\ell|} D^\ell \delta_{y'}$. Since $|\ell| \leqslant n$, we have

$$\left\| D^\ell \delta_y - D^\ell \delta_{y'} \right\|_{-(n+\alpha')} \leqslant |y - y'|^{\alpha'}.$$

To prove the first estimate, we can apply Corollary 4.4.6 with

$$\tilde{V}_t = D_p \tilde{H}(\cdot, D\tilde{u}_t), \quad \tilde{V}'_t = D_p \tilde{H}(\cdot, D\tilde{u}'_t),$$
$$\Gamma_t = D^2_{pp} \tilde{H}_t(\cdot, D\tilde{u}_t), \quad \Gamma'_t = D^2_{pp} \tilde{H}_t(\cdot, D\tilde{u}'_t),$$

and $(\tilde{b}^0_t)_{t \in [0,T]} \equiv 0$, $(\tilde{f}^0_t)_{t \in [0,T]} \equiv 0$, $\tilde{g}^0_T \equiv 0$ and $\vartheta = 0$ in (4.38), so that, following the proof of Proposition 5.2.4,

$$\|\tilde{V}_t - \tilde{V}'_t\|_{n+\alpha} + \|\Gamma_t - \Gamma'_t\|_0 \leqslant C \|\tilde{u}_t - \tilde{u}'_t\|_{n+1+\alpha}.$$

Now, the first estimate in the statement follows from the combination of Theorem 4.3.1 and Corollary 4.4.6.

The second estimate is a straightforward consequence of the first one. $\square$

Proposition 5.2.6. *Propositions 5.2.4 and 5.2.5 easily extend to any initial time* $t_0 \in [0,T]$. *Then, for any* $\alpha' \in (0, \alpha)$, *any* $t_0 \in [0,T]$, *and* $m_0 \in \mathcal{P}(\mathbb{T}^d)$

$$\lim_{h \to 0} \sup_{\ell \in \{0, \dots, n\}^d, |\ell| \leqslant n} \left\| D_y^\ell \frac{\delta U}{\delta m}(t_0 + h, \cdot, m_0, \cdot) - D_y^\ell \frac{\delta U}{\delta m}(t_0, \cdot, m_0, \cdot) \right\|_{n+1+\alpha', \alpha'} = 0.$$

Proof. Given two probability measures $m, m' \in \mathcal{P}(\mathbb{T}^d)$, we know from Proposition 5.2.4 that, for any $t \in [0, T]$,

$$
U(t, \cdot, m') - U(t, \cdot, m)
$$
$$
= \int_{\mathbb{T}^d} \frac{\delta U}{\delta m}(t, \cdot, m, y) d(m' - m)(y) + O(\mathbf{d}_1^2(m, m')), \qquad (5.9)
$$

the equality holding true in $\mathcal{C}^{n+1+\alpha}(\mathbb{T}^d)$ and the Landau notation $O(\cdot)$ being uniform in t_0 and m (the constant C in the statement of Proposition 5.2.4 being explicitly quantified by means of Proposition 4.4.5, related to the stability of solutions to the linear equation).

From Proposition 5.2.5, we deduce that the set of functions $([\mathbb{T}^d]^2 \ni (x, y) \mapsto (\delta U / \delta m)(t, x, m, y))_{t \in [0, T]}$ is relatively compact in $\mathcal{C}^{n+1+\alpha'}(\mathbb{T}^d) \times \mathcal{C}^{n+\alpha'}(\mathbb{T}^d)$, for any $\alpha' \in (0, \alpha)$. Any limit $\Phi : [\mathbb{T}^d]^2 \to \mathbb{R}$ obtained by letting t tend to t_0 in (5.9) must satisfy (use Corollary 5.1.2 to pass to the limit in the left-hand side):

$$
U(t_0, \cdot, m') - U(t_0, \cdot, m) = \int_{\mathbb{T}^d} \Phi(\cdot, y) d(m' - m)(y) + O(\mathbf{d}_1^2(m, m')),
$$

the equality holding true in $\mathcal{C}^0(\mathbb{T}^d)$. This proves that, for any $x \in \mathbb{T}^d$,

$$
\int_{\mathbb{T}^d} \frac{\delta U}{\delta m}(t_0, x, m, y) d(m' - m)(y) = \int_{\mathbb{T}^d} \Phi(x, y) d(m' - m)(y).
$$

Choosing m' as the solution at time h of the Fokker–Planck equation

$$
\partial_t m_t = -\mathrm{div}(b m_t), \quad t \geqslant 0,
$$

for a smooth field b and with $m_0 = m$ as initial condition, and then letting h tend to 0, we deduce that

$$
\int_{\mathbb{T}^d} D_m U(t_0, x, m, y) \cdot b(y) dm(y) = \int_{\mathbb{T}^d} D_y \Phi(x, y) \cdot b(y) dm(y).
$$

When m has full support, this proves that

$$
\Phi(x, y) = \frac{\delta U}{\delta m}(t_0, x, m, y) + c(x), \quad x, y \in \mathbb{T}^d.
$$

Since both sides have a zero integral in y with respect to m, $c(x)$ must be zero.

When the support of m does not cover $\mathbb{T}^d$, we can approximate m by a sequence $(m_n)_{n \geqslant 1}$ of measures with full supports. By Proposition 5.2.5, we know that, for any $\alpha' \in (0, \alpha)$,

$$
\lim_{n \to \infty} \sup_{t \in [0, T]} \left\| \frac{\delta U}{\delta m}(t, \cdot, m_n, \cdot) - \frac{\delta U}{\delta m}(t, \cdot, m, \cdot) \right\|_{n+1+\alpha', \alpha'} = 0,
$$

so that, in $\mathcal{C}^{n+1+\alpha'}(\mathbb{T}^d) \times \mathcal{C}^{\alpha'}(\mathbb{T}^d)$,

$$\lim_{t \to t_0} \frac{\delta U}{\delta m}(t, \cdot, m, \cdot) = \lim_{n \to \infty} \lim_{t \to t_0} \frac{\delta U}{\delta m}(t, \cdot, m_n, \cdot) = \frac{\delta U}{\delta m}(t_0, \cdot, m, \cdot).$$

We easily complete the proof when $|\ell| = 0$. Since the set of functions $([\mathbb{T}^d]^2 \ni (x,y) \mapsto (D_y^\ell \delta U/\delta m)(t,x,m,y))_{t \in [0,T]}$ is relatively compact in $\mathcal{C}^{n+1+\alpha'}(\mathbb{T}^d) \times \mathcal{C}^{\alpha'}(\mathbb{T}^d)$, any limit as t tends to t_0 must coincide with the derivative of index ℓ in y of the limit of $[\mathbb{T}^d]^2 \ni (x,y) \mapsto [\delta U/\delta m](t,x,m,y)$ as t tends to t_0.   □

5.3   SECOND-ORDER DIFFERENTIABILITY

Assumption. Throughout the section, we assume that F, G, and H satisfy (2.4) and (2.5) in Subsection 2.3 and that, for some integer $n \geqslant 2$ and some $\alpha \in (0,1)$, **(HF2(n))** and **(HG2($n+1$))** hold true.

To complete the analysis of the master equation, we need to investigate the second-order differentiability in the direction of the measure, on the same model as for the first-order derivatives.

As for the first order, the idea is to write the second-order derivative of U in the direction m as the initial value of the backward component of a linearized system of the type (4.37), which is referred next to as the *second-order linearized system*. Basically, the *second-order linearized system* is obtained by differentiating one more time the *first-order linearized system* (5.5). Recalling that (5.5) has the form

$$d_t \tilde{z}_t = \left\{ -\Delta \tilde{z}_t + \langle D_p \tilde{H}_t(\cdot, D\tilde{u}_t), D\tilde{z}_t \rangle - \frac{\delta \tilde{F}_t}{\delta m}(\cdot, m_t)(\rho_t) \right\} dt + d\tilde{M}_t,$$

$$\partial_t \tilde{\rho}_t - \Delta \tilde{\rho}_t - \text{div}\big(\tilde{\rho}_t D_p \tilde{H}_t(\cdot, D\tilde{u}_t)\big) - \text{div}\big(\tilde{m}_t D_{pp}^2 \tilde{H}_t(\cdot, D\tilde{u}_t) D\tilde{z}_t\big) = 0, \tag{5.10}$$

with the boundary condition

$$\tilde{z}_T = \frac{\delta \tilde{G}}{\delta m}(\cdot, m_T)(\rho_T),$$

the procedure is to differentiate the pair $(\tilde{\rho}_t, \tilde{z}_t)_{t \in [0,T]}$ with respect to the initial condition m_0 of $(\tilde{m}_t, \tilde{u}_t)_{t \in [0,T]}$, the initial condition of $(\tilde{\rho}_t, \tilde{z}_t)_{t \in [0,T]}$ being kept frozen.

Above, $(\tilde{m}_t, \tilde{u}_t)_{0 \leqslant t \leqslant T}$ is indeed chosen as the solution of the system (4.7), for a given initial distribution $m_0 \in \mathcal{P}(\mathbb{T}^d)$, and $(\tilde{\rho}_t, \tilde{z}_t)_{t \in [0,T]}$ as the solution of the system (5.10) with an initial condition $\rho_0 \in (\mathcal{C}^{n+\alpha'}(\mathbb{T}^d))'$, for some $\alpha' < \alpha$. Implicitly, the initial condition ρ_0 is understood as some $m_0' - m_0$ for another $m_0' \in \mathcal{P}(\mathbb{T}^d)$, in which case we know from Proposition 5.2.4 that $(\tilde{\rho}_t, \tilde{z}_t)_{t \in [0,T]}$ reads as the derivative, at $\varepsilon = 0$, of the solution to (4.7) when initialized with the measure $m_0 + \varepsilon(m'_0 - m_0)$. However, following the strategy used in the analysis of the first-order derivatives of U, it is much more convenient, in order to investigate the second-order derivatives of U, to distinguish the initial condition of $(\tilde{\rho}_t)_{t\in[0,T]}$ from the direction $m'_0 - m_0$ used to differentiate the system (4.7). This says that, in (5.10), we should allow $(\tilde{\rho}_t, \tilde{z}_t)_{t\in[0,T]}$ to be driven by an arbitrary initial condition $\rho_0 \in (C^{n+\alpha'}(\mathbb{T}^d))'$.

Now, when (5.10) is driven by an arbitrary initial condition ρ_0 and m_0 is perturbed in the direction $m'_0 - m_0$ for another $m'_0 \in \mathcal{P}(\mathbb{T}^d)$ (that is m_0 is changed into $m_0 + \varepsilon(m'_0 - m_0)$ for some small ε), the system obtained by differentiating (5.10) (at $\varepsilon = 0$) takes the form

$$
d_t \tilde{z}_t^{(2)} = \Big\{ -\Delta \tilde{z}_t^{(2)} + \big\langle D_p \tilde{H}_t(\cdot, D\tilde{u}_t), D\tilde{z}_t^{(2)} \big\rangle - \frac{\delta \tilde{F}_t}{\delta m}(\cdot, m_t)(\rho_t^{(2)})
$$
$$
+ \big\langle D_{pp}^2 \tilde{H}_t(\cdot, D\tilde{u}_t), D\tilde{z}_t \otimes D\partial_m \tilde{u}_t \big\rangle - \frac{\delta^2 \tilde{F}_t}{\delta m^2}(\cdot, m_t)(\rho_t, \partial_m m_t) \Big\} dt
$$
$$
+ d\tilde{M}_t,
$$
$$
\partial_t \tilde{\rho}_t^{(2)} - \Delta \tilde{\rho}_t^{(2)} - \mathrm{div}\Big(\tilde{\rho}_t^{(2)} D_p \tilde{H}_t(\cdot, D\tilde{u}_t)\Big) - \mathrm{div}\Big(\tilde{m}_t D_{pp}^2 \tilde{H}_t(\cdot, D\tilde{u}_t) D\tilde{z}_t^{(2)}\Big)
$$
$$
- \mathrm{div}\Big(\tilde{\rho}_t D_{pp}^2 \tilde{H}_t(\cdot, D\tilde{u}_t) D\partial_m \tilde{u}_t\Big) - \mathrm{div}\Big(\partial_m \tilde{m}_t D_{pp}^2 \tilde{H}_t(\cdot, D\tilde{u}_t) D\tilde{z}_t\Big)
$$
$$
- \mathrm{div}\Big(\tilde{m}_t D_{ppp}^3 \tilde{H}_t(\cdot, D\tilde{u}_t) D\tilde{z}_t \otimes D\partial_m \tilde{u}_t\Big) = 0, \tag{5.11}
$$

with a terminal boundary condition of the form

$$
\tilde{z}_T^{(2)} = \frac{\delta \tilde{G}}{\delta m}(\cdot, m_T)(\rho_T^{(2)}) + \frac{\delta^2 \tilde{G}}{\delta m^2}(\cdot, m_T)(\rho_T, \partial_m m_T),
$$

where we have denoted by $(\partial_m \tilde{m}_t, \partial_m \tilde{u}_t)_{t\in[0,T]}$ the derivative of $(\tilde{m}_t, \tilde{u}_t)_{t\in[0,T]}$ when the initial condition is differentiated in the direction $m'_0 - m_0$ at point m_0, for another $m'_0 \in \mathcal{P}(\mathbb{T}^d)$. In (5.11), the pair $(\tilde{\rho}_t^{(2)}, \tilde{z}_t^{(2)})_{t\in[0,T]}$ is then understood as the derivative of the solution $(\tilde{\rho}_t, \tilde{z}_t)_{t\in[0,T]}$ to (5.10).

Now, using the same philosophy as in the analysis of the first-order derivatives, we can choose freely the initial condition ρ_0 of (5.10). Generally speaking, we will choose $\rho_0 = (-1)^{|\ell|} D^\ell \delta_y$, for some multi-index $\ell \in \{0, \ldots, n-1\}^d$ with $|\ell| \leqslant n - 1$ and some $y \in \mathbb{T}^d$. Since ρ_0 is expected to be insensitive to any perturbation that could apply to m_0, it then makes sense to let $\rho_0^{(2)} = 0$. As said earlier, the initial condition $\partial_m m_0$ of $(\partial_m \tilde{m}_t)_{0 \leqslant t \leqslant T}$ is expected to have the form $m'_0 - m_0$ for another probability measure $m'_0 \in \mathcal{P}(\mathbb{T}^d)$. Anyhow, by the same linearity argument as in the analysis of the first-order derivative, we can start with the case when $\partial_m m_0$ is the derivative of a Dirac mass, namely $\partial_m m_0 = (-1)^{|k|} D^k \delta_\zeta$, for another multi-index $k \in \{0, \ldots, n-1\}^d$, and another $\zeta \in \mathbb{T}^d$, in which case $(\partial_m \tilde{m}_t, \partial_m \tilde{u}_t)_{0 \leqslant t \leqslant T}$ is another solution to (5.10), but with

$\partial_m m_0 = (-1)^{|k|} D^k \delta_\zeta$ as the initial condition. Given these initial conditions, we then let

$$v^{(\ell,k)}(\cdot, m_0, y, \zeta) = \tilde{z}_0^{(2)},$$

provided that (5.11) has a unique solution.

To check that existence and uniqueness hold true, we may proceed as follows. The system (5.11) is of the type (4.37), with

$$
\begin{aligned}
\tilde{V}_t &= D_p \tilde{H}_t(\cdot, D\tilde{u}_t), \quad \Gamma_t = D_{pp}^2 \tilde{H}_t(\cdot, D\tilde{u}_t), \\
\tilde{b}_t^0 &= \tilde{\rho}_t D_{pp}^2 \tilde{H}_t(\cdot, D\tilde{u}_t) D\partial_m \tilde{u}_t \\
&\quad + \partial_m \tilde{m}_t D_{pp}^2 \tilde{H}_t(\cdot, D\tilde{u}_t) D\tilde{z}_t + \tilde{m}_t D_{ppp}^3 \tilde{H}_t(\cdot, D\tilde{u}_t) D\tilde{z}_t \otimes D\partial_m \tilde{u}_t, \\
\tilde{f}_t^0 &= \langle D_{pp}^2 \tilde{H}_t(\cdot, D\tilde{u}_t), D\tilde{z}_t \otimes D\partial_m \tilde{u}_t \rangle - \frac{\delta^2 \tilde{F}_t}{\delta m^2}(\cdot, m_t)(\rho_t, \partial_m m_t), \\
\tilde{g}_T^0 &= \frac{\delta^2 \tilde{G}}{\delta m^2}(\cdot, m_T)(\rho_T, \partial_m m_T).
\end{aligned}
\tag{5.12}
$$

Recall from Theorem 4.3.1 and Lemma 4.3.7 on the one hand and from Corollary 4.4.6 on the other hand that we can find a constant C (the value of which is allowed to increase from line to line), independent of m_0, y, ζ, ℓ, and k, such that

$$
\begin{aligned}
&\operatorname{essup}_{\omega \in \Omega} \sup_{t \in [0,T]} \|\tilde{u}_t\|_{n+1+\alpha} \leqslant C, \\
&\operatorname{essup}_{\omega \in \Omega} \Big[\sup_{t \in [0,T]} \big(\|\tilde{z}_t\|_{n+1+\alpha} + \|\partial_m \tilde{u}_t\|_{n+1+\alpha} \\
&\qquad\qquad\qquad + \|\tilde{\rho}_t\|_{-(n+\alpha')} + \|\partial_m \tilde{m}_t\|_{-(n+\alpha')} \big) \Big] \leqslant C.
\end{aligned}
\tag{5.13}
$$

Since $|\ell|, |k| \leqslant n-1$, we can apply Corollary 4.4.6 with n replaced by $n-1$ (notice that $n-1$ satisfies the assumption of Section 5.2), so that

$$
\operatorname{essup}_{\omega \in \Omega} \Big[\sup_{t \in [0,T]} \big(\|\tilde{\rho}_t\|_{-(n+\alpha'-1)} + \|\partial_m \tilde{m}_t\|_{-(n+\alpha'-1)} \big) \Big] \leqslant C.
\tag{5.14}
$$

Therefore, we deduce that

$$
\operatorname{essup}_{\omega \in \Omega} \sup_{t \in [0,T]} \|\tilde{b}_t^0\|_{-(n+\alpha'-1)} \leqslant C.
$$

Similarly,

$$
\operatorname{essup}_{\omega \in \Omega} \sup_{t \in [0,T]} \|\tilde{f}_t^0\|_{n+\alpha} + \operatorname{essup}_{\omega \in \Omega} \|\tilde{g}_T^0\|_{n+1+\alpha} \leqslant C.
$$

From Theorem 4.4.2, we deduce that, with the prescribed initial conditions, (5.11) has a unique solution. Moreover, by Corollary 4.4.6,

$$\text{essup}_{\omega \in \Omega} \sup_{t \in [0,T]} \|\tilde{z}_t^{(2)}\|_{n+1+\alpha} + \text{essup}_{\omega \in \Omega} \sup_{t \in [0,T]} \|\tilde{\rho}_t^{(2)}\|_{-(n+\alpha')} \leqslant C, \qquad (5.15)$$

where C is independent of m_0, y, ζ, ℓ and k.

On the model of Lemma 5.2.2, we claim:

Lemma 5.3.1. *The function*

$$[\mathbb{T}^d]^3 \ni (x, y, \zeta) \mapsto v^{(0,0)}(x, m_0, y, \zeta)$$

admits continuous crossed derivatives in (y, ζ), up to the order $n-1$ in y and to the order $n-1$ in ζ, the derivative

$$D_y^\ell D_\zeta^k v^{(0,0)}(\cdot, m_0, y, \zeta) : \mathbb{T}^d \ni x \mapsto D_y^\ell D_\zeta^k v^{(0,0)}(x, m_0, y, \zeta),$$

for $|\ell|, |k| \leqslant n-1$, belonging to $\mathcal{C}^{n+1+\alpha}(\mathbb{T}^d)$. Moreover, writing

$$v^{(\ell,k)}(x, m_0, y, \zeta) = D_y^\ell D_\zeta^k v^{(0,0)}(x, m_0, y, \zeta), \quad x, y, \zeta \in \mathbb{T}^d,$$

there exists, for any $\alpha' \in (0, \alpha)$, a constant C such that, for any multi-indices ℓ, k with $|\ell|, |k| \leqslant n-1$; any $y, y', \zeta, \zeta' \in \mathbb{T}^d$; and any $m_0 \in \mathcal{P}(\mathbb{T}^d)$,

$$\left\| v^{(\ell,k)}(\cdot, m_0, y, \zeta) \right\|_{n+1+\alpha} \leqslant C,$$
$$\left\| v^{(\ell,k)}(\cdot, m_0, y, \zeta) - v^{(\ell,k)}(\cdot, m_0, y', \zeta') \right\|_{n+1+\alpha} \leqslant C \big(|y - y'|^{\alpha'} + |\zeta - \zeta'|^{\alpha'} \big).$$

Proof. With the same notations as in the statement of Lemma 5.2.2, we denote by $(\tilde{\rho}_t^{k,\zeta}, \tilde{z}_t^{k,\zeta})_{t \in [0,T]}$ the solution to (5.5) with $(-1)^{|k|} D^k \delta_\zeta$ as the initial condition and by $(\tilde{\rho}_t^{\ell,y}, \tilde{z}_t^{\ell,y})_{t \in [0,T]}$ the solution to (5.5) with $(-1)^{|\ell|} D^\ell \delta_y$ as the initial condition.

By Proposition 5.2.5 (applied with both $n-1$ and n), we have, for any $y, y' \in \mathbb{T}^d$ and any $\zeta, \zeta' \in \mathbb{T}^d$,

$$\text{essup}_{\omega \in \Omega} \left[\sup_{t \in [0,T]} \|\tilde{z}_t^{k,\zeta} - \tilde{z}_t^{k,\zeta'}\|_{n+1+\alpha} + \sup_{t \in [0,T]} \|\tilde{\rho}_t^{k,\zeta} - \tilde{\rho}_t^{k,\zeta'}\|_{-(n+\alpha'-1)} \right]$$
$$\leqslant C|\zeta - \zeta'|^{\alpha'},$$
$$\text{essup}_{\omega \in \Omega} \left[\sup_{t \in [0,T]} \|\tilde{z}_t^{\ell,y} - \tilde{z}_t^{\ell,y'}\|_{n+1+\alpha} + \sup_{t \in [0,T]} \|\tilde{\rho}_t^{\ell,y} - \tilde{\rho}_t^{\ell,y'}\|_{-(n+\alpha'-1)} \right]$$
$$\leqslant C|y - y'|^{\alpha'}. \qquad (5.16)$$

Denote now by $(\tilde{b}_t^{\ell,k,y,\varsigma})_{t\in[0,T]}$ the process $(\tilde{b}_t^0)_{t\in[0,T]}$ in (5.12) when $(\tilde{\rho}_t,\tilde{z}_t)_{t\in[0,T]}$ stands for the process $(\tilde{\rho}_t^{\ell,y},\tilde{z}_t^{\ell,y})_{t\in[0,T]}$ and $(\partial_m\tilde{m}_t,\partial_m\tilde{u}_t)_{t\in[0,T]}$ is replaced by $(\tilde{\rho}_t^{k,\varsigma},\tilde{z}_t^{k,\varsigma})_{t\in[0,T]}$. Define in a similar way $(\tilde{f}_t^{\ell,k,y,\varsigma})_{t\in[0,T]}$ and $\tilde{g}_T^{\ell,k,y,\varsigma}$. Then, combining (5.16) with (5.13) and (5.14),

$$\operatorname{essup}_\omega\Big[\big\|\tilde{g}_T^{\ell,k,y',\varsigma'}-\tilde{g}_T^{\ell,k,y,\varsigma}\big\|_{n+1+\alpha}$$
$$+\sup_{t\in[0,T]}\Big(\big\|\tilde{b}_t^{\ell,k,y',\varsigma'}-\tilde{b}_t^{\ell,k,y,\varsigma}\big\|_{-(n+\alpha'-1)}+\big\|\tilde{f}_t^{\ell,k,y',\varsigma'}-\tilde{f}_t^{\ell,k,y,\varsigma}\big\|_{n+\alpha}\Big)\Big]$$
$$\leqslant C\big(|y-y'|^{\alpha'}+|z-z'|^{\alpha'}\big).$$

By Proposition 4.4.5, we deduce that

$$\big\|v^{(\ell,k)}(\cdot,m_0,y,\varsigma)-v^{(\ell,k)}(\cdot,m_0,y',\varsigma')\big\|_{n+1+\alpha}\leqslant C\big(|y-y'|^{\alpha'}+|\varsigma-\varsigma'|^{\alpha'}\big),\quad(5.17)$$

which provides the last claim in the statement (the L^∞ bound following from (5.15)).

Now, by Lemma 5.2.2 (applied with both n and $n-1$), we know that, for $|k|\leqslant n-2$ and $j\in\{1,\ldots,d\}$,

$$\lim_{\mathbb{R}\backslash\{0\}\ni h\to0}\operatorname{essup}_{\omega\in\Omega}\Big[\sup_{t\in[0,T]}\Big(\big\|\tfrac{1}{h}\big(\tilde{\rho}_t^{\varsigma+he_j,k}-\tilde{\rho}_t^{\varsigma,k}\big)-\tilde{\rho}_t^{\varsigma,k+e_j}\big\|_{-(n+\alpha'-1)}$$
$$+\big\|\tfrac{1}{h}\big(\tilde{z}_t^{\varsigma+he_j,k}-\tilde{z}_t^{\varsigma,k}\big)-\tilde{z}_t^{\varsigma,k+e_j}\big\|_{n+1+\alpha}\Big)\Big]=0,$$

where e_j denotes the jth vector of the canonical basis of $\mathbb{R}^d$. Therefore, by (5.13),

$$\lim_{\mathbb{R}\backslash\{0\}\ni h\to0}\operatorname{essup}_{\omega\in\Omega}\Big[\sup_{t\in[0,T]}\Big(\big\|\tfrac{1}{h}\big(\tilde{b}_t^{\ell,k,y,\varsigma+he_j}-\tilde{b}_t^{\ell,k,y,\varsigma}\big)-\tilde{b}_t^{\ell,k+e_j,y,\varsigma}\big\|_{-(n+\alpha'-1)}$$
$$+\big\|\tfrac{1}{h}\big(\tilde{f}_t^{\ell,k,y,\varsigma+he_j}-\tilde{f}_t^{\ell,k,y,\varsigma}\big)-\tilde{f}_t^{\ell,k+e_j,y,\varsigma}\big\|_{n+\alpha}\Big)$$
$$+\big\|\tfrac{1}{h}\big(\tilde{g}_T^{\ell,k,y,\varsigma+he_j}-\tilde{g}_T^{\ell,k,y,\varsigma}\big)-\tilde{g}_T^{\ell,k+e_j,y,\varsigma}\big\|_{n+1+\alpha}\Big)\Big]=0.$$

By Proposition 4.4.5,

$$\lim_{h\to0}\Big\|\tfrac{1}{h}\big(v^{(\ell,k)}(\cdot,m_0,y,\varsigma+he_j)-v^{(\ell,k)}(\cdot,m_0,y,\varsigma)\big)$$
$$-v^{(\ell,k+e_j)}(\cdot,m_0,y,\varsigma)\Big\|_{n+1+\alpha}=0,$$

which proves, by induction, that

$$D_\varsigma^k v^{(\ell,0)}(x,m_0,y,\varsigma)=v^{(\ell,k)}(x,m_0,y,\varsigma),\quad x,y,\varsigma\in\mathbb{T}^d.$$

Similarly, we can prove that

$$D_y^\ell v^{(0,k)}(x, m_0, y, \zeta) = v^{(\ell,k)}(x, m_0, y, \zeta), \quad x, y, \zeta \in \mathbb{T}^d.$$

Together with the continuity property (5.17), we complete the proof. □

We claim that

Proposition 5.3.2. *We can find a constant C such that, for any $m_0, m_0' \in$*
$\mathcal{P}(\mathbb{T}^d)$, any $y, \zeta \in \mathbb{T}^d$, and any multi-indices ℓ, k with $|\ell|, |k| \leqslant n - 1$,

$$\left\| v^{(\ell,k)}(\cdot, m_0, y, \zeta) - v^{(\ell,k)}(\cdot, m_0', y, \zeta) \right\|_{n+1+\alpha} \leqslant C \mathbf{d}_1(m_0, m_0').$$

Proof. The proof consists of a new application of Proposition 4.4.5. Given

- the solutions $(\tilde{m}_t, \tilde{u}_t)_{t \in [0,T]}$ and $(\tilde{m}_t', \tilde{u}_t')_{t \in [0,T]}$ to (4.7) with $\tilde{m}_0 = m_0$ and $\tilde{m}_0' = m_0'$ as the respective initial conditions;
- the solutions $(\partial_m \tilde{m}_t, \partial_m \tilde{u}_t)_{t \in [0,T]}$ and $(\partial_m \tilde{m}_t', \partial_m \tilde{u}_t')_{t \in [0,T]}$ to (5.10), with $(\tilde{m}_t, \tilde{u}_t)_{t \in [0,T]}$ and $(\tilde{m}_t', \tilde{u}_t')_{t \in [0,T]}$ as the respective input and $\partial_m \tilde{m}_0 = \partial_m \tilde{m}_0' = (-1)^{|k|} D^k \delta_\zeta$ as the initial condition, for some multi-index k with $|k| \leqslant n - 1$ and for some $\zeta \in \mathbb{T}^d$;
- the solutions $(\tilde{\rho}_t, \tilde{z}_t)_{t \in [0,T]}$ and $(\tilde{\rho}_t', \tilde{z}_t')_{t \in [0,T]}$ to (5.10), with $(\tilde{m}_t, \tilde{u}_t)_{t \in [0,T]}$ and $(\tilde{m}_t', \tilde{u}_t')_{t \in [0,T]}$ as the respective input and $(-1)^{|\ell|} D^\ell \delta_y$ as the initial condition, for some multi-index ℓ with $|\ell| \leqslant n - 1$ and for some $y \in \mathbb{T}^d$;
- the solutions $(\tilde{\rho}_t^{(2)}, \tilde{z}_t^{(2)})_{t \in [0,T]}$ and $(\tilde{\rho}_t^{(2)'}, \tilde{z}_t^{(2)'})_{t \in [0,T]}$ to the second-order linearized system (5.11) with the tuples $(\tilde{m}_t, \tilde{u}_t, \tilde{\rho}_t, \tilde{z}_t, \partial_m \tilde{m}_t, \partial_m \tilde{u}_t)_{t \in [0,T]}$ and $(\tilde{m}_t', \tilde{u}_t', \tilde{\rho}_t', \tilde{z}_t', \partial_m \tilde{m}_t', \partial_m \tilde{u}_t')_{t \in [0,T]}$ as the respective input and with 0 as the initial condition.

Notice from (5.7) that $\tilde{z}_0 = v^{(\ell)}(\cdot, m_0, y)$ and $\tilde{z}_0' = v^{(\ell)}(\cdot, m_0', y)$.

With each of the two aforementioned tuples $(\tilde{m}_t, \tilde{u}_t, \tilde{\rho}_t, \tilde{z}_t, \partial_m \tilde{m}_t, \partial_m \tilde{u}_t)_{t \in [0,T]}$ and $(\tilde{m}_t', \tilde{u}_t', \tilde{\rho}_t', \tilde{z}_t', \partial_m \tilde{m}_t', \partial_m \tilde{u}_t')_{t \in [0,T]}$, we can associate the same coefficients as in (5.12), labeling with a prime the coefficients associated with the input $(\tilde{m}_t', \tilde{u}_t', \tilde{\rho}_t', \tilde{z}_t', \partial_m \tilde{m}_t', \partial_m \tilde{u}_t')_{t \in [0,T]}$. Combining (5.13) and (5.14) together with (5.12) and the version of (5.12) associated with the "primed" coefficients and the "primed" inputs, we obtain

$$\|\tilde{V}_t - \tilde{V}_t'\|_{n+\alpha} + \|\Gamma_t - \Gamma_t'\|_0$$
$$+ \|b_t^0 - b_t^{0'}\|_{-(n+\alpha'-1)} + \|f_t^0 - f_t^{0'}\|_{n+\alpha} + \|g_T^0 - g_T^{0'}\|_{n+1+\alpha}$$
$$\leqslant C \Big(\|\tilde{u}_t - \tilde{u}_t'\|_{n+1+\alpha} + \|\tilde{z}_t - \tilde{z}_t'\|_{n+1+\alpha} + \|\partial_m \tilde{u}_t - \partial_m \tilde{u}_t'\|_{n+1+\alpha}$$
$$+ \mathbf{d}_1(\tilde{m}_t, \tilde{m}_t') + \|\tilde{\rho}_t - \tilde{\rho}_t'\|_{-(n+\alpha'-1)} + \|\partial_m \tilde{m}_t - \partial_m \tilde{m}_t'\|_{-(n+\alpha'-1)} \Big).$$

By Propositions 4.4.5 and 5.2.5 (applied with both n and $n - 1$), we complete the proof. □

On the model of Lemma 5.2.3, we have

Lemma 5.3.3.   *Given a finite measure μ on $\mathbb{T}^d$, the solution $\tilde{z}^{(2)}$ to (5.5), with 0 as the initial condition, when $(\tilde{m}_t)_{0 \leqslant t \leqslant T}$ is initialized with m_0, $(\tilde{\rho}_t)_{0 \leqslant t \leqslant T}$ is initialized with $(-1)^{|\ell|} D^\ell \delta_y$, for $|\ell| \leqslant n-1$ and $y \in \mathbb{T}^d$, and $(\partial_m \tilde{m}_t)_{0 \leqslant t \leqslant T}$ is initialized with μ, reads, when taken at time 0,*

$$\tilde{z}_0^{(2)} : \mathbb{R}^d \ni x \mapsto \tilde{z}_0^{(2)}(x) = \int_{\mathbb{T}^d} v^{(\ell,0)}(x, m_0, y, \zeta) d\mu(\zeta).$$

Now,

Proposition 5.3.4.   *We can find a constant C such that, for any multi-index ℓ with $|\ell| \leqslant n-1$, any $m_0, m_0' \in \mathcal{P}(\mathbb{T}^d)$ and any $y \in \mathbb{T}^d$,*

$$\left\| v^{(\ell)}(\cdot, m_0', y) - v^{(\ell)}(\cdot, m_0, y) - \int_{\mathbb{T}^d} v^{(\ell,0)}(\cdot, m_0, y, \zeta) d(m_0' - m_0)(\zeta) \right\|_{n+1+\alpha}$$
$$\leqslant C \mathbf{d}_1^2(m_0, m_0').$$

Proof.   We follow the lines of the proof of Proposition 5.2.4. Given two initial conditions $m_0, m_0' \in \mathcal{P}(\mathbb{T}^d)$, we consider

- the solutions $(\tilde{m}_t, \tilde{u}_t)_{t \in [0,T]}$ and $(\tilde{m}_t', \tilde{u}_t')_{t \in [0,T]}$ to (4.7) with $\tilde{m}_0 = m_0$ and $\tilde{m}_0' = m_0'$ as the respective initial conditions;
- the solution $(\partial_m \tilde{m}_t, \partial_m \tilde{u}_t)_{t \in [0,T]}$ to the system (5.5), when driven by the input $(\tilde{m}_t, \tilde{u}_t)_{t \in [0,T]}$ and by the initial condition $\partial_m \tilde{m}_0 = m_0' - m_0$;
- the solutions $(\tilde{\rho}_t, \tilde{z}_t)_{t \in [0,T]}$ and $(\tilde{\rho}_t', \tilde{z}_t')_{t \in [0,T]}$ to (5.10), with $(\tilde{m}_t, \tilde{u}_t)_{t \in [0,T]}$ and $(\tilde{m}_t', \tilde{u}_t')_{t \in [0,T]}$ as the respective input and $(-1)^{|\ell|} D^\ell \delta_y$ as the initial condition, for some multi-index ℓ with $|\ell| \leqslant n-1$ and for some $y \in \mathbb{T}^d$;
- the solution $(\tilde{\rho}_t^{(2)}, \tilde{z}_t^{(2)})_{t \in [0,T]}$ to (5.11) with $(\tilde{m}_t, \tilde{u}_t, \tilde{\rho}_t, \tilde{z}_t, \partial_m \tilde{m}_t, \partial_m \tilde{u}_t)_{t \in [0,T]}$ as input and 0 as the initial condition.

Then, we let

$$\delta \tilde{\rho}_t^{(2)} = \tilde{\rho}_t' - \tilde{\rho}_t - \tilde{\rho}_t^{(2)}, \quad \delta \tilde{z}_t^{(2)} = \tilde{z}_t' - \tilde{z}_t - \tilde{z}_t^{(2)}, \quad t \in [0,T].$$

We have

$$d_t(\delta \tilde{z}_t^{(2)}) = \left\{ -\Delta(\delta \tilde{z}_t^{(2)}) + \langle D_p \tilde{H}_t(\cdot, D\tilde{u}_t), D(\delta \tilde{z}_t^{(2)}) \rangle - \frac{\delta \tilde{F}_t}{\delta m}(\cdot, m_t)(\delta \rho_t^{(2)}) + \tilde{f}_t \right\} dt$$
$$+ d\tilde{M}_t,$$

$$\partial_t(\delta \tilde{\rho}_t^{(2)}) - \Delta(\delta \tilde{\rho}_t^{(2)}) - \mathrm{div}\left[(\delta \tilde{\rho}_t^{(2)}) D_p \tilde{H}_t(\cdot, D\tilde{u}_t)\right]$$
$$- \mathrm{div}\left[\tilde{m}_t D_{pp}^2 \tilde{H}_t(\cdot, D\tilde{u}_t)(D\delta \tilde{z}_t^{(2)}) + \tilde{b}_t\right] = 0,$$

with a boundary condition of the form

$$\delta\tilde{z}_T^{(2)} = \frac{\delta\tilde{G}}{\delta m}(\cdot, m_T)\big(\delta\rho_T^{(2)}\big) + \tilde{g}_T,$$

where

$$
\begin{aligned}
\tilde{b}_t = {}& \tilde{\rho}_t'\Big(D_p\tilde{H}_t(\cdot, D\tilde{u}_t') - D_p\tilde{H}_t(\cdot, D\tilde{u}_t)\Big) \\
& + \Big(\tilde{m}_t' D_{pp}^2\tilde{H}_t(\cdot, D\tilde{u}_t') - \tilde{m}_t D_{pp}^2\tilde{H}_t(\cdot, D\tilde{u}_t)\Big)D\tilde{z}_t' \\
& - \partial_m\tilde{m}_t D_{pp}^2\tilde{H}_t(\cdot, D\tilde{u}_t)D\tilde{z}_t - \tilde{\rho}_t D_{pp}^2\tilde{H}_t(\cdot, D\tilde{u}_t)D\partial_m\tilde{u}_t \\
& - \tilde{m}_t D_{ppp}^3\tilde{H}_t(\cdot, D\tilde{u}_t)D\tilde{z}_t \otimes D\partial_m\tilde{u}_t,
\end{aligned}
$$

and,

$$
\begin{aligned}
\tilde{f}_t = {}& \big\langle D_p\tilde{H}_t(\cdot, D\tilde{u}_t') - D_p\tilde{H}_t(\cdot, D\tilde{u}_t), D\tilde{z}_t'\big\rangle \\
& - \big\langle D_{pp}^2\tilde{H}_t(\cdot, D\tilde{u}_t), D\tilde{z}_t \otimes D\partial_m\tilde{u}_t\big\rangle \\
& - \Big(\frac{\delta\tilde{F}_t}{\delta m}(\cdot, m_t')(\rho_t') - \frac{\delta\tilde{F}_t}{\delta m}(\cdot, m_t)(\rho_t') - \frac{\delta^2\tilde{F}_t}{\delta m^2}(\cdot, m_t)(\rho_t, \partial_m m_t)\Big), \\
\tilde{g}_T = {}& \frac{\delta\tilde{G}}{\delta m}(\cdot, m_T')(\rho_T') - \frac{\delta\tilde{G}}{\delta m}(\cdot, m_T)(\rho_T') - \frac{\delta^2\tilde{G}}{\delta m^2}(\cdot, m_T)(\rho_T, \partial_m m_T),
\end{aligned}
$$

and where $(\tilde{M}_t)_{t\in[0,T]}$ is a square integrable martingale as in (4.37).

Therefore,

$$
\begin{aligned}
\tilde{b}_t = {}& \big(\tilde{\rho}_t' - \tilde{\rho}_t\big)\Big(D_p\tilde{H}_t(\cdot, D\tilde{u}_t') - D_p\tilde{H}_t(\cdot, D\tilde{u}_t)\Big) \\
& + \tilde{\rho}_t\Big(D_p\tilde{H}_t(\cdot, D\tilde{u}_t') - D_p\tilde{H}_t(\cdot, D\tilde{u}_t) - D_{pp}^2\tilde{H}_t(\cdot, D\tilde{u}_t)D\partial_m\tilde{u}_t\Big) \\
& + \Big(\tilde{m}_t' D_{pp}^2\tilde{H}_t(\cdot, D\tilde{u}_t') - \tilde{m}_t D_{pp}^2\tilde{H}_t(\cdot, D\tilde{u}_t)\Big)\big(D\tilde{z}_t' - D\tilde{z}_t\big) \\
& + \Big(\tilde{m}_t' - \tilde{m}_t\Big)\Big(D_{pp}^2\tilde{H}_t(\cdot, D\tilde{u}_t') - D_{pp}^2\tilde{H}_t(\cdot, D\tilde{u}_t)\Big)D\tilde{z}_t \\
& + \Big(\tilde{m}_t' - \tilde{m}_t - \partial_m\tilde{m}_t\Big)D_{pp}^2\tilde{H}_t(\cdot, D\tilde{u}_t)D\tilde{z}_t \\
& + \tilde{m}_t\Big(D_{pp}^2\tilde{H}_t(\cdot, D\tilde{u}_t') - D_{pp}^2\tilde{H}_t(\cdot, D\tilde{u}_t) - D_{ppp}^3\tilde{H}_t(\cdot, D\tilde{u}_t)D\partial_m\tilde{u}_t\Big)D\tilde{z}_t,
\end{aligned}
$$

and

$$
\begin{aligned}
\tilde{f}_t = {}& \Big\langle D_p\tilde{H}_t(\cdot, D\tilde{u}_t') - D_p\tilde{H}_t(\cdot, D\tilde{u}_t), D\tilde{z}_t' - D\tilde{z}_t\Big\rangle \\
& + \Big\langle D_p\tilde{H}_t(\cdot, D\tilde{u}_t') - D_p\tilde{H}_t(\cdot, D\tilde{u}_t) - D_{pp}^2\tilde{H}_t(\cdot, D\tilde{u}_t)D\partial_m\tilde{u}_t, D\tilde{z}_t\Big\rangle \\
& + \Big(\frac{\delta\tilde{F}_t}{\delta m}(\cdot, m_t') - \frac{\delta\tilde{F}_t}{\delta m}(\cdot, m_t)\Big)(\rho_t' - \rho_t) \\
& + \Big(\frac{\delta\tilde{F}_t}{\delta m}(\cdot, m_t')(\rho_t) - \frac{\delta\tilde{F}_t}{\delta m}(\cdot, m_t)(\rho_t) - \frac{\delta^2\tilde{F}_t}{\delta m^2}(\cdot, m_t)(\rho_t, \partial_m m_t)\Big),
\end{aligned}
$$

Similarly,

$$\tilde{g}_T = \left(\frac{\delta\tilde{G}}{\delta m}(\cdot,m_T') - \frac{\delta\tilde{G}}{\delta m}(\cdot,m_T)\right)(\rho_T' - \rho_T)$$
$$+ \left(\frac{\delta\tilde{G}}{\delta m}(\cdot,m_t')(\rho_T) - \frac{\delta\tilde{G}}{\delta m}(\cdot,m_t)(\rho_T) - \frac{\delta^2\tilde{G}}{\delta m^2}(\cdot,m_T)(\rho_T,\partial_m m_T)\right).$$

Applying Theorem 4.3.1, Lemma 4.3.7, Propositions 5.2.4 and 5.2.5, and (5.13) and (5.14) and using the same kind of Taylor expansion as in the proof of Proposition 5.2.4, we deduce that

$$\operatorname{essup}_{\omega\in\Omega}\sup_{t\in[0,T]}\left[\|\tilde{b}_t\|_{-(n+\alpha'-1)} + \|\tilde{f}_t\|_{n+\alpha} + \|\tilde{g}_T\|_{n+1+\alpha}\right] \leqslant C\mathbf{d}_1^2(m_0,m_0').$$

By Proposition 4.4.5, we complete the proof. $\square$

We thus deduce:

Proposition 5.3.5. *For any $x \in \mathbb{T}^d$, the function $\mathcal{P}(\mathbb{T}^d) \ni m \mapsto U(0,x,m)$ is twice differentiable in the direction m and the second-order derivatives read, for any $m \in \mathcal{P}(\mathbb{T}^d)$*

$$\frac{\delta^2 U}{\delta m^2}(0,x,m,y,y') = v^{(0,0)}(x,m,y,y'), \quad y,y' \in \mathbb{T}^d.$$

In particular, for any $\alpha' \in (0,\alpha)$, $t \in [0,T]$ and $m \in \mathcal{P}(\mathbb{T}^d)$, the function $[\delta^2 U/\delta m^2](0,\cdot,m,\cdot,\cdot)$ belongs to $\mathcal{C}^{n+1+\alpha'}(\mathbb{T}^d) \times \mathcal{C}^{n-1+\alpha'}(\mathbb{T}^d) \times \mathcal{C}^{n-1+\alpha'}(\mathbb{T}^d)$ and the mapping

$$\mathcal{P}(\mathbb{T}^d) \ni m \mapsto \frac{\delta^2 U}{\delta m^2}(0,\cdot,m,\cdot,\cdot) \in \mathcal{C}^{n+1+\alpha'}(\mathbb{T}^d) \times \mathcal{C}^{n-1+\alpha'}(\mathbb{T}^d) \times \mathcal{C}^{n-1+\alpha'}(\mathbb{T}^d)$$

is continuous (with respect to $\mathbf{d}_1$). The derivatives in y and y' read

$$D_y^\ell D_{y'}^k \frac{\delta^2 U}{\delta m^2}(0,x,m,y,y') = v^{(\ell,k)}(x,m,y,y'), \quad y,y' \in \mathbb{T}^d, \quad |k|,|\ell| \leqslant n-1.$$

Proof. By Proposition 5.3.4, we indeed know that, for any multi-index ℓ with $|\ell| \leqslant n-1$ and any $x,y \in \mathbb{T}^d$, the map $\mathcal{P}(\mathbb{T}^d) \ni m \mapsto D_y^\ell[\delta U/\delta m](0,x,m,y)$ is differentiable with respect to m, the derivative writing, for any $m \in \mathcal{P}(\mathbb{T}^d)$,

$$\frac{\delta}{\delta m}\left[D_y^\ell \frac{\delta U}{\delta m}\right](0,x,m,y,y') = v^{(\ell,0)}(0,x,m,y,y'), \quad y,y' \in \mathbb{T}^d.$$

By Lemma 5.3.1, $[\delta/\delta m][D_y^\ell[\delta U/\delta m]](0,x,m,y,y')$ is $n-1$ times differentiable with respect to y' and, together with Proposition 5.3.2, the derivatives are continuous in all the parameters. Making use of Schwarz' Lemma 2.2.4 (when $\ell = 1$

and iterating the argument to handle the case $\ell \geqslant 2$), the proof is easily completed.   □

Following Proposition 5.2.6, we finally claim:

Proposition 5.3.6. *Proposition 5.3.5 easily extends to any initial time $t_0 \in [0, T]$. Then, for any $\alpha' \in (0, \alpha)$, any $t_0 \in [0, T]$, and $m_0 \in \mathcal{P}(\mathbb{T}^d)$*

$$
\lim_{h \to 0} \sup_{|k| \leqslant n-1} \sup_{|\ell| \leqslant n-1} \left\| D_y^\ell D_{y'}^k \frac{\delta^2 U}{\delta m^2} (t_0 + h, \cdot, m_0, \cdot) \right.
$$

$$
\left. - D_y^\ell D_{y'}^k \frac{\delta^2 U}{\delta m^2} (t_0, \cdot, m_0, \cdot) \right\|_{n+1+\alpha', \alpha', \alpha'} = 0.
$$

5.4 DERIVATION OF THE MASTER EQUATION

We now prove Theorem 2.4.5. Of course the key point is to prove that U, as constructed in the previous section, is a solution of the master equation (2.13).

5.4.1 Regularity Properties of the Solution

The regularity properties of U follow from Sections 5.1, 5.2, and 5.3, see in particular Propositions 5.3.5 and 5.3.6 (take note that, in the statements of Theorem 2.4.5 and of Proposition 5.3.5, the indices of regularity in y and y' are not exactly the same; this comes from the fact that, in the statement of Theorem 2.4.5, we require **(HF2)** and **(HG2)** to hold at one more rank).

5.4.2 Derivation of the Master Equation

We now have all the necessary ingredients to derive the master equation satisfied by U. The first point is to recall that, whenever the forward component $(\tilde{m}_t)_{t \in [t_0, T]}$ in (5.1) is initialized with $m_0 \in \mathcal{P}(\mathbb{T}^d)$ at time t_0, then

$$
U(t_0, x, m_0) = \tilde{u}_{t_0}(x), \quad x \in \mathbb{T}^d,
$$

$(\tilde{u}_t)_{t \in [t_0, T]}$ denoting the backward component in (5.1). Moreover, by Lemma 5.1.1, for any $h \in [T - t_0]$,

$$
\tilde{u}_{t_0 + h}(x) = U\big(t_0 + h, x + \sqrt{2}(W_{t_0 + h} - W_{t_0}), m_{t_0, t_0 + h}\big), \quad x \in \mathbb{T}^d,
$$

where $m_{t_0, t}$ is the image of $\tilde{m}_t$ by the random mapping $\mathbb{T}^d \ni x \mapsto x + \sqrt{2}(W_t - W_{t_0})$, that is, $m_{t_0, t} = [id + \sqrt{2}(W_t - W_{t_0})] \sharp \tilde{m}_t$. In particular, we can write

$$\frac{U(t_0 + h, x, m_0) - U(t_0, x, m_0)}{h}$$

$$= \frac{\mathbb{E}\big[U\big(t_0 + h, x + \sqrt{2}(W_{t_0+h} - W_{t_0}), m_{t_0,t_0+h}\big)\big] - U(t_0, x, m_0)}{h}$$

$$+ \frac{U(t_0 + h, x, m_0) - \mathbb{E}\big[U\big(t_0 + h, x + \sqrt{2}(W_{t_0+h} - W_{t_0}), m_{t_0,t_0+h}\big)\big]}{h} \qquad (5.18)$$

$$= \frac{\mathbb{E}[\tilde{u}_{t_0+h}(x)] - \tilde{u}_{t_0}(x)}{h}$$

$$+ \frac{U(t_0 + h, x, m_0) - \mathbb{E}\big[U\big(t_0 + h, x + \sqrt{2}(W_{t_0+h} - W_{t_0}), m_{t_0,t_0+h}\big)\big]}{h}.$$

We start with the first term in the right-hand side of (5.18). Following (4.12), we deduce from the backward equation in (5.1) that, for any $x \in \mathbb{T}^d$,

$$d_t\big[\mathbb{E}(\tilde{u}_t(x))\big] = \mathbb{E}\Big[\big\{-\Delta\tilde{u}_t + \tilde{H}_{t_0,t}(\cdot, D\tilde{u}_t) - \tilde{F}_{t_0,t}(\cdot, m_{t_0,t})\big\}(x)\Big]dt,$$

where the coefficients $\tilde{F}_{t_0,t}$ and $\tilde{H}_{t_0,t}$ are given by (5.2). In particular, thanks to the regularity property in Corollary 5.1.2, we deduce that

$$\lim_{h \searrow 0} \frac{\mathbb{E}[\tilde{u}_{t_0+h}(x)] - \tilde{u}_{t_0}(x)}{h} \qquad (5.19)$$

$$= -\Delta_x U(t_0, m_0, x) + H\big(x, D_x U(t_0, m_0, x)\big) - F\big(x, m_0\big).$$

To pass to the limit in the last term in (5.18), we need a specific form of Itô's formula. The precise version is given in the Appendix; see Lemma A.3.1. Applied to the current setting, with

$$\beta_t(\cdot) = D_p H\big(\cdot, D_x U(t, \cdot, m_{t_0,t})\big),$$

it says that

$$\lim_{h \searrow 0} \frac{1}{h} \mathbb{E}\Big[U\big(t_0 + h, x + \sqrt{2}(W_{t_0+h} - W_{t_0}), m_{t_0,t_0+h}\big)$$

$$- U\big(t_0 + h, x, m_0\big)\Big]$$

$$= \Delta_x U(t_0, x, m_0)$$

$$+ 2\int_{\mathbb{T}^d} \text{div}_y\big[D_m U\big](t_0, x, m_0, y)dm_0(y)$$

$$- \int_{\mathbb{T}^d} D_m U\big(t_0, x, m_0, y\big)D_p H\big(y, DU(t_0, y, m_0)\big)dm_0(y) \qquad (5.20)$$

$$+ 2\int_{\mathbb{T}^d} \text{div}_x\big[D_m U\big](t_0, x, m_0, y)dm_0(y)$$

$$+ \int_{[\mathbb{T}^d]^2} \text{Tr}\Big[D^2_{mm}U\big(t_0, x, m_0, y, y'\big)\Big]dm_0(y)dm_0(y').$$

From (5.19) and (5.20), we deduce that, for any $(x, m_0) \in \mathbb{T}^d \times \mathcal{P}(\mathbb{T}^d)$, the mapping $[0, T] \ni t \mapsto U(t, x, m_0)$ is right-differentiable and, for any $t_0 \in [0, T)$,

$$
\begin{aligned}
\lim_{h \searrow 0} & \frac{U(t_0 + h, x, m_0) - U(t_0, x, m_0)}{h} \\
= & -2\Delta_x U(t_0, x, m_0) + H(x, D_x U(t_0, x, m_0)) - F(x, m_0) \\
& - 2 \int_{\mathbb{T}^d} \mathrm{div}_y [D_m U](t_0, x, m_0, y) \, dm_0(y) \\
& + \int_{\mathbb{T}^d} D_m U(t_0, x, m_0, y) D_p H(y, DU(t_0, y, m_0)) \, dm_0(y) \\
& - 2 \int_{\mathbb{T}^d} \mathrm{div}_x [D_m U](t_0, x, m_0, y) \, dm_0(y) \\
& - \int_{[\mathbb{T}^d]^2} \mathrm{Tr} [D_{mm}^2 U(t_0, x, m_0, y, y')] \, dm_0(y) dm_0(y').
\end{aligned}
$$

Since the right-hand side is continuous in (t_0, x, m_0), we deduce that U is continuously differentiable in time and satisfies the master equation (2.13) (with $\beta = 1$).

5.4.3   Uniqueness

It now remains to prove uniqueness. Considering a solution V to the master equation (2.13) (with $\beta = 1$) along the lines of Definition 2.4.4, the strategy is to expand

$$
\tilde{u}_t' = V(t, x + \sqrt{2} W_t, m_t'), \quad t \in [0, T],
$$

where, for a given initial condition $m_0 \in \mathcal{P}(\mathbb{T}^d)$, m_t' is the image of $\tilde{m}_t'$ by the mapping $\mathbb{T}^d \ni x \mapsto x + \sqrt{2} W_t$, $(\tilde{m}_t')_{t \in [0, T]}$ denoting the solution of the Fokker–Planck equation

$$
d_t \tilde{m}_t' = \left\{ \Delta \tilde{m}_t' + \mathrm{div}(\tilde{m}_t' D_p \tilde{H}_t(\cdot, D_x V(t, x + \sqrt{2} W_t, \tilde{m}_t'))) \right\} dt,
$$

which reads, for almost every realization of $(W_t)_{t \in [0, T]}$, as the flow of conditional marginal distributions (given $(W_t)_{t \in [0, T]}$) of the McKean–Vlasov process

$$
\begin{aligned}
dX_t = & -D_p \tilde{H}_t(X_t, D_x V(t, x + \sqrt{2} W_t, \mathcal{L}(X_t | W))) dt \\
& + \sqrt{2} dB_t, \quad t \in [0, T],
\end{aligned}
\tag{5.21}
$$

X_0 having m_0 as distribution. Notice that the above equation is uniquely solvable since $D_x V$ is Lipschitz continuous in the space and measure arguments (by the simple fact that $D_x^2 V$ and $D_m D_x V$ are continuous functions on a compact set). We refer to [92] for standard solvability results for McKean–Vlasov SDEs (which may be easily extended to the current setting).

Of course, the key point is to prove that the pair $(\tilde{m}'_t, \tilde{u}'_t)_{t \in [0,T]}$ solves the same forward–backward system (4.7) as $(\tilde{m}_t, \tilde{u}_t)_{t \in [0,T]}$, in which case it will follow that $V(0, x, m_0) = \tilde{u}'_0 = \tilde{u}_0 = U(0, x, m_0)$. (The same argument may be repeated for any other initial condition with another initial time.)

The strategy consists of a suitable application of Lemma A.3.1 below. Given $0 \leqslant t \leqslant t + h \leqslant T$, we have to expand the difference

$$
\begin{aligned}
&\mathbb{E}\big[V\big(t + h, x + \sqrt{2}W_{t+h}, m'_{t+h}\big)|\mathcal{F}_t\big] - V\big(t, x + \sqrt{2}W_t, m'_t\big) \\
&= \mathbb{E}\big[V\big(t + h, x + \sqrt{2}W_{t+h}, m'_{t+h}\big)|\mathcal{F}_t\big] - V\big(t + h, x + \sqrt{2}W_t, m'_t\big) \\
&\quad + V\big(t + h, x + \sqrt{2}W_t, m'_t\big) - V\big(t, x + \sqrt{2}W_t, m'_t\big) \\
&= S^1_{t,h} + S^2_{t,h}.
\end{aligned}
\tag{5.22}
$$

By Lemma A.3.1, with

$$
\beta_t(\cdot) = D_p H\big(\cdot, D_x V(t, \cdot, m'_t)\big), \quad t \in [0, T],
$$

it holds that

$$
\begin{aligned}
&S^1_{t,h} \\
&= h\bigg[\Delta_x V\big(t, x + \sqrt{2}W_t, m'_t\big) \\
&\quad + 2\int_{\mathbb{T}^d} \mathrm{div}_y\big[D_m V\big]\big(t, x + \sqrt{2}W_t, m'_t, y\big)dm'_t(y) \\
&\quad - \int_{\mathbb{T}^d} D_m V\big(t, x + \sqrt{2}W_t, m'_t, y\big) \cdot D_p H\big(y, D_x V(t, y, m'_t)\big)dm'_t(y) \\
&\quad + 2\int_{\mathbb{T}^d} \mathrm{div}_x\big[D_m V\big]\big(t, x + \sqrt{2}W_t, m'_t, y\big)dm'_t(y) \\
&\quad + \int_{[\mathbb{T}^d]^2} \mathrm{Tr}\big[D^2_{mm} V\big]\big(t, x + \sqrt{2}W_t, m'_t, y, y'\big)dm'_t(y)dm'_t(y') \\
&\quad + \varepsilon_{t,t+h}\bigg],
\end{aligned}
\tag{5.23}
$$

where $(\varepsilon_{s,t})_{s,t \in [0,T]:s \leqslant t}$ is a family of real-valued random variables such that

$$
\lim_{h \searrow 0} \sup_{s,t \in [0,T]:|s-t| \leqslant h} \mathbb{E}\big[|\varepsilon_{s,t}|\big] = 0.
$$

Expand now $S^2_{t,h}$ in (5.22) to the first order in t and use the fact that $\partial_t V$ is uniformly continuous on the compact set $[0, T] \times \mathbb{T}^d \times \mathcal{P}_2(\mathbb{T}^d)$. Combining (5.22),

(5.23), and the master PDE (2.13) satisfied by V, we deduce that

$$
\mathbb{E}\big[V\big(t+h, x+\sqrt{2}W_{t+h}, m'_{t+h}\big)|\mathcal{F}_t\big] - V\big(t, x+\sqrt{2}W_t, m'_t\big)
$$
$$
= -h\Big[\Delta_x V\big(t, x+\sqrt{2}W_t, m'_t\big) - H\big(x+\sqrt{2}W_t, D_x V\big(t, x+\sqrt{2}W_t, m'_t\big)\big)
$$
$$
+ F\big(x+\sqrt{2}W_t, m'_t\big) + \varepsilon_{t,t+h}\Big].
$$

Considering a partition $t = t_0 < t_1 < \cdots < t_N = T$ of $[t, T]$ of step size h, the above identity yields

$$
\mathbb{E}\big[G\big(x+\sqrt{2}W_T, m'_T\big) - V\big(t, x+\sqrt{2}W_t, m'_t\big)|\mathcal{F}_t\big]
$$
$$
= -h\sum_{i=0}^{N-1}\Big[\Delta_x V\big(t_i, x+\sqrt{2}W_{t_i}, m'_{t_i}\big)
$$
$$
- H\big(x+\sqrt{2}W_{t_i}, D_x V\big(t_i, x+\sqrt{2}W_{t_i}, m'_{t_i}\big)\big) + F\big(x+\sqrt{2}W_{t_i}, m'_{t_i}\big)\Big]
$$
$$
+ h\sum_{i=0}^{N-1}\mathbb{E}\big[\varepsilon_{t_i,t_{i+1}}|\mathcal{F}_t\big].
$$

Since

$$
\limsup_{h\searrow 0}\ \sup_{r,s\in[0,T]:|r-s|\leqslant h}\ \mathbb{E}\Big[\big|\mathbb{E}\big[\varepsilon_{r,s}|\mathcal{F}_t\big]\big|\Big] \leqslant \limsup_{h\searrow 0}\ \sup_{r,s\in[0,T]:|r-s|\leqslant h}\ \mathbb{E}\big[|\varepsilon_{r,s}|\big] = 0,
$$

we can easily replace each $\mathbb{E}[\varepsilon_{t_i,t_{i+1}}|\mathcal{F}_t]$ by $\varepsilon_{t_i,t_{i+1}}$ itself, allowing for a modification of $\varepsilon_{t_i,t_{i+1}}$. Moreover, here and below, we use the fact that, for a random process $(\gamma_t)_{t\in[0,T]}$, with paths in $\mathcal{C}^0([0,T],\mathbb{R})$, satisfying

$$
\mathrm{essup}_{\omega\in\Omega}\ \sup_{t\in[0,T]}|\gamma_t| < \infty, \tag{5.24}
$$

it must hold that

$$
\lim_{h\searrow 0}\ \sup_{s,t\in[0,T]:|s-t|\leqslant h}\ \mathbb{E}\big[|\eta_{s,t}|\big] = 0, \quad \eta_{s,t} = \frac{1}{|s-t|}\int_s^t (\gamma_r - \gamma_s)dr. \tag{5.25}
$$

The proof just consists in bounding $|\eta_{s,t}|$ by $w_\gamma(h)$, where w_γ stands for the pathwise modulus of continuity of $(\gamma_t)_{t\in[0,T]}$, which satisfies, thanks to (5.24) and Lebesgue's dominated convergence theorem,

$$
\lim_{h\searrow 0}\mathbb{E}\big[w_\gamma(h)\big] = 0.
$$

Therefore, allowing for a new modification of the random variables $\varepsilon_{t_i, t_{i+1}}$, for $i = 0, \ldots, N-1$, we deduce that

$$
\mathbb{E}\big[G\big(x + \sqrt{2}W_T, m_T'\big) - V\big(t, x + \sqrt{2}W_t, m_t'\big)|\mathcal{F}_t\big]
$$
$$
= -\int_t^T \Big[\Delta_x V\big(s, x + \sqrt{2}W_s, m_s'\big) - H\big(x + \sqrt{2}W_s, D_x V(s, x + \sqrt{2}W_s, m_s')\big)
$$
$$
+ F\big(x + \sqrt{2}W_o, m_s'\big)\Big]ds + h \sum_{i=0}^{N-1} \varepsilon_{t_i, t_{i+1}}.
$$

Letting, for any $x \in \mathbb{T}^d$,

$$
\tilde{M}_t'(x) = V\big(t, x + \sqrt{2}W_t, m_t'\big) + \int_0^t \Big[\Delta_x V\big(s, x + \sqrt{2}W_s, m_s'\big)
$$
$$
- H\big(x + \sqrt{2}W_s, D_x V(s, x + \sqrt{2}W_s, m_s')\big) + F\big(x + \sqrt{2}W_s, m_s'\big)\Big]ds,
$$

we deduce that
$$
\mathbb{E}\big[\tilde{M}_T'(x) - \tilde{M}_t'(x)|\mathcal{F}_t\big] = h \sum_{i=0}^{N-1} \varepsilon_{t_i, t_{i+1}}.
$$

Now, letting h tend to 0, we deduce that $(\tilde{M}_t'(x))_{t \in [0,T]}$ is a martingale. Thanks to the regularity properties of V and its derivatives, it is bounded.

Letting
$$
\tilde{v}_t(x) = V\big(t, x + \sqrt{2}W_t, m_t'\big), \quad t \in [0, T],
$$

we finally notice that

$$
\tilde{v}_t(x) = \tilde{G}_T(x, m_T') + \int_t^T \Big[\Delta_x \tilde{v}_s(x) - \tilde{H}_s\big(x, D\tilde{v}_s(x)\big) + \tilde{F}(x, m_s')\Big]ds
$$
$$
- \big(\tilde{M}_T' - \tilde{M}_t'\big)(x), \quad t \in [0, T],
$$

which proves that $(\tilde{m}_t', \tilde{v}_t, \tilde{M}_t')_{t \in [0,T]}$ solves (4.7).

5.5   WELL-POSEDNESS OF THE STOCHASTIC MFG SYSTEM

We are now ready to come back to the well-posedness of the stochastic MFG system

$$
\begin{cases}
d_t u_t = \{-2\Delta u_t + H(x, Du_t) - F(x, m_t) - \sqrt{2}\,\mathrm{div}(v_t)\}dt \\
\qquad\qquad\qquad\qquad\qquad\qquad\qquad + v_t \cdot dW_t, \\
d_t m_t = \big[2\Delta m_t + \mathrm{div}\big(m_t D_p H(x, Du_t)\big)\big]dt - \sqrt{2}\,\mathrm{div}(m_t dW_t), \\
\qquad\qquad\qquad\qquad\qquad\qquad\qquad \text{in } [t_0, T] \times \mathbb{T}^d, \\
m_{t_0} = m_0, \ u_T(x) = G(x, m_T) \quad \text{in } \mathbb{T}^d.
\end{cases}
\tag{5.26}
$$

For simplicity of notation, we prove the existence and uniqueness of the solution for $t_0 = 0$.

First step. Existence of a solution. We start with the solution, denoted by $(\tilde{u}_t, \tilde{m}_t, \tilde{M}_t)_{t \in [0,T]}$, to the system

$$\begin{cases} d_t \tilde{m}_t = \left\{ \Delta \tilde{m}_t + \text{div}\left(\tilde{m}_t D_p \tilde{H}_t(\cdot, D\tilde{u}_t) \right) \right\} dt, \\ d_t \tilde{u}_t = \left\{ -\Delta \tilde{u}_t + \tilde{H}_t(\cdot, D\tilde{u}_t) - \tilde{F}_t(\cdot, m_t) \right\} dt + d\tilde{M}_t, \\ \tilde{m}_0 = m_0, \ \tilde{u}_T(x) = \tilde{G}(x, m_T) \quad \text{in } \mathbb{T}^d. \end{cases} \quad (5.27)$$

where $\tilde{H}_t(x, p) = H(x + \sqrt{2}W_t, p)$, $\tilde{F}_t(x, m) = F(x + \sqrt{2}W_t, m)$ and $\tilde{G}(x, m) = G(x + \sqrt{2}W_T, m)$. The existence and uniqueness of a solution $(\tilde{u}_t, \tilde{m}_t, \tilde{M}_t)_{t \in [0,T]}$ to (5.27) are ensured by Theorem 4.3.1. Given such a solution, we let

$$u_t(x) = \tilde{u}_t(x - \sqrt{2}W_t), \quad x \in \mathbb{T}^d ; \qquad m_t = (\text{id} + \sqrt{2}W_t) \sharp \tilde{m}_t, \quad t \in [0, T],$$

and claim that the pair $(u_t, m_t)_{t \in [0,T]}$ thus defined satisfies (5.26) (for a suitable $(v_t)_{t \in [0,T]}$).

The dynamics satisfied by $(m_t)_{t \in [0,T]}$ are given by the so-called Itô–Wentzell formula for Schwartz distribution-valued processes (see [67, Theorem 1.1]), the proof of which works as follows: for any test function $\phi \in \mathcal{C}^4(\mathbb{T}^d)$ and any $z \in \mathbb{R}^d$, we have $\int_{\mathbb{T}^d} \phi(x) dm_t(x) = \int_{\mathbb{T}^d} \phi(x + \sqrt{2}W_t) d\tilde{m}_t(x)$; expanding the variation of $(\int_{\mathbb{T}^d} \phi(x + z) d\tilde{m}_t(x))_{t \in [0,T]}$ by means of the Fokker–Planck equation satisfied by $(\tilde{m}_t)_{t \in [0,T]}$ and then replacing z by $\sqrt{2}W_t$, we then obtain the semi-martingale expansion of $(\int_{\mathbb{T}^d} \phi(x + \sqrt{2}W_t) d\tilde{m}_t(x))_{t \in [0,T]}$ by applying the standard Itô–Wentzell formula, as given in Subsection A.3.1. Once again we refer to [67, Theorem 1.1] for a complete account.

Applying this strategy to our framework (with the formal writing $(m_t(x) = \tilde{m}_t(x - \sqrt{2}W_t))_{t \in [0,T]}$), this shows exactly that $(m_t)_{t \in [0,T]}$ solves

$$d_t m_t = \left\{ 2\Delta m_t + \text{div}\left(D_p \tilde{H}_t \left(x - \sqrt{2}W_t, D\tilde{u}_t(x - \sqrt{2}W_t) \right) \right) \right\} dt$$
$$- \sqrt{2}\text{div}(m_t dW_t) \quad (5.28)$$
$$= \left\{ 2\Delta m_t + \text{div}\left(D_p H \left(x, Du_t(x) \right) \right) \right\} dt - \sqrt{2}\text{div}(m_t dW_t).$$

Next we consider the equation satisfied by $(u_t)_{t \in [0,T]}$. Generally speaking, the strategy is similar. Intuitively, it consists in applying Itô–Wenztell formula again, but to $(u_t(x) = \tilde{u}_t(x - \sqrt{2}W_t))_{t \in [0,T]}$. In any case, to apply Itô–Wentzell formula, we need first to identify the martingale part in $(\tilde{u}_t(x))_{t \in [0,T]}$ (namely $(\tilde{M}_t(x))_{t \in [0,T]}$). Recalling from Lemma 5.1.1 the formula

$$\tilde{u}_t(x) = U\left(t, x + \sqrt{2}W_t, m_t \right), \quad t \in [0, T],$$

we understand that the martingale part of $(\tilde{u}_t(x))_{t\in[0,T]}$ should be given by the first-order expansion of the above right-hand side (using an appropriate version of Itô's formula for functionals defined on $[0,T]\times\mathbb{T}^d\times\mathcal{P}(\mathbb{T}^d)$). For our purpose, it is simpler to express $u_t(x)$ in terms of U directly:

$$u_t(x) = U(t, x, m_t), \quad t \in [0, T].$$

The trick is then to expand the above right-hand side by taking advantage of the master equation satisfied by U and from a new Itô's formula for functionals of a measure argument; see Theorem A.1 in the Appendix.

As consequence of Theorem A.1, we get that

$$
\begin{aligned}
d_t u_t(x) = \Big\{ & \partial_t U(t, x, m_t) \\
& + \int_{\mathbb{T}^d} 2\mathrm{div}_y\big[D_m U\big](t, x, m_t, y)dm_t(y) \\
& - \int_{\mathbb{T}^d} D_m U(t, x, m_t, y)\cdot D_p H_t\big(y, Du_t(y)\big)dm_t(y) \\
& + \int_{\mathbb{T}^d}\int_{\mathbb{T}^d} \mathrm{Tr}\big[D_{mm}^2 U\big](t, x, m_t, y, y')dm_t(y)dm_t(y') \Big\}dt \\
& + \sqrt{2}\Big(\int_{\mathbb{T}^d} D_m U(t, x, m_t, y)dm_t(y)\Big)\cdot dW_t.
\end{aligned}
$$

Letting

$$v_t(x) = \int_{\mathbb{T}^d} D_m U(t, x, m_t, y)dm_t(y), \quad t \in [0, T],\ x \in \mathbb{T}^d,$$

and using the master equation satisfied by U, we obtain therefore

$$
\begin{aligned}
d_t u_t(x) = & -2\Delta u_t(x) + H\big(x, Du_t(x)\big) - F(x, m_t) - \sqrt{2}\mathrm{div}\big(v_t(x)\big)dt \\
& + v_t(x)\cdot dW_t.
\end{aligned}
$$

Together with (5.28), this completes the proof of the existence of a solution to (5.26).

Second step. Uniqueness of the solution. We now prove uniqueness of the solution to (5.26). Given a solution $(u_t, m_t)_{t\in[0,T]}$ (with some $(v_t)_{t\in[0,T]}$) to (5.26), we let

$$\tilde{u}_t(x) = u_t(x + \sqrt{2}W_t), \quad x \in \mathbb{T}^d, \quad \tilde{m}_t = (id - \sqrt{2}W_t)\sharp m_t, \quad t \in [0, T].$$

To prove uniqueness, it suffices to show that $(\tilde{u}_t, \tilde{m}_t)_{t\in[0,T]}$ is a solution to (5.27) (for some martingale $(\tilde{M}_t)_{t\in[0,T]}$).

We first investigate the dynamics of $(\tilde{m}_t)_{t\in[0,T]}$. As in the first step (*existence of a solution*), we may apply Itô–Wenztell formula to the process $(\int_{\mathbb{T}^d} \phi(x - \sqrt{2}W_t)dm_t(x))_{t\in[0,T]}$ for a smooth function $\phi \in \mathcal{C}^4(\mathbb{T}^d)$, see Subsection A.3.1. Equivalently, we may apply the Itô–Wenztell formula for Schwartz distribution-valued processes as given in [67, Theorem 1.1]. We get exactly that $(\tilde{m}_t)_{t\in[0,T]}$ satisfies the first equation in (5.27).

To prove the second equation in (5.27), we apply the Itô–Wentzell formula for real-valued processes to $(\tilde{u}_t(x) = u_t(x + \sqrt{2}W_t))_{t\in[0,T]}$, see again Subsection A.3.1.

Chapter Six

Convergence of the Nash System

IN THIS CHAPTER, we consider, for an integer $N \geqslant 2$, a classical solution $(v^{N,i})_{i \in \{1,\dots,N\}}$ of the Nash system with a common noise:

$$
\begin{cases}
-\partial_t v^{N,i}(t, \boldsymbol{x}) - \sum_{j=1}^{N} \Delta_{x_j} v^{N,i}(t, \boldsymbol{x}) - \beta \sum_{j,k=1}^{N} \mathrm{Tr} D^2_{x_j, x_k} v^{N,i}(t, \boldsymbol{x}) \\
\quad + H(x_i, D_{x_i} v^{N,i}(t, \boldsymbol{x})) + \sum_{j \neq i} D_p H(x_j, D_{x_j} v^{N,j}(t, \boldsymbol{x})) \cdot D_{x_j} v^{N,i}(t, \boldsymbol{x}) \qquad (6.1) \\
\quad = F(x_i, m_{\boldsymbol{x}}^{N,i}) \qquad\qquad \text{in } [0, T] \times (\mathbb{T}^d)^N, \\
v^{N,i}(T, \boldsymbol{x}) = G(x_i, m_{\boldsymbol{x}}^{N,i}) \qquad \text{in } (\mathbb{T}^d)^N,
\end{cases}
$$

where we set, for $\boldsymbol{x} = (x_1, \dots, x_N) \in (\mathbb{T}^d)^N$, $m_{\boldsymbol{x}}^{N,i} = \dfrac{1}{N-1} \sum_{j \neq i} \delta_{x_j}$.

Our aim is to prove Theorem 2.4.8, which says that $(v^{N,i})_{i \in \{1,\dots,N\}}$ converges, in a suitable sense, to the solution of the second-order master equation and Theorem 2.4.9, which claims that the optimal trajectories also converge.

Throughout this part we assume that H, F, and G satisfy the assumption of Theorem 2.4.5 with $n \geqslant 2$. This allows us to define $U = U(t, x, m)$ the solution of the second-order master equation

$$
\begin{cases}
-\partial_t U - (1 + \beta) \Delta_x U + H(x, D_x U) - (1 + \beta) \displaystyle\int_{\mathbb{T}^d} \mathrm{div}_y [D_m U] \, dm(y) \\
\quad + \displaystyle\int_{\mathbb{T}^d} D_m U \cdot D_p H\big(y, D_x U(\cdot, y, \cdot)\big) \, dm(y) \\
\quad - 2\beta \displaystyle\int_{\mathbb{T}^d} \mathrm{div}_x [D_m U] \, dm(y) - \beta \displaystyle\int_{\mathbb{T}^{2d}} \mathrm{Tr} \left[D^2_{mm} U \right] dm^{\otimes 2}(y, y') \qquad (6.2) \\
\quad = F(x, m) \qquad\qquad \text{in } [0, T] \times \mathbb{T}^d \times \mathcal{P}(\mathbb{T}^d), \\
U(T, x, m) = G(x, m) \qquad \text{in } \mathbb{T}^d \times \mathcal{P}(\mathbb{T}^d),
\end{cases}
$$

where $\partial_t U$, $D_x U$ and $\Delta_x U$ are understood as $\partial_t U(t, x, m)$, $D_x U(t, x, m)$ and $\Delta_x U(t, x, m)$, $D_x U(\cdot, y, \cdot)$ is understood as $D_x U(t, y, m)$, and $D_m U$ and $D^2_{mm} U$ are understood as $D_m(t, x, m, y)$ and $D^2_{mm} U(t, x, m, y, y')$. Above, $\beta \geqslant 0$ is a parameter for the common noise. For $\alpha' \in (0, \alpha)$, we have for any

$(t, x) \in [0, T] \times \mathbb{T}^d$, $m, m' \in \mathcal{P}(\mathbb{T}^d)$

$$\|U(t, \cdot, m)\|_{n+2+\alpha'} + \left\|\frac{\delta U}{\delta m}(t, \cdot, m, \cdot)\right\|_{(n+2+\alpha', n+1+\alpha')}$$

$$+ \left\|\frac{\delta^2 U}{\delta m^2}(t, \cdot, m, \cdot, \cdot)\right\|_{(n+2+\alpha', n+\alpha', n+\alpha')} \tag{6.3}$$

$$\leqslant C_0,$$

and the mapping

$$[0, T] \times \mathcal{P}(\mathbb{T}^d) \ni (t, m) \mapsto \frac{\delta^2 U}{\delta m^2}(t, \cdot, m, \cdot, \cdot) \in \mathcal{C}^{n+2+\alpha'}(\mathbb{T}^d) \times \left[\mathcal{C}^{n+\alpha'}(\mathbb{T}^d)\right]^2 \tag{6.4}$$

is continuous. As previously stated, a solution of (6.2) satisfying the above properties has been built into Theorem 2.4.5. When $\beta = 0$, one just needs to replace the above assumptions by those of Theorem 2.4.2, which do not require the second-order differentiability of F and G with respect to m.

The main idea for proving the convergence of the $(v^{N,i})_{i \in \{1,\dots,N\}}$ toward U is to use the fact that suitable finite dimensional projections of U are nearly solutions to the Nash equilibrium equation. Actually, as we already alluded to at the end of Chapter 2, this strategy works under weaker assumptions than that required in the statement of Theorem 2.4.5. What is really needed is that H and $D_p H$ are globally Lipschitz continuous, that the coefficient diffusion driving the independent noises $((B_t^i)_{t \in [0,T]})_{i=1,\dots,N}$ in (1.3) is nondegenerate, and that the master equation has a classical solution satisfying the conclusion of Theorem 2.4.5 (or Theorem 2.4.2 if $\beta = 0$). In particular, the monotonicity properties of F and G have no role in the proof of the convergence of the N-Nash system. We refer to Remarks 6.1.5 and 6.2.2 and we let the interested reader reformulate the statements of Theorems 2.4.8 and 2.4.9 accordingly.

6.1   FINITE DIMENSIONAL PROJECTIONS OF U

For $N \geqslant 2$ and $i \in \{1, \dots, N\}$ we set

$$u^{N,i}(t, \boldsymbol{x}) = U(t, x_i, m_{\boldsymbol{x}}^{N,i}) \quad \text{where } \boldsymbol{x} = (x_1, \dots, x_N) \in (\mathbb{T}^d)^N,$$

$$m_{\boldsymbol{x}}^{N,i} = \frac{1}{N-1} \sum_{j \neq i} \delta_{x_j}.$$

Note that the $u^{N,i}$ are at least $\mathcal{C}^2$ with respect to the x_i variable because so is U. Moreover, $\partial_t u^{N,i}$ exists and is continuous because of the regularity of U. The next statement says that $u^{N,i}$ is actually globally $\mathcal{C}^2$ in the space variables:

Proposition 6.1.1. *Assume* (6.3) *and* (6.4). *For any* $N \geqslant 2$, $i \in \{1, \ldots, N\}$, $u^{N,i}$ *is of class* $\mathcal{C}^2$ *in the space variables, with*

$$D_{x_j} u^{N,i}(t, \boldsymbol{x}) = \frac{1}{N-1} D_m U(t, x_i, m_{\boldsymbol{x}}^{N,i}, x_j) \qquad (j \neq i),$$

$$D_{x_i, x_j}^2 u^{N,i}(t, \boldsymbol{x}) = \frac{1}{N-1} D_x D_m U(t, x_i, m_{\boldsymbol{x}}^{N,i}, x_j) \qquad (j \neq i),$$

$$D_{x_j, x_j}^2 u^{N,i}(t, \boldsymbol{x}) = \frac{1}{N-1} D_y \left[D_m U \right] (t, x_i, m_{\boldsymbol{x}}^{N,i}, x_j)$$

$$+ \frac{1}{(N-1)^2} D_{mm}^2 U(t, x_i, m_{\boldsymbol{x}}^{N,i}, x_j, x_j) \qquad (j \neq i),$$

$$D_{x_j, x_k}^2 u^{N,i}(t, \boldsymbol{x}) = \frac{1}{(N-1)^2} D_{mm}^2 U(t, x_i, m_{\boldsymbol{x}}^{N,i}, x_j, x_k) \qquad (i, j, k \text{ distinct}).$$

Remark 6.1.2. If, instead of assumptions (6.3) and (6.4), one only requires the assumptions of Theorem 2.4.2 (no second-order differentiability for F and G), then, for any $N \geqslant 2$, $i \in \{1, \ldots, N\}$, $u^{N,i}$ is of class $\mathcal{C}^1$ in the space variables, with uniformly Lipschitz continuous space derivatives and with

$$D_{x_j} u^{N,i}(t, \boldsymbol{x}) = \frac{1}{N-1} D_m U(t, x_i, m_{\boldsymbol{x}}^{N,i}, x_j) \qquad (j \neq i),$$

$$D_{x_i, x_j}^2 u^{N,i}(t, \boldsymbol{x}) = \frac{1}{N-1} D_x D_m U(t, x_i, m_{\boldsymbol{x}}^{N,i}, x_j) \qquad (j \neq i),$$

$$|D_{x_j, x_j}^2 u^{N,i}(t, \mathbf{x}) - \frac{1}{N-1} D_y \left[D_m U \right] (t, x_i, m_{\boldsymbol{x}}^{N,i}, x_j)| \leqslant \frac{C}{N} \qquad (j \neq i),$$

where the two equalities hold for any $\mathbf{x} \in (\mathbb{T}^d)^N$ while the inequality only holds for a.e. $\mathbf{x} \in (\mathbb{T}^d)^N$, the constant C depending on the Lipschitz regularity of $D_m U$ in the m variable. The proof is the same as for Proposition 6.1.1 except that one uses Proposition A.2.1 instead of Proposition A.2.3 in the argument that follows.

Proof. For $\boldsymbol{x} = (x_j)_{j \in \{1, \ldots, N\}}$ such that $x_j \neq x_k$ for any $j \neq k$, let $\epsilon = \min_{j \neq k} |x_j - x_k|$. For $\boldsymbol{v} = (v_j) \in (\mathbb{R}^d)^N$ with $v_i = 0$ (the value of $i \in \{1, \ldots, N\}$ being fixed), we consider a smooth vector field ϕ such that

$$\phi(x) = v_j \qquad \text{if } x \in B(x_j, \epsilon/4),$$

where $B(x_j, \epsilon/4)$ is the ball of center x_j and of radius $\epsilon/4$. Then, in view of our assumptions (6.3) and (6.4) on U, Propositions A.2.3 and A.2.4 in the Appendix imply that

$$\left| U\big(t, x_i, (id + \phi)\sharp m_{\boldsymbol{x}}^{N,i}\big) - U\big(t, x_i, m_{\boldsymbol{x}}^{N,i}\big) \right.$$

$$- \int_{\mathbb{T}^d} D_m U\big(t, x_i, m_{\boldsymbol{x}}^{N,i}, y\big) \cdot \phi(y)\, dm_{\boldsymbol{x}}^{N,i}(y)$$

$$- \frac{1}{2} \int_{\mathbb{T}^d} D_y \big[D_m U \big]\big(t, x_i, m_{\boldsymbol{x}}^{N,i}, y\big) \phi(y) \cdot \phi(y)\, dm_{\boldsymbol{x}}^{N,i}(y)$$

$$\left. - \frac{1}{2} \int_{\mathbb{T}^d} \int_{\mathbb{T}^d} D_{mm}^2 U\big(t, x_i, m_{\boldsymbol{x}}^{N,i}, y, y'\big) \phi(y) \cdot \phi(y')\, dm_{\boldsymbol{x}}^{N,i}(y) dm_{\boldsymbol{x}}^{N,i}(y') \right|$$

$$\leqslant \|\phi\|_{L^3_{m_{\boldsymbol{x}}^{N,i}}}^2 \, \omega\big(\|\phi\|_{L^3_{m_{\boldsymbol{x}}^{N,i}}}\big),$$

for some modulus ω such that $\omega(s) \to 0$ as $s \to 0^+$. Therefore (recalling that $v_i = 0$),

$$u^{N,i}(t, \boldsymbol{x} + \boldsymbol{v}) - u^{N,i}(t, \boldsymbol{x})$$

$$= U\big(t, x_i, (id + \phi)\sharp m_{\boldsymbol{x}}^{N,i}\big) - U\big(t, x_i, m_{\boldsymbol{x}}^{N,i}\big)$$

$$= \int_{\mathbb{T}^d} D_m U\big(t, x_i, m_{\boldsymbol{x}}^{N,i}, y\big) \cdot \phi(y)\, dm_{\boldsymbol{x}}^{N,i}(y)$$

$$+ \frac{1}{2} \int_{\mathbb{T}^d} D_y \big[D_m U \big]\big(t, x_i, m_{\boldsymbol{x}}^{N,i}, y\big) \phi(y) \cdot \phi(y) dm_{\boldsymbol{x}}^{N,i}(y)$$

$$+ \frac{1}{2} \int_{\mathbb{T}^d} \int_{\mathbb{T}^d} D_{mm}^2 U\big(t, x_i, m_{\boldsymbol{x}}^{N,i}, y, z\big) \phi(y) \cdot \phi(z) dm_{\boldsymbol{x}}^{N,i}(y) dm_{\boldsymbol{x}}^{N,i}(z)$$

$$+ O\Big(\|\phi\|_{L^3(m_{\boldsymbol{x}}^{N,i})}^2 \omega\big(\|\phi\|_{L^3(m_{\boldsymbol{x}}^{N,i})}\big) \Big)$$

$$= \frac{1}{N-1} \sum_{j \neq i} D_m U\big(t, x_i, m_{\boldsymbol{x}}^{N,i}, x_j\big) \cdot v_j$$

$$+ \frac{1}{2(N-1)} \sum_{j \neq i} D_y \big[D_m U \big]\big(t, x_i, m_{\boldsymbol{x}}^{N,i}, x_j\big) v_j \cdot v_j$$

$$+ \frac{1}{2(N-1)^2} \sum_{j,k \neq i} D_{mm}^2 U\big(t, x_i, m_{\boldsymbol{x}}^{N,i}, x_j, x_k\big) v_j \cdot v_k + O\big(|\boldsymbol{v}|^2 \omega(|\boldsymbol{v}|)\big),$$

where $|O(r)| \leqslant C|r|$ for a constant C independent of $(t, \boldsymbol{x})$, N and i and where ω is allowed to vary from line to line as long as $\omega(s) \to 0$ as $s \to 0^+$.

This shows that $u^{N,i}$ has a second-order expansion at $\boldsymbol{x}$ with respect to the variables $(x_j)_{j \neq i}$ and that

$$D_{x_j} u^{N,i}(t, \boldsymbol{x}) = \frac{1}{N-1} D_m U\big(t, x_i, m_{\boldsymbol{x}}^{N,i}, x_j\big) \qquad\qquad (j \neq i),$$

$$D_{x_j, x_j}^2 u^{N,i}(t, \boldsymbol{x}) = \frac{1}{N-1} D_y \big[D_m U \big]\big(t, x_i, m_{\boldsymbol{x}}^{N,i}, x_j\big)$$

$$+ \frac{1}{(N-1)^2} D_{mm}^2 U\big(t, x_i, m_{\boldsymbol{x}}^{N,i}, x_j, x_j\big) \qquad (j \neq i),$$

$$D^2_{x_j,x_k} u^{N,i}(t, \boldsymbol{x}) = \frac{1}{(N-1)^2} D^2_{mm} U\left(t, x_i, m_{\boldsymbol{x}}^{N,i}, x_j, x_k\right) \qquad (i, j, k \text{ distinct}).$$

Observing that the existence of the derivatives in the direction x_i is obvious, we have thus proved the existence of first- and second-order space derivatives of U in the open subset of $[0, T] \times (\mathbb{T}^d)^N$ consisting of the points $(t, \boldsymbol{x}) = (t, x_1, \cdots x_N)$ such that $x_i \neq x_j$ for any $i \neq j$. As $D_m U$, $D_y [D_m U]$ and $D^2_{mm} U$ are continuous, these first- and second-order derivatives can be continuously extended to the whole space $[0, T] \times (\mathbb{T}^d)^N$, and therefore $u^{N,i}$ is $\mathcal{C}^2$ with respect to the space variables in $[0, T] \times \mathbb{T}^{Nd}$. $\qquad\qquad\qquad\qquad\qquad\qquad\qquad\qquad\qquad\qquad\qquad\square$

We now show that $(u^{N,i})_{i\in\{1,\dots,N\}}$ is "almost" a solution to the Nash system (6.1):

Proposition 6.1.3. *Assume (6.3) and (6.4). One has, for any $i \in \{1, \dots, N\}$,*

$$\begin{cases} -\partial_t u^{N,i}(t, \boldsymbol{x}) - \displaystyle\sum_{j=1}^{N} \Delta_{x_j} u^{N,i}(t, \boldsymbol{x}) - \beta \displaystyle\sum_{j,k=1}^{N} \text{Tr} D^2_{x_j, x_k} u^{N,i}(t, \boldsymbol{x}) \\ + H(x_i, D_{x_i} u^{N,i}(t, \boldsymbol{x})) + \displaystyle\sum_{j\neq i} D_p H(x_j, D_{x_j} u^{N,j}(t, \boldsymbol{x})) \cdot D_{x_j} u^{N,i}(t, \boldsymbol{x}) \qquad (6.5) \\ = F(x_i, m_{\boldsymbol{x}}^{N,i}) + r^{N,i}(t, \boldsymbol{x}) \qquad \text{in } [0, T] \times (\mathbb{T}^d)^N, \\ v^{N,i}(T, \boldsymbol{x}) = G(x_i, m_{\boldsymbol{x}}^{N,i}) \qquad \text{in } (\mathbb{T}^d)^N, \end{cases}$$

where $r^{N,i} \in \mathcal{C}^0([0, T] \times \mathbb{T}^{Nd})$ with

$$\|r^{N,i}\|_\infty \leqslant \frac{C}{N}.$$

Remark 6.1.4. When $\beta = 0$, assumptions (6.3) and (6.4) can be replaced by those of Theorem 2.4.2 (no second-order differentiability for F and G). In this case, equation (6.5) only holds a.e. with $r^{N,i} \in L^\infty$ still satisfying $\|r^{N,i}\|_\infty \leqslant C/N$. The proof is the same as for Proposition 6.1.3, except that one ignores of course the terms involving β and one replaces the conclusion of Proposition 6.1.1 by those of Remark 6.1.2.

Proof. As U solves (6.2), one has at a point $(t, x_i, m_{\boldsymbol{x}}^{N,i})$:

$$- \partial_t U - (1 + \beta)\Delta_x U + H(x_i, D_x U)$$
$$- (1 + \beta) \int_{\mathbb{T}^d} \text{div}_y [D_m U](t, x_i, m_{\boldsymbol{x}}^{N,i}, y) dm_{\boldsymbol{x}}^{N,i}(y)$$
$$+ \int_{\mathbb{T}^d} D_m U(t, x_i, m_{\boldsymbol{x}}^{N,i}, y) \cdot D_p H(y, D_x U(t, y, m_{\boldsymbol{x}}^{N,i})) dm_{\boldsymbol{x}}^{N,i}(y)$$
$$- 2\beta \int_{\mathbb{T}^d} \text{div}_x [D_m U](t, x_i, m_{\boldsymbol{x}}^{N,i}, y) dm_{\boldsymbol{x}}^{N,i}(y)$$
$$- \beta \int_{\mathbb{T}^d} \text{Tr} D^2_{mm} U(t, x_i, m_{\boldsymbol{x}}^{N,i}, y, z) dm_{\boldsymbol{x}}^{N,i}(y) dm_{\boldsymbol{x}}^{N,i}(z) = F(x_i, m_{\boldsymbol{x}}^{N,i}).$$

So $u^{N,i}$ satisfies

$$
\begin{aligned}
&-\partial_t u^{N,i} - (1+\beta)\Delta_{x_i} u^{N,i} + H(x_i, D_{x_i} u^{N,i}) \\
&\quad - (1+\beta) \int_{\mathbb{T}^d} \operatorname{div}_y [D_m U] (t, x_i, m_{\boldsymbol{x}}^{N,i}, y) dm_{\boldsymbol{x}}^{N,i}(y) \\
&\quad + \frac{1}{N-1} \sum_{j \neq i} D_m U (t, x_i, m_{\boldsymbol{x}}^{N,i}, x_j) \cdot D_p H (x_j, D_x U(t, x_j, m_{\boldsymbol{x}}^{N,i})) \\
&\quad - 2\beta \int_{\mathbb{T}^d} \operatorname{div}_x [D_m U] (t, x_i, m_{\boldsymbol{x}}^{N,i}, y) dm_{\boldsymbol{x}}^{N,i}(y) \\
&\quad - \beta \int_{\mathbb{T}^d} \operatorname{Tr} D_{mm}^2 U (t, x_i, m_{\boldsymbol{x}}^{N,i}, y, z) dm_{\boldsymbol{x}}^{N,i}(y) dm_{\boldsymbol{x}}^{N,i}(z) = F(x_i, m_{\boldsymbol{x}}^{N,i}).
\end{aligned}
$$

Note that, by Proposition 6.1.1,

$$
\frac{1}{N-1} D_m U (t, x_i, m_{\boldsymbol{x}}^{N,i}, x_j) = D_{x_j} u^{N,i}(t, \boldsymbol{x}).
$$

In particular,

$$
\|D_{x_j} u^{N,i}\|_\infty \leq \frac{C}{N}. \tag{6.6}
$$

By the Lipschitz continuity of $D_x U$ with respect to m, we have

$$
\left| D_x U(t, x_j, m_{\boldsymbol{x}}^{N,i}) - D_x U(t, x_j, m_{\boldsymbol{x}}^{N,j}) \right| \leq C \mathbf{d}_1(m_{\boldsymbol{x}}^{N,i}, m_{\boldsymbol{x}}^{N,j}) \leq \frac{C}{N-1},
$$

so that, by Lipschitz continuity of $D_p H$,

$$
\left| D_p H (x_j, D_x U(t, x_j, m_{\boldsymbol{x}}^{N,i})) - D_p H (x_j, D_{x_j} u^{N,j}(t, \boldsymbol{x})) \right| \leq \frac{C}{N}. \tag{6.7}
$$

Collecting the above relations, we obtain

$$
\begin{aligned}
&\frac{1}{N-1} \sum_{j \neq i} D_m U (t, x_i, m_{\boldsymbol{x}}^{N,i}, x_j) \cdot D_p H (x_j, D_x U(t, x_j, m_{\boldsymbol{x}}^{N,i})) \\
&\quad = \sum_{j \neq i} D_{x_j} u^{N,i}(t, \boldsymbol{x}) \cdot D_p H (x_j, D_x U(t, x_j, m_{\boldsymbol{x}}^{N,i})) \\
&\quad = \sum_{j \neq i} D_{x_j} u^{N,i}(t, \boldsymbol{x}) \cdot D_p H (x_j, D_{x_j} u^{N,j}(t, \boldsymbol{x})) + O\left(\frac{1}{N}\right),
\end{aligned}
$$

where we used (6.6) in the last inequality. On the other hand,

$$\sum_{j=1}^{N} \Delta_{x_j} u^{N,i} + \beta \sum_{j,k=1}^{N} \mathrm{Tr} D_{x_j,x_k}^2 u^{N,i}$$

$$= (1+\beta)\Delta_{x_i} u^{N,i} + (1+\beta)\sum_{j \neq i} \Delta_{x_j} u^{N,i}$$

$$+ 2\beta \sum_{j \neq i} \mathrm{Tr} D_{x_i,x_j}^2 u^{N,i} + \beta \sum_{j,k,i \text{ distinct}} \mathrm{Tr} D_{x_j,x_k}^2 u^{N,i},$$

where, using Proposition 6.1.1,

$$\sum_{j \neq i} \Delta_{x_j} u^{N,i}(t, \boldsymbol{x}) = \int_{\mathbb{T}^d} \mathrm{div}_y \left[D_m U \right] \left(t, x_i, m_{\boldsymbol{x}}^{N,i}, y \right) dm_{\boldsymbol{x}}^{N,i}(y)$$

$$+ \frac{1}{N-1} \int_{\mathbb{T}^d} \mathrm{Tr} \left[D_{mm}^2 U \right] \left(t, x_i, m_{\boldsymbol{x}}^{N,i}, y, y \right) dm_{\boldsymbol{x}}^{N,i}(y)$$

$$\sum_{j \neq i} \mathrm{Tr} D_{x_i,x_j}^2 u^{N,i}(t, \boldsymbol{x}) = \int_{\mathbb{T}^d} \mathrm{div}_x \left[D_m U \right] \left(t, x_i, m_{\boldsymbol{x}}^{N,i}, y \right) dm_{\boldsymbol{x}}^{N,i}(y)$$

$$\sum_{j,k,i \text{ distinct}} \mathrm{Tr} D_{x_j,x_k}^2 u^{N,i}(t, \boldsymbol{x})$$

$$= \int_{\mathbb{T}^d} \int_{\mathbb{T}^d} \mathrm{Tr} \left[D_{mm}^2 U \right] \left(t, x_i, m_{\boldsymbol{x}}^{N,i}, y, z \right) dm_{\boldsymbol{x}}^{N,i}(y) dm_{\boldsymbol{x}}^{N,i}(z).$$

Therefore

$$- \partial_t u^{N,i}(t, \boldsymbol{x}) - \sum_{j=1}^{N} \Delta_{x_j} u^{N,i}(t, \boldsymbol{x}) - \beta \sum_{j,k=1}^{N} \mathrm{Tr} D_{x_j,x_k}^2 u^{N,i}(t, \boldsymbol{x})$$

$$+ H\left(x_i, D_{x_i} u^{N,i}(t, \boldsymbol{x}) \right) + \sum_{j \neq i} D_{x_j} u^{N,i}(t, \boldsymbol{x}) \cdot D_p H\left(x_j, D_{x_j} u^{N,j}(t, \boldsymbol{x}) \right)$$

$$+ \frac{1+\beta}{N-1} \int_{\mathbb{T}^d} \mathrm{Tr} D_{mm}^2 U\left(t, x_i, m_{\boldsymbol{x}}^{N,i}, y, y \right) dm_{\boldsymbol{x}}^{N,i}(y)$$

$$= F(x_i, m_{\boldsymbol{x}}^{N,i}) + O\left(\frac{1}{N}\right),$$

which shows the result. $\qquad\qquad\qquad\qquad\qquad\qquad\qquad\qquad\qquad\qquad$ $\square$

Remark 6.1.5. *The reader may observe that, in addition to the existence of a classical solution U (to the master equation) satisfying the conclusion of Theorem 2.4.5, only the global Lipschitz property of $D_p H$ is used in the proof; see (6.7).*

6.2   CONVERGENCE

We now turn to the proof of Theorem 2.4.8. For this, we consider the solution $(v^{N,i})_{i \in \{1,\dots,N\}}$ of the Nash system (6.1). By uniqueness of the solution, the $(v^{N,i})_{i \in \{1,\dots,N\}}$ must be symmetrical. By symmetrical, we mean that, for any $\boldsymbol{x} = (x_l)_{l \in \{1,\dots,N\}} \in \mathbb{T}^{Nd}$ and for any indices $j \neq k$, if $\tilde{\boldsymbol{x}} = (\tilde{x}_l)_{l \in \{1,\dots,N\}}$ is the N-tuple obtained from $\boldsymbol{x}$ by permuting the j and k vectors (i.e., $\tilde{x}_l = x_l$ for $l \notin \{j, k\}$, $\tilde{x}_j = x_k$, $\tilde{x}_k = x_j$), then

$$v^{N,i}(t, \tilde{\boldsymbol{x}}) = v^{N,i}(t, \boldsymbol{x}) \text{ if } i \notin \{j, k\}, \text{ while } v^{N,i}(t, \tilde{\boldsymbol{x}}) = v^{N,k}(t, \boldsymbol{x}) \text{ if } i = j,$$

which may be reformulated as follows: There exists a function $V^N : \mathbb{T}^d \times [\mathbb{T}^d]^{N-1} \to \mathbb{R}$ such that, for any $x \in \mathbb{T}^d$, the function $[\mathbb{T}^d]^{N-1} \ni (y_1, \dots, y_{N-1}) \mapsto V^N(x, (y_1, \dots, y_{N-1}))$ is invariant under permutation, and

$$\forall i \in \{1, \dots, N\}, \ \boldsymbol{x} \in [\mathbb{T}^d]^N, \quad v^{N,i}(t, \boldsymbol{x}) = V^N\big(x_i, (x_1, \dots, x_{i-1}, x_{i+1}, \dots, x_N)\big).$$

Note that the $(u^{N,i})_{i \in \{1,\dots,N\}}$ are also symmetrical.

The proof of Theorem 2.4.8 consists in comparing the "optimal trajectories" for $v^{N,i}$ and for $u^{N,i}$, for any $i \in \{1, \dots, N\}$. For this, let us fix $t_0 \in [0, T)$, $m_0 \in \mathcal{P}(\mathbb{T}^d)$ and let $(Z_i)_{i \in \{1,\dots,N\}}$ be an i.i.d family of N random variables of law m_0. We set $\boldsymbol{Z} = (Z_i)_{i \in \{1,\dots,N\}}$. Let also $((B_t^i)_{t \in [0,T]})_{i \in \{1,\dots,N\}}$ be a family of N independent d-dimensional Brownian motions that is also independent of $(Z_i)_{i \in \{1,\dots,N\}}$ and let W be a d-dimensional Brownian motion independent of the family $((B_t^i)_{t \in [0,T]}, Z_i)_{i \in \{1,\dots,N\}}$. We consider the systems of stochastic differential equations (SDEs) with variables $(\boldsymbol{X}_t = (X_{i,t})_{i \in \{1,\dots,N\}})_{t \in [0,T]}$ and $(\boldsymbol{Y}_t = (Y_{i,t})_{i \in \{1,\dots,N\}})_{t \in [0,T]}$ (the SDEs being set on $\mathbb{R}^d$ with periodic coefficients):

$$\begin{cases} dX_{i,t} = -D_p H\big(X_{i,t}, D_{x_i} u^{N,i}(t, \boldsymbol{X}_t)\big) dt \\ \qquad\qquad\qquad + \sqrt{2} dB_t^i + \sqrt{2\beta} dW_t, \quad t \in [t_0, T], \\ X_{i,t_0} = Z_i, \end{cases} \tag{6.8}$$

and

$$\begin{cases} dY_{i,t} = -D_p H\big(Y_{i,t}, D_{x_i} v^{N,i}(t, \boldsymbol{Y}_t)\big) dt \\ \qquad\qquad\qquad + \sqrt{2} dB_t^i + \sqrt{2\beta} dW_t, \quad t \in [t_0, T], \\ Y_{i,t_0} = Z_i. \end{cases} \tag{6.9}$$

Since the $(u^{N,i})_{i \in \{1,\dots,N\}}$ are symmetrical, the processes $((X_{i,t})_{t \in [t_0,T]})_{i \in \{1,\dots,N\}}$ are exchangeable. The same holds for the $((Y_{i,t})_{t \in [t_0,T]})_{i \in \{1,\dots,N\}}$ and, actually, the N $\mathbb{R}^{2d}$-valued processes $((X_{i,t}, Y_{i,t})_{t \in [t_0,T]})_{i \in \{1,\dots,N\}}$ are also exchangeable.

Theorem 6.2.1. *Assume (6.3) and (6.4). Then we have, for any $i \in \{1, \ldots, N\}$,*

$$\mathbb{E}\Big[\sup_{t \in [t_0, T]} |Y_{i,t} - X_{i,t}| \Big] \leqslant \frac{C}{N}, \qquad \forall t \in [t_0, T], \tag{6.10}$$

$$\mathbb{E}\Big[\int_{t_0}^T |D_{x_i} v^{N,i}(t, \boldsymbol{Y}_t) - D_{x_i} u^{N,i}(t, \boldsymbol{Y}_t)|^2 dt \Big] \leqslant \frac{C}{N^2}, \tag{6.11}$$

and, $\mathbb{P}$–almost surely, for all $i = 1, \ldots, N$,

$$|u^{N,i}(t_0, \boldsymbol{Z}) - v^{N,i}(t_0, \boldsymbol{Z})| \leqslant \frac{C}{N}, \tag{6.12}$$

where C is a (deterministic) constant that does not depend on t_0, m_0, and N.

Proof of Theorem 6.2.1. *First step.* We start with the proof of (6.11). For simplicity, we work with $t_0 = 0$. Let us first introduce new notations:

$$U_t^{N,i} = u^{N,i}(t, \boldsymbol{Y}_t), \quad V_t^{N,i} = v^{N,i}(t, \boldsymbol{Y}_t),$$

$$DU_t^{N,i,j} = D_{x_j} u^{N,i}(t, \boldsymbol{Y}_t), \quad DV_t^{N,i,j} = D_{x_j} v^{N,i}(t, \boldsymbol{Y}_t), \quad t \in [0, T].$$

Using equation (6.1) satisfied by the $(v^{N,i})_{i \in \{1, \ldots, N\}}$, we deduce from Itô's formula that, for any $i \in \{1, \ldots, N\}$,

$$
\begin{aligned}
dV_t^{N,i} &= \Big[\partial_t v^{N,i}(t, \boldsymbol{Y}_t) - \sum_{j=1}^N D_{x_j} v^{N,i}(t, \boldsymbol{Y}_t) \cdot D_p H\big(Y_{j,t}, D_{x_j} v^{N,i}(t, \boldsymbol{Y}_t)\big) \\
&\quad + \sum_{j=1}^N \Delta_{x_j} v^{N,i}(t, \boldsymbol{Y}_t) + \beta \sum_{j,k=1}^N \mathrm{Tr} D^2_{x_j, x_k} v^{N,i}(t, \boldsymbol{Y}_t) \Big] dt \\
&\quad + \sqrt{2} \sum_{j=1}^N D_{x_j} v^{N,i}(t, \boldsymbol{Y}_t) \cdot dB_t^j + \sqrt{2\beta} \sum_{j=1}^N D_{x_j} v^{N,i}(t, \boldsymbol{Y}_t) \cdot dW_t \\
&= \Big[H\big(Y_{i,t}, D_{x_i} v^{N,i}(t, \boldsymbol{Y}_t)\big) - D_{x_i} v^{N,i}(t, \boldsymbol{Y}_t) \cdot D_p H\big(Y_{i,t}, D_{x_i} v^{N,i}(t, \boldsymbol{Y}_t)\big) \\
&\quad - F\big(Y_{i,t}, m_{\boldsymbol{Y}_t}^{N,i}\big) \Big] dt + \sqrt{2} \sum_{j=1}^N D_{x_j} v^{N,i}(t, \boldsymbol{Y}_t) \cdot dB_t^j \\
&\quad + \sqrt{2\beta} \sum_{j=1}^N D_{x_j} v^{N,i}(t, \boldsymbol{Y}_t) \cdot dW_t.
\end{aligned}
\tag{6.13}
$$

Similarly, as $(u^{N,i})_{i\in\{1,\dots,N\}}$ satisfies (6.5), we have by standard computation:

$$
\begin{aligned}
dU_t^{N,i} = \Big[& H\big(Y_{i,t}, D_{x_i} u^{N,i}(t, \boldsymbol{Y}_t)\big) - D_{x_i} u^{N,i}(t, \boldsymbol{Y}_t) \\
& \cdot D_p H\big(Y_{i,t}, D_{x_i} u^{N,i}(t, \boldsymbol{Y}_t)\big) - F\big(Y_{i,t}, m_{\boldsymbol{Y}_t}^{N,i}\big) - r^{N,i}(t, \boldsymbol{Y}_t) \Big] dt \\
& - \sum_{j=1}^{N} D_{x_j} u^{N,i}(t, \boldsymbol{Y}_t) \cdot \Big(D_p H\big(Y_{j,t}, D_{x_j} v^{N,j}(t, \boldsymbol{Y}_t)\big) \\
& - D_p H\big(Y_{j,t}, D_{x_j} u^{N,j}(t, \boldsymbol{Y}_t)\big) \Big) dt + \sqrt{2} \sum_{j=1}^{N} D_{x_j} u^{N,i}(t, \boldsymbol{Y}_t) \\
& \cdot dB_t^j + \sqrt{2\beta} \sum_{j=1}^{N} D_{x_j} u^{N,i}(t, \boldsymbol{Y}_t) \cdot dW_t.
\end{aligned}
\tag{6.14}
$$

Compute the difference between (6.13) and (6.14), take the square, and apply Itô's formula again:

$$
\begin{aligned}
d\big[U_t^{N,i} - V_t^{N,i}\big]^2 = \Big[& 2(U_t^{N,i} - V_t^{N,i})\Big(H\big(Y_{i,t}, DU_t^{N,i,i}\big) \\
& - H\big(Y_{i,t}, DV_t^{N,i,i}\big)\Big) - 2(U_t^{N,i} - V_t^{N,i})\Big(DU_t^{N,i,i} \cdot [D_p H\big(Y_{i,t}, DU_t^{N,i,i}\big) \\
& - D_p H\big(Y_{i,t}, DV_t^{N,i,i}\big)]\Big) - 2(U_t^{N,i} - V_t^{N,i})\Big([DU_t^{N,i,i} - DV_t^{N,i,i}] \\
& \cdot D_p H\big(Y_{i,t}, DV_t^{N,i,i}\big)\Big) - 2(U_t^{N,i} - V_t^{N,i}) r^{N,i}(t, \boldsymbol{Y}_t) \Big] dt \\
& - 2(U_t^{N,i} - V_t^{N,i}) \sum_{j=1}^{N} DU_t^{N,i,j} \cdot \Big(D_p H\big(Y_{j,t}, DV_t^{N,j,j}\big) \\
& - D_p H\big(Y_{j,t}, DU_t^{N,j,j}\big)\Big) dt + \Big[2 \sum_{j=1}^{N} |DU_t^{N,i,j} - DV_t^{N,i,j}|^2 \\
& + 2\beta \Big|\sum_{j=1}^{N} (DU_t^{N,i,j} - DV_t^{N,i,j})\Big|^2 \Big] dt + \sqrt{2}(U_t^{N,i} - V_t^{N,i}) \\
& \sum_{j=1}^{N} \Big[(DU_t^{N,i,j} - DV_t^{N,i,j}) \cdot dB_t^j + \sqrt{\beta}(DU_t^{N,i,j} - DV_t^{N,i,j}) \cdot dW_t\Big].
\end{aligned}
\tag{6.15}
$$

Recall now that H and $D_p H$ are Lipschitz continuous in the variable p. Recall also that $DU_t^{N,i,i} = D_x U(t, Y_{i,t}, m_{\boldsymbol{Y}_t}^{N,i})$ is bounded, independently of i, N, and t, and that $DU_t^{N,i,j} = (1/(N-1)) D_m U(t, Y_{i,t}, m_{\boldsymbol{Y}_t}^{N,i})$ is bounded by C/N when

$i \neq j$, for C independent of i, j, N, and t. Recall finally from Proposition 6.1.3 that $r^{N,i}$ is bounded by C/N. Integrating from t to T in the above formula and taking the conditional expectation given $\boldsymbol{Z}$ (with the shortened notation $\mathbb{E}^{\boldsymbol{Z}}[\cdot] = \mathbb{E}[\cdot|\boldsymbol{Z}]$), we deduce:

$$\mathbb{E}^{\boldsymbol{Z}}\left[|U_t^{N,i} - V_t^{N,i}|^2\right] + 2\sum_{j=1}^{N}\mathbb{E}^{\boldsymbol{Z}}\left[\int_t^T |DU_s^{N,i,j} - DV_s^{N,i,j}|^2 ds\right]$$

$$+ 2\beta\mathbb{E}^{\boldsymbol{Z}}\left[\int_t^T \left|\sum_{j=1}^{N}(DU_s^{N,i,j} - DV_t^{N,i,j})\right|^2 ds\right]$$

$$\leqslant \mathbb{E}^{\boldsymbol{Z}}\left[|U_T^{N,i} - V_T^{N,i}|^2\right] + \frac{C}{N}\int_t^T \mathbb{E}^{\boldsymbol{Z}}\left[|U_s^{N,i} - V_s^{N,i}|\right]ds \qquad (6.16)$$

$$+ C\int_t^T \mathbb{E}^{\boldsymbol{Z}}\left[|U_s^{N,i} - V_s^{N,i}| \cdot |DU_s^{N,i,i} - DV_s^{N,i,i}|\right]ds$$

$$+ \frac{C}{N}\sum_{j \neq i}\int_t^T \mathbb{E}^{\boldsymbol{Z}}\left[|U_s^{N,i} - V_s^{N,i}| \cdot |DU_s^{N,j,j} - DV_s^{N,j,j}|\right]ds.$$

Note that the boundary condition $U_T^{N,i} - V_T^{N,i}$ is 0. By a standard convexity argument, we get

$$\mathbb{E}^{\boldsymbol{Z}}\left[|U_t^{N,i} - V_t^{N,i}|^2\right] + \mathbb{E}^{\boldsymbol{Z}}\left[\int_t^T |DU_s^{N,i,i} - DV_s^{N,i,i}|^2 ds\right]$$

$$\leqslant \frac{C}{N^2} + C\int_t^T \mathbb{E}^{\boldsymbol{Z}}\left[|U_s^{N,i} - V_s^{N,i}|^2\right]ds \qquad (6.17)$$

$$+ \frac{1}{2N}\sum_{j=1}^{N}\mathbb{E}^{\boldsymbol{Z}}\left[\int_t^T |DU_s^{N,j,j} - DV_s^{N,j,j}|^2 ds\right].$$

By taking the mean over $i \in \{1, \dots, N\}$, we can get rid of the second term in the left-hand side. By Gronwall's lemma, we obtain (allowing the constant C to increase from line to line):

$$\sup_{0 \leqslant t \leqslant T}\left[\frac{1}{N}\sum_{i=1}^{N}\mathbb{E}^{\boldsymbol{Z}}\left[|U_t^{N,i} - V_t^{N,i}|^2\right]\right] \leqslant \frac{C}{N^2}. \qquad (6.18)$$

Plugging (6.18) into (6.17), we deduce that

$$\frac{1}{N}\sum_{j=1}^{N}\mathbb{E}^{\boldsymbol{Z}}\left[\int_0^T |DU_s^{N,j,j} - DV_s^{N,j,j}|^2 ds\right] \leqslant \frac{C}{N^2}.$$

Inserting this bound on the right-hand side of (6.17) and applying Gronwall's lemma once again, we finally end up with:

$$\sup_{t\in[0,T]} \mathbb{E}^{\mathbf{Z}}\left[|U_t^{N,i} - V_t^{N,i}|^2\right] + \mathbb{E}^{\mathbf{Z}}\left[\int_0^T |DU_s^{N,i,i} - DV_s^{N,i,i}|^2 ds\right] \leqslant \frac{C}{N^2}. \quad (6.19)$$

This proves (6.11).

Second step. We now derive (6.10) and (6.12). We start with (6.12). Noticing that $U_0^{N,i} - V_0^{N,i} = u^{N,i}(0, \mathbf{Z}) - v^{N,i}(0, \mathbf{Z})$, we deduce from (6.19) that, with probability 1 under $\mathbb{P}$, for all $i \in \{1, \dots, N\}$,

$$|u^{N,i}(0, \mathbf{Z}) - v^{N,i}(0, \mathbf{Z})| \leqslant \frac{C}{N},$$

which is exactly (6.12).

We are now ready to estimate the difference $X_{i,t} - Y_{i,t}$, for $t \in [0, T]$ and $i \in \{1, \dots, N\}$. In view of the equation satisfied by the processes $(X_{i,t})_{t\in[0,T]}$ and by $(Y_{i,t})_{t\in[0,T]}$, we have

$$|X_{i,t} - Y_{i,t}| \leqslant \int_0^t \Big|D_p H\big(X_{i,s}, D_{x_i} u^{N,i}(s, \mathbf{X}_s)\big)$$
$$- D_p H\big(Y_{i,s}, D_{x_i} v^{N,i}(s, \mathbf{Y}_s)\big)\Big| ds$$

Using the Lipschitz regularity of $D_p H$, the regularity of U and Proposition 6.1.1, we obtain

$$|X_{i,t} - Y_{i,t}| \leqslant C \int_0^t \left(|X_{i,s} - Y_{i,s}| + \frac{1}{N}\sum_{j\neq i}|X_{j,s} - Y_{j,s}|\right) ds$$

$$+ \int_0^t \Big|D_p H\big(Y_{i,s}, D_{x_i} u^{N,i}(s, \mathbf{Y}_s)\big) \qquad (6.20)$$

$$- D_p H\big(Y_{i,s}, D_{x_i} v^{N,i}(s, \mathbf{Y}_s)\big)\Big| ds.$$

Taking the sup over $t \in [0, \tau]$ (for $\tau \in [0, T]$) and the conditional expectation with respect to $\mathbf{Z}$ we find

$$\mathbb{E}^{\mathbf{Z}}\Big[\sup_{t\in[0,\tau]} |X_{i,t} - Y_{i,t}|\Big] \leqslant C \int_0^\tau \Big(\mathbb{E}^{\mathbf{Z}}\Big[\sup_{t\in[0,s]}|X_{i,t} - Y_{i,t}|\Big]$$

$$+ \frac{1}{N}\sum_{j\neq i} \mathbb{E}^{\mathbf{Z}}\Big[\sup_{t\in[0,s]}|X_{j,t} - Y_{j,t}|\Big]\Big) ds \qquad (6.21)$$

$$+ \mathbb{E}^{\mathbf{Z}}\left[\int_0^T |DU_s^{N,i,i} - DV_s^{N,i,i}| ds\right].$$

Summing over $i \in \{1, \ldots, N\}$ and using (6.19), we derive by Gronwall's inequality:

$$\sum_{i=1}^{N} \mathbb{E}^{\mathbf{Z}} \Big[\sup_{t \in [0,T]} |X_{i,t} - Y_{i,t}| \Big] ds \leqslant C.$$

Inserting the above inequality into (6.21) and using once again Gronwall's lemma, we obtain (6.10). □

Remark 6.2.2. *The reader may observe that, in addition to the existence of a classical solution U (to the master equation) satisfying the conclusion of Theorem 2.4.5, only the global Lipschitz properties of H and $D_p H$ and the ellipticity of σ are used in the proof; see (6.16) and (6.20).*

Remark 6.2.3. *If $\beta = 0$, the conclusion of Theorem 6.2.1 also holds under the assumptions of Theorem 2.4.2 (no second-order differentiability for F and G) instead of (6.3) and (6.4). The proof is the same as above because one can still apply Itô's formula to the maps $u^{N,i}$ which are C^1 in time and $C^{1,1}$ in space and the diffusions driving the processes are non degenerate.*

Proof of Theorem 2.4.8. For part (i), let us choose $m_0 \equiv 1$ and apply (6.12):

$$\Big| U(t_0, Z_i, m_{\mathbf{Z}}^{N,i}) - v^{N,i}(t_0, \mathbf{Z}) \Big| \leqslant \frac{C}{N} \qquad \text{a.e.,} \qquad i \in \{1, \ldots, N\},$$

where $\mathbf{Z} = (Z_1, \ldots, Z_N)$ with $Z_1, \ldots, Z_N$ i.i.d. random variables with uniform density on $\mathbb{T}^d$. The support of $\mathbf{Z}$ being $(\mathbb{T}^d)^N$, we derive from the continuity of U and of the $(v^{N,i})_{i \in \{1,\ldots,N\}}$ that the above inequality holds for any $\mathbf{x} \in (\mathbb{T}^d)^N$:

$$\Big| U(t_0, x_i, m_{\mathbf{x}}^{N,i}) - v^{N,i}(t_0, \mathbf{x}) \Big| \leqslant \frac{C}{N} \qquad \forall \mathbf{x} \in (\mathbb{T}^d)^N, \qquad i \in \{1, \ldots, N\},$$

for all $i \in \{1, \ldots, N\}$. Then we use the Lipschitz continuity of U with respect to m to replace $U(t_0, x_i, m_{\mathbf{x}}^{N,i})$ by $U(t_0, x_i, m_{\mathbf{x}}^{N})$ in the above inequality, the additional error term being of order $1/N$.

To prove (ii), we use the Lipschitz continuity of U and a result by Fournier and Guillin [42] (see also Dereich, Scheutzow, and Schottstedt [34] when $d \geqslant 3$ and Ajtai, Komlos, and Tusndy [9] when $d = 2$), to deduce that for any $x_i \in \mathbb{T}^d$:

$$\int_{\mathbb{T}^{d(N-1)}} |u^{N,i}(t, \mathbf{x}) - U(t, x_i, m_0)| \prod_{j \neq i} m_0(dx_j)$$

$$= \int_{\mathbb{T}^{d(N-1)}} |U(t, x_i, m_{\mathbf{x}}^{N,i}) - U(t, x_i, m_0)| \prod_{j \neq i} m_0(dx_j)$$

$$\leqslant C \int_{\mathbb{T}^{d(N-1)}} \mathbf{d}_1(m_{\mathbf{x}}^{N,i}, m_0) \prod_{j \neq i} m_0(dx_j) \leqslant C \varepsilon_N,$$

where

$$
\varepsilon_N = \begin{cases} N^{-1/d} & \text{if } d \geqslant 3, \\ N^{-1/2}\ln(N) & \text{if } d = 2, \\ N^{-1/2} & \text{if } d = 1. \end{cases}
$$

Combining Theorem 6.2.1 with the above inequality, we therefore obtain

$$
\left\| w^{N,i}(t_0, \cdot, m_0) - U(t_0, \cdot, m_0) \right\|_{L^1(m_0)}
$$

$$
= \int_{\mathbb{T}^d} \left| \int_{\mathbb{T}^{d(N-1)}} v^{N,i}\big(t, (x_j)\big) \prod_{j \neq i} m_0(dx_j) - U(t, x_i, m_0) \right| dm_0(x_i)
$$

$$
\leqslant \mathbb{E}\big[|v^{N,i}(t, \mathbf{Z}) - u^{N,i}(t, \mathbf{Z})|\big] + \int_{\mathbb{T}^{dN}} |u^{N,i}(t, \mathbf{x}) - U(t, x_i, m_0)| \prod_{j=1}^N m_0(dx_j)
$$

$$
\leqslant CN^{-1} + C\varepsilon_N \leqslant C\varepsilon_N.
$$

This shows part (ii) of the theorem.   □

6.3   PROPAGATION OF CHAOS

We now prove Theorem 2.4.9. Let us recall the notation. Throughout this part, $(v^{N,i})_{i\in\{1,\dots,N\}}$ is the solution of the Nash system (6.1) and the processes $((Y_{i,t})_{t\in[t_0,T]})_{i\in\{1,\dots,N\}}$ are "optimal trajectories" for this system, i.e., solve (6.9) with $Y_{i,t_0} = Z_i$ as the initial condition at time t_0. Our aim is to understand the behavior of the $((Y_{i,t})_{t\in[t_0,T]})_{i\in\{1,\dots,N\}}$ for a large number of players N.

For any $i \in \{1, \dots, N\}$, let $(\tilde{X}_{i,t})_{t\in[t_0,T]}$ be the solution of the SDE of McKean–Vlasov type:

$$
d\tilde{X}_{i,t} = -D_pH\left(\tilde{X}_{i,t}, D_xU(t, \tilde{X}_{i,t}, \mathcal{L}(\tilde{X}_{i,t}|W))\right) dt + \sqrt{2}dB_t^i + \sqrt{2\beta}dW_t,
$$

$$
\tilde{X}_{i,t_0} = Z_i.
$$

Recall that, for any $i \in \{1, \dots, N\}$, the conditional law $\mathcal{L}(\tilde{X}_{i,t}|W)$ is equal to m_t, where $(u_t, m_t)_{t\in[0,T]}$ is the solution of the mean field game (MFG) system with common noise given by (5.1)–(5.2) (see Section 5.4.3). Solvability of the McKean–Vlasov equation may be discussed on the model of (5.21).

Our aim is to show that

$$
\mathbb{E}\left[\sup_{t\in[t_0,T]} |Y_{i,t} - \tilde{X}_{i,t}| \right] \leqslant C\varepsilon_N,
$$

for some $C > 0$, where, as before,

$$\varepsilon_N = \begin{cases} N^{-1/d} & \text{if } d \geqslant 3, \\ N^{-1/2}\ln(N) & \text{if } d = 2, \\ N^{-1/2} & \text{if } d = 1. \end{cases}$$

Proof of Theorem 2.4.9. The proof is a direct application of Theorem 6.2.1 combined with the following estimate on the distance between $(\tilde{X}_{i,t})_{t\in[t_0,T]}$ and the solution $(X_{i,t})_{t\in[t_0,T]}$ of (6.8):

$$\mathbb{E}\left[\sup_{t\in[t_0,T]}\left|X_{i,t} - \tilde{X}_{i,t}\right|\right] \leqslant C\varepsilon_N. \tag{6.22}$$

Indeed, by the triangle inequality, we have, provided that (6.22) holds true:

$$\mathbb{E}\left[\sup_{t\in[t_0,T]}\left|Y_{i,t} - \tilde{X}_{i,t}\right|\right] \leqslant \mathbb{E}\left[\sup_{t\in[t_0,T]}\left|Y_{i,t} - X_{i,t}\right|\right] + \mathbb{E}\left[\sup_{t\in[t_0,T]}\left|X_{i,t} - \tilde{X}_{i,t}\right|\right]$$

$$\leqslant C(N^{-1} + \varepsilon_N),$$

where we used (6.10) to pass from the first to the second line.

It now remains to check (6.22). For this, we fix $i \in \{1, \ldots, N\}$ and let

$$\rho(t) = \mathbb{E}\left[\sup_{s\in[t_0,t]}\left|X_{i,s} - \tilde{X}_{i,s}\right|\right].$$

Then, for any $s \in [t_0, t]$, we have

$$\begin{aligned} \left|X_{i,s} - \tilde{X}_{i,s}\right| &\leqslant \int_{t_0}^s \left|-D_pH(X_{i,r}, D_{x_i}u^{N,i}(r, \boldsymbol{X}_r))\right. \\ &\qquad\qquad \left. + D_pH(\tilde{X}_{i,r}, D_xU(r, \tilde{X}_{i,r}, m_r))\right|dr \\ &\leqslant \int_{t_0}^s \left|-D_pH(X_{i,r}, D_xU(r, X_{i,r}, m_{\boldsymbol{X}_r}^{N,i}))\right. \\ &\qquad\qquad \left. + D_pH(\tilde{X}_{i,r}, D_xU(r, \tilde{X}_{i,r}, m_{\tilde{\boldsymbol{X}}_r}^{N,i}))\right|dr \\ &\qquad + \int_{t_0}^s \left|-D_pH(\tilde{X}_{i,r}, D_xU(r, \tilde{X}_{i,r}, m_{\tilde{\boldsymbol{X}}_r}^{N,i}))\right. \\ &\qquad\qquad \left. + D_pH(\tilde{X}_{i,r}, D_xU(r, \tilde{X}_{i,r}, m_r))\right|dr. \end{aligned}$$

As $(x, m) \mapsto D_xU(t, x, m)$ and $(x, z) \mapsto D_pH(x, z)$ are uniformly Lipschitz continuous, we get

$$\left|X_{i,s} - \tilde{X}_{i,s}\right| \leqslant C\int_{t_0}^s \left(\left|X_{i,r} - \tilde{X}_{i,r}\right| + \mathbf{d}_1(m_{\boldsymbol{X}_r}^{N,i}, m_{\tilde{\boldsymbol{X}}_r}^{N,i}) + \mathbf{d}_1(m_{\tilde{\boldsymbol{X}}_r}^{N,i}, m_r)\right)dr,$$

where

$$\mathbf{d}_1\big(m_{X_t}^{N,i}, m_{\tilde{X}_t}^{N,i}\big) \leqslant \frac{1}{N-1} \sum_{j \neq i} |X_{j,s} - \tilde{X}_{j,s}|. \tag{6.23}$$

Hence,

$$\begin{aligned} |X_{i,s} - \tilde{X}_{i,s}| &\leqslant C \int_{t_0}^{s} \Big(|X_{i,r} - \tilde{X}_{i,r}| \\ &\quad + \frac{1}{N-1} \sum_{j \neq i} |X_{j,r} - \tilde{X}_{j,r}| + \mathbf{d}_1(m_{\tilde{X}_r}^{N,i}, m_r) \Big) dr. \end{aligned}$$

Taking the supremum over $s \in [t_0, t]$ and then the expectation, we have, recalling that the random variables $(X_{j,r} - \tilde{X}_{j,r})_{j \in \{1,\dots,N\}}$ have the same law:

$$\begin{aligned} \rho(t) &= \mathbb{E}\Big[\sup_{s \in [t_0,t]} |X_{i,s} - \tilde{X}_{i,s}| \Big] \\ &\leqslant C \int_{t_0}^{t} \Big(\mathbb{E}\big[|X_{i,s} - \tilde{X}_{i,s}| \big] + \frac{1}{N-1} \sum_{j \neq i} \mathbb{E}\big[|X_{j,s} - \tilde{X}_{j,s}| \big] \Big) ds \\ &\quad + C \int_{t_0}^{t} \mathbb{E}\Big[\mathbf{d}_1\big(m_{\tilde{X}_s}^{N,i}, m_s\big) \Big] ds \\ &\leqslant C \int_{t_0}^{t} \rho(s) ds + C \varepsilon_N, \end{aligned} \tag{6.24}$$

where we used, as in the proof of Theorem 2.4.8, the result by Fournier and Guillin [42]. Namely, since the variables $(\tilde{X}_{i,s})_{i=1,\dots,N}$ are, conditional on W, i.i.d. random variables with law m_t, we have

$$\mathbb{E}\Big[\mathbf{d}_1\big(m_{\tilde{X}_s}^{N,i}, m_s\big) | (W_r)_{r \in [t_0,T]} \Big] \leqslant c \mathbb{E}\Big[|\tilde{X}_{1,t}|^4 \,|\, (W_r)_{r \in [t_0,T]} \Big]^{1/4} \varepsilon_N,$$

for a constant c only depending on the dimension d. Taking the expectation, we easily derive the last line in (6.24). By Gronwall's inequality, the proof is easily completed. $\qquad \square$

Appendix

We now provide several basic results on the notion of differentiability on the space of probability measures used in the book, including a short comparison with the derivative on the set of random variables.

A.1 LINK WITH THE DERIVATIVE ON THE SPACE OF RANDOM VARIABLES

As a first step, we discuss the connection between the derivative $\delta U/\delta m$ in Definition 2.2.1 and the derivative introduced by Lions in [76] and used (among others) in [24, 31].

The notion introduced in [76] consists in lifting up functionals defined on the space of probability measures into functionals defined on the set of random variables. When the underlying probability measures are defined on a (finite dimensional) vector space E (so that the random variables that are distributed along these probability measures also take values in E), this permits us to take advantage of the standard differential calculus on the Hilbert space formed by the square-integrable random variables with values in E.

Here the setting is slightly different, as the probability measures that are considered throughout the book are defined on the torus. Some care is thus needed in the definition of the linear structure underpinning the argument.

Throughout the Appendix, we call $d_{\mathbb{T}^d}$ the distance on the d-dimensional torus, which is defined as follows:

$$d_{\mathbb{T}^d}(x, y) = \inf_{c \in \mathbb{Z}^d} |\hat{x} - \hat{y} + c|, \quad x, y \in \mathbb{T}^d,$$

where $\hat{x}$ and $\hat{y}$ are the only representatives of x and y in $[0, 1)^d$. For clarity, we call ϕ the mapping $\mathbb{T}^d \ni x \mapsto \phi(x) := \hat{x}$. We observe that, for any coordinate $i \in \{1, \cdots, d\}$, the function ϕ_i is lower-semicontinuous and thus measurable. Above, $|\cdot|$ is the Euclidean norm on $\mathbb{R}^d$.

A.1.1 First-Order Expansion with Respect to Torus-Valued Random Variables

On the torus $\mathbb{T}^d$, we may consider the group of translations $(\tau_y)_{y \in \mathbb{R}^d}$, parameterized by elements y of $\mathbb{R}^d$. For any $y \in \mathbb{R}^d$, τ_y maps $\mathbb{T}^d$ into itself. The mapping

$\mathbb{R}^d \ni y \mapsto \tau_y(0)$ being obviously continuous allows us to define, for any square integrable random variable $\tilde{X} \in L^2(\Omega, \mathcal{A}, \mathbb{P}; \mathbb{R}^d)$ (where $(\Omega, \mathcal{A}, \mathbb{P})$ is an atomless probability space), the random variable $\tau_{\tilde{X}}(0)$, which takes values in $\mathbb{T}^d$. Given a mapping $U : \mathcal{P}(\mathbb{T}^d) \to \mathbb{R}$, we may define its *lifted* version as

$$\tilde{U} : L^2(\Omega, \mathcal{A}, \mathbb{P}; \mathbb{R}^d) \ni \tilde{X} \mapsto \tilde{U}(\tilde{X}) = U\big(\mathcal{L}(\tau_{\tilde{X}}(0))\big), \tag{A.1}$$

where the argument in the right-hand side denotes the law of $\tau_{\tilde{X}}(0)$ (seen as a $\mathbb{T}^d$-valued random variable). Quite obviously, $\mathcal{L}(\tau_{\tilde{X}}(0))$ only depends on the law of $\tilde{X}$.

Assume now that the mapping $\tilde{U}$ is continuously Fréchet differentiable on $L^2(\Omega, \mathcal{A}, \mathbb{P}; \mathbb{R}^d)$. What [76] says is that, for any $\tilde{X} \in L^2(\Omega, \mathcal{A}, \mathbb{P}; \mathbb{R}^d)$, the Fréchet derivative has the form

$$D\tilde{U}(\tilde{X}) = \widetilde{\partial_\mu U}\big(\mathcal{L}(\tilde{X})\big)(\tilde{X}), \quad \mathbb{P} \text{ almost surely}, \tag{A.2}$$

for a mapping $\{\widetilde{\partial_\mu U}(\mathcal{L}(\tilde{X})) : \mathbb{R}^d \ni y \mapsto \widetilde{\partial_\mu U}(\mathcal{L}(\tilde{X}))(y) \in \mathbb{R}^d\} \in L^2(\mathbb{R}^d, \mathcal{L}(\tilde{X}); \mathbb{R}^d)$. This relationship is fundamental. Another key observation is that, for any random variables $\tilde{X}$ and $\tilde{Y}$ with values in $\mathbb{R}^d$ and $\tilde{\xi}$ with values in $\mathbb{Z}^d$, it holds that

$$\lim_{\varepsilon \to 0} \frac{1}{\varepsilon} \Big[\tilde{U}\big(\tilde{X} + \tilde{\xi} + \varepsilon \tilde{Y}\big) - \tilde{U}(\tilde{X}) \Big] = \mathbb{E}\Big[D\tilde{U}\big(\tilde{X} + \tilde{\xi}\big) \cdot \tilde{Y} \Big],$$

which, by the simple fact that $\tau_{\tilde{X}+\tilde{\xi}}(0) = \tau_{\tilde{X}}(0)$, is also equal to

$$\lim_{\varepsilon \to 0} \frac{1}{\varepsilon} \Big[\tilde{U}\big(\tilde{X} + \varepsilon \tilde{Y}\big) - \tilde{U}(\tilde{X}) \Big] = \mathbb{E}\Big[D\tilde{U}(\tilde{X}) \cdot \tilde{Y} \Big],$$

proving that

$$D\tilde{U}(\tilde{X}) = D\tilde{U}\big(\tilde{X} + \tilde{\xi}\big). \tag{A.3}$$

Consider now a random variable X from Ω with values into $\mathbb{T}^d$. With X, we may associate the random variable $\hat{X} = \phi(X)$, with values in $[0,1)^d$, given (pointwise) as the only representative of X in $[0,1)^d$. We observe that the law of $\hat{X}$ is uniquely determined by the law of X. Moreover, for any Borel function $h : \mathbb{T}^d \to \mathbb{R}$,

$$\mathbb{E}[h(X)] = \mathbb{E}[\hat{h}(\hat{X})],$$

where $\hat{h}$ is the identification of h as a function from $[0,1)^d$ to $\mathbb{R}$, namely $\hat{h}(x) = h(\tau_x(0))$, for $x \in [0,1)^d$.

Then, we deduce from (A.2) that

$$D\tilde{U}(\hat{X}) = \widetilde{\partial_\mu U}\big(\mathcal{L}(\hat{X})\big)(\hat{X}), \quad \mathbb{P} \text{ almost surely}.$$

Moreover, from (A.3), we also have, for any random variable $\tilde{\xi}$ with values in $\mathbb{Z}^d$,

$$D\tilde{U}(\hat{X} + \tilde{\xi}) = \widetilde{\partial_\mu U}\big(\mathcal{L}(\hat{X})\big)(\hat{X}), \quad \mathbb{P} \text{ almost surely.}$$

Since $\partial_\mu U(\mathcal{L}(\hat{X}))(\cdot)$ is in $L^2(\mathbb{R}^d, \mathcal{L}(\hat{X}); \mathbb{R}^d)$ and $\hat{X}$ takes values in $[0,1)^d$, we can identify $\partial_\mu U(\mathcal{L}(\hat{X}))(\cdot)$ with a function in $L^2(\mathbb{T}^d, \mathcal{L}(X); \mathbb{R}^d)$. Without any ambiguity, we may denote this function (up to a choice of a version) by

$$\mathbb{T}^d \ni y \mapsto \partial_\mu U\big(\mathcal{L}(X)\big)(y) := \partial_\mu U\big(\mathcal{L}(\phi(X))\big)\big(\phi(y)\big).$$

As an application we have that, for any random variables X and Y with values in $\mathbb{T}^d$,

$$\begin{aligned}
U\big(\mathcal{L}(Y)\big) - U\big(\mathcal{L}(X)\big) &= U\big(\mathcal{L}(\tau_{\hat{Y}}(0))\big) - U\big(\mathcal{L}(\tau_{\hat{X}}(0))\big) \\
&= \tilde{U}(\hat{Y}) - \tilde{U}(\hat{X}) \\
&= \mathbb{E} \int_0^1 D\tilde{U}\big(\mathcal{L}(\lambda\hat{Y} + (1-\lambda)\hat{X})\big) \cdot (\hat{Y} - \hat{X}) \, d\lambda.
\end{aligned}$$

Now, we can write

$$\lambda\hat{Y} + (1-\lambda)\hat{X} = \hat{X} + \lambda(\hat{Y} - \hat{X}) = \hat{Z}, \quad \text{with } Z = \tau_{\lambda(\hat{Y} - \hat{X})}(X).$$

Noticing that Z is a random variable with values in $\mathbb{T}^d$, we deduce that

$$\begin{aligned}
&U\big(\mathcal{L}(Y)\big) - U\big(\mathcal{L}(X)\big) \\
&= \mathbb{E} \int_0^1 \partial_\mu U\big(\mathcal{L}(\tau_{\lambda(\hat{Y}-\hat{X})}(X))\big)\big(\tau_{\lambda(\hat{Y}-\hat{X})}(X)\big) \cdot (\hat{Y} - \hat{X}) \, d\lambda.
\end{aligned}$$

Similarly, for any random variable $\tilde{\xi}$ with values in $\mathbb{Z}^d$,

$$\begin{aligned}
U\big(\mathcal{L}(Y)\big) - U\big(\mathcal{L}(X)\big) &= \tilde{U}(\hat{Y} + \tilde{\xi}) - \tilde{U}(\hat{X}) \\
&= \mathbb{E} \int_0^1 D\tilde{U}\big(\hat{X} + \lambda(\hat{Y} + \tilde{\xi} - \hat{X})\big) \cdot (\hat{Y} + \tilde{\xi} - \hat{X}) \, d\lambda.
\end{aligned}$$

Now, $\hat{X} + \lambda(\hat{Y} + \tilde{\xi} - \hat{X})$ writes $\hat{Z} + \tilde{\zeta}$, where $\tilde{\zeta}$ is a random variable with values in $\mathbb{Z}^d$ and $\hat{Z}$ is associated with the $\mathbb{T}^d$-valued random variable $Z = \tau_{\lambda(\hat{Y}+\tilde{\xi}-\hat{X})}(X)$, so that

$$\begin{aligned}
&U\big(\mathcal{L}(Y)\big) - U\big(\mathcal{L}(X)\big) \\
&= \mathbb{E} \int_0^1 D\tilde{U}(\hat{Z}) \cdot (\hat{Y} + \tilde{\xi} - \hat{X}) \, d\lambda \\
&= \mathbb{E} \int_0^1 \partial_\mu U\big(\mathcal{L}(\tau_{\lambda(\hat{Y}+\tilde{\xi}-\hat{X})}(X))\big)\big(\tau_{\lambda(\hat{Y}+\tilde{\xi}-\hat{X})}(X)\big) \cdot (\hat{Y} + \tilde{\xi} - \hat{X}) \, d\lambda.
\end{aligned} \tag{A.4}$$

The fact that $\tilde{\xi}$ can be chosen in a completely arbitrary way says that the choice of the representatives of X and Y in (A.4) does not matter. Of course, this is a consequence of the periodic structure underpinning the whole analysis. Precisely, for any representatives $\bar{X}$ and $\bar{Y}$ (with values in $\mathbb{R}^d$) of X and Y, we can write

$$
\begin{aligned}
U\big(\mathcal{L}(Y)\big) &- U\big(\mathcal{L}(X)\big) \\
&= \mathbb{E}\int_0^1 \partial_\mu U\big(\mathcal{L}(\tau_{\lambda(\bar{Y}-\bar{X})}(X))\big)\big(\tau_{\lambda(\bar{Y}-\bar{X})}(X)\big)\cdot(\bar{Y}-\bar{X})\,d\lambda.
\end{aligned}
\tag{A.5}
$$

Formula (A.5) gives a rule for expanding, along torus-valued random variables, functionals depending on torus-supported probability measures. It is the analogue of the *differentiation rule* defined in [76] on the space of probability measures on $\mathbb{R}^d$ through the differential calculus in $L^2(\Omega, \mathcal{A}, \mathbb{P}; \mathbb{R}^d)$.

In particular, if $\tilde{U}$ is continuously differentiable, with (say) $D\tilde{U}$ being Lipschitz continuous on $L^2(\Omega, \mathcal{A}, \mathbb{P}; \mathbb{R}^d)$, then (with the same notations as in (A.4))

$$
\begin{aligned}
\mathbb{E}\big[|D\tilde{U}(\hat{Y}) - D\tilde{U}(\hat{X})|^2\big] &= \mathbb{E}\big[|D\tilde{U}(\hat{Y}+\tilde{\xi}) - D\tilde{U}(\hat{X})|^2\big] \\
&\leqslant C\mathbb{E}\big[|\hat{Y}+\tilde{\xi}-\hat{X}|^2\big].
\end{aligned}
\tag{A.6}
$$

Now, for two random variables X and Y with values in the torus, one may find a random variable $\tilde{\xi}$, with values in $\mathbb{Z}^d$, such that, pointwise,

$$
\tilde{\xi} = \operatorname{argmin}_{c\in\mathbb{Z}^d}|\tau_c(\hat{Y}) - \hat{X}|,
$$

the right-hand side being the distance $d_{\mathbb{T}^d}(X,Y)$ between X and Y on the torus. Stated differently, we may choose $\tilde{\xi}$ such that $|\hat{Y}+\tilde{\xi}-\hat{X}| = d_{\mathbb{T}^d}(X,Y)$. Plugged into (A.6), this shows that the Lipschitz property of $D\tilde{U}$ (on $L^2(\Omega, \mathcal{A}, \mathbb{P}; \mathbb{R}^d)$) reads as a Lipschitz property with respect to torus-valued random variables. Namely,

$$
\mathbb{E}\big[|D\tilde{U}(\hat{Y}) - D\tilde{U}(\hat{X})|^2\big] \leqslant C\mathbb{E}\big[\big(d_{\mathbb{T}^d}(X,Y)\big)^2\big],
$$

which may be rewritten as

$$
\mathbb{E}\Big[\big|\partial_\mu U\big(\mathcal{L}(Y)\big)(Y) - \partial_\mu U\big(\mathcal{L}(X)\big)(X)\big|^2\Big] \leqslant C\mathbb{E}\Big[\big(d_{\mathbb{T}^d}(X,Y)\big)^2\Big].
$$

Taking the infimum over all the pairs of $\mathbb{T}^d$-valued random variables (X,Y) with prescribed marginal laws, we also have

$$
W_2\Big(\partial_\mu U(m)(\cdot)\sharp m, \partial_\mu U(m')(\cdot)\sharp m'\Big) \leqslant C^{1/2}\mathbf{d}_2(m,m'),
$$

for all $m, m' \in \mathcal{P}(\mathbb{T}^d)$. Here as well as throughout, for any $p \geqslant 1$, W_p is the standard p-Wasserstein distance on the space $\mathcal{P}_p(\mathbb{R}^d)$ of probability measures on $\mathbb{R}^d$ with a finite p-order moment and $\mathbf{d}_p$ is the p-Wasserstein distance on the space $\mathcal{P}(\mathbb{T}^d)$, whose definition is as follows:

$$\mathbf{d}_p(m, m') = \inf\left(\int_{[\mathbb{T}^d]^2} d^p_{\mathbb{T}^d}(x, y) d\pi(x, y) \right)^{1/p},$$

the infimum being taken over all the probability measures π on $[\mathbb{T}^d]^2$ with m and m' as marginal laws.

In this regard, we recall that the Kantorovich–Rubinstein duality formula (see, for instance, Chapter 6 in [95]), asserts that, when $p = 1$, the above definition of $\mathbf{d}_1$ is consistent with the definition we gave in Section 2.1, namely

$$\mathbf{d}_1(m, m') = \sup_{\phi} \int_{\mathbb{T}^d} \phi(y) d(m - m')(y),$$

the supremum being taken over all Lipschitz-continuous functions $\phi : \mathbb{T}^d \to \mathbb{R}$ with a Lipschitz constant less than 1.

A.1.2   From Differentiability Along Random Variables to Differentiability in m

We now address the connection between the mapping $\mathcal{P}(\mathbb{T}^d) \times \mathbb{T}^d \ni (m, y) \mapsto \partial_\mu U(m)(y) \in \mathbb{R}^d$ and the derivative $\mathcal{P}(\mathbb{T}^d) \times \mathbb{T}^d \ni (m, y) \mapsto [\delta U/\delta m](m, y) \in \mathbb{R}^d$ described in Definition 2.2.1.

Proposition A.1.1. *Assume that the function U is differentiable in the sense explained in Subsection A.1.1 and thus satisfies the expansion formula (A.5). Assume moreover that there exists a continuous version of the mapping $\partial_\mu U : \mathcal{P}(\mathbb{T}^d) \times \mathbb{T}^d \ni (m, y) \mapsto \partial_\mu U(m, y) \in \mathbb{R}^d$.*

Then, U is differentiable in the sense of Definition 2.2.1. Moreover, $\delta U/\delta m$ is continuously differentiable with respect to the second variable and

$$D_m U(m, y) = \partial_\mu U(m)(y), \quad m \in \mathcal{P}(\mathbb{T}^d), \ y \in \mathbb{T}^d.$$

Proof.   *First step.* The first step is to prove that, for any $m \in \mathcal{P}(\mathbb{T}^d)$, there exists a continuously differentiable map $V(m, \cdot) : \mathbb{T}^d \ni y \mapsto V(m, y) \in \mathbb{R}$ such that

$$\partial_\mu U(m)(y) = D_y V(m, y), \quad y \in \mathbb{T}^d.$$

The strategy is to prove that $\partial_\mu U(m) : \mathbb{T}^d \ni y \mapsto \partial_\mu U(m)(y)$ is orthogonal (in $L^2(\mathbb{T}^d, dy; \mathbb{R}^d)$) to divergence-free vector fields. It suffices to prove that, for any smooth divergence-free vector field $b : \mathbb{T}^d \to \mathbb{R}^d$,

$$\int_{\mathbb{T}^d} \partial_\mu U(m)(y) \cdot b(y) dy = 0.$$

Since $\partial_\mu U$ is jointly continuous in (m, y), it is enough to prove the above identity for any m with a positive smooth density. When m is not smooth, we may indeed approximate it by $m \star \rho$, where $\star$ denotes the convolution and ρ a smooth kernel on $\mathbb{R}^d$ with full support.

With such an m and such a b, we consider the ordinary differential equation (set on $\mathbb{R}^d$ but driven by periodic coefficients)

$$dX_t = \frac{b(X_t)}{m(X_t)} \, dt, \quad t \geqslant 0,$$

the initial condition X_0 being $[0, 1)^d$-valued and distributed according to some $m \in \mathcal{P}(\mathbb{T}^d)$ (identifying m with the probability measure $\phi \sharp m$ on $[0, 1)^d$). By periodicity of b and m, $(X_t)_{t \geqslant 0}$ generates on $\mathbb{T}^d$ a flow of probability measures $(m_t)_{t \geqslant 0}$ satisfying the Fokker–Planck equation

$$\partial_t m_t = -\mathrm{div}(\frac{b}{m} m_t), \quad t \geqslant 0, \quad m_0 = m.$$

Since b is divergence free, we get that $m_t = m$ for all $t \geqslant 0$. Indeed, m is an obvious solution of the equation and it is the unique one within the class of continuous weak solutions from $[0, \infty)$ into $\mathcal{P}(\mathbb{T}^d)$, which shows in particular that m is an invariant measure of $(X_t)_{t \geqslant 0}$. Then, for all $t \geqslant 0$,

$$U(m_t) - U(m_0) = 0,$$

so that, with the same notation as in (A.1), $\lim_{t \searrow 0}[(\tilde{U}(X_t) - \tilde{U}(X_0))/t] = 0$. Now, choosing $\bar{Y} = X_t$ and $\bar{X} = X_0$ in (A.5), dividing by t and letting $t \searrow 0$, we get

$$\int_{\mathbb{T}^d} \partial_\mu U(m)(y) \cdot b(y) dy = 0.$$

We easily deduce that $\partial_\mu U(m)$ reads as a gradient, that is,

$$\partial_\mu U(m)(y) = \partial_y V(m, y).$$

It is given as a solution of the Poisson equation

$$\Delta_y V(m, y) = \mathrm{div}_y \partial_\mu U(m)(y)$$

Of course, $V(m, \cdot)$ is uniquely defined up to an additive constant. We can choose it in such a way that

$$\int_{\mathbb{T}^d} V(m, y) dm(y) = 0.$$

Using the representation of the solution of the Poisson equation by means of the Poisson kernel, we easily deduce that the function V is jointly continuous. Indeed, under the above centering condition,

$$V(m, y) = \int_0^{+\infty} \mathbb{E}\big[\operatorname{div}_y \partial_\mu U(m)(y + B_t)\big]\, dt,$$

where $(B_t)_{t \geqslant 0}$ is a d-dimensional Brownian motion. Recalling from a standard spectral gap argument that, for some $\rho > 0$,

$$\sup_{y \in \mathbb{T}^d} \big|\mathbb{E}\big[\operatorname{div}_y \partial_\mu U(m)(y + B_t)\big]\big| \leqslant e^{-\rho t} \sup_{y \in \mathbb{T}^d} \big|\operatorname{div}_y \partial_\mu U(m)(y)\big|, \quad t \geqslant 0,$$

continuity easily follows.

Second step. The second step of the proof is to check that Definition 2.2.1 holds true. Let us consider two measures of the form m_X^N and m_Y^N, where $N \in \mathbb{N}^*$, $X = (x_1, \ldots, x_N) \in (\mathbb{T}^d)^N$ is such that $x_i \neq x_j$ and $Y = (y_1, \ldots, y_N) \in (\mathbb{T}^d)^N$. Without loss of generality we assume that the indices for Y are such that

$$\mathbf{d}_1(m_X^N, m_Y^N) = \frac{1}{N} \sum_{i=1}^N d_{\mathbb{T}^d}(x_i, y_i) = \frac{1}{N} \sum_{i=1}^N |\bar{x}_i - \bar{y}_i|, \tag{A.7}$$

where $\bar{x}_1, \ldots, \bar{x}_N$ and $\bar{y}_1, \ldots, \bar{y}_N$ are well-chosen representatives, in $\mathbb{R}^d$, of the points $x_1, \ldots, x_N$ and $y_1, \ldots, y_N$ in $\mathbb{T}^d$ ($d_{\mathbb{T}^d}$ denoting the distance on the torus). Let $\bar{X}$ be a random variable such that $\mathbb{P}(\bar{X} = \bar{x}_i) = 1/N$ and $\bar{Y}$ be the random variable defined by $\bar{Y} = \bar{y}_i$ if $\bar{X} = \bar{x}_i$. Then, with the same notations as in (A.1), $\mathbb{P}_{\mathcal{L}(\tau_{\bar{X}}(0))} = m_X^N$ and $\mathbb{P}_{\mathcal{L}(\tau_{\bar{Y}}(0))} = m_Y^N$.

Thanks to (A.5), we get

$$U(m_Y^N) - U(m_X^N)$$
$$= \int_0^1 \mathbb{E}\Big[\partial_\mu U\big(\mathcal{L}(\tau_{\lambda \bar{Y} + (1-\lambda)\bar{X}}(0))\big)\big(\tau_{\lambda \bar{Y} + (1-\lambda)\bar{X}}(0)\big) \cdot (\bar{Y} - \bar{X})\Big] d\lambda.$$

So, if w is a modulus of continuity of the map $\partial_\mu U$ on the compact set $\mathcal{P}(\mathbb{T}^d) \times \mathbb{T}^d$, we obtain by (A.7):

$$\Big| U(m_Y^N) - U(m_X^N) - \int_0^1 \mathbb{E}\big[\partial_\mu U(m_X^N)\big(\tau_{\lambda \bar{Y} + (1-\lambda)\bar{X}}(0)\big) \cdot (\bar{Y} - \bar{X})\big] d\lambda \Big|$$
$$\leqslant \mathbb{E}\big[|\bar{Y} - \bar{X}|\big] w\big(\mathbf{d}_1(m_X^N, m_Y^N)\big) \tag{A.8}$$
$$= \mathbf{d}_1(m_X^N, m_Y^N) w\big(\mathbf{d}_1(m_X^N, m_Y^N)\big).$$

Moreover, since $D_y V(m, y) = \partial_\mu U(m)(y)$, we have

$$\int_0^1 \mathbb{E}\Big[\partial_\mu U(m_X^N)\big(\tau_{\lambda \bar{Y} + (1-\lambda)\bar{X}}(0)\big)\big(\tau_{\lambda \bar{Y} + (1-\lambda)\bar{X}}\big) \cdot (\bar{Y} - \bar{X})\Big] d\lambda$$

$$= \frac{1}{N} \sum_{i=1}^N \int_0^1 D_y V\big(m_X^N, \tau_{\lambda \bar{y}_i + (1-\lambda)\bar{x}_i}(0)\big) \cdot (\bar{y}_i - \bar{x}_i)\, d\lambda$$

$$= \frac{1}{N} \sum_{i=1}^N \int_0^1 D_y V\big(m_X^N, \lambda \bar{y}_i + (1-\lambda)\bar{x}_i\big) \cdot (\bar{y}_i - \bar{x}_i)\, d\lambda,$$

where we saw $D_y V(m_X^N, \cdot)$ as a periodic function defined on the whole $\mathbb{R}^d$. Then,

$$\int_0^1 \mathbb{E}\Big[\partial_\mu U(m_X^N)\big(\tau_{\lambda \bar{Y} + (1-\lambda)\bar{X}}(0)\big) \cdot (\bar{Y} - \bar{X})\Big] d\lambda$$

$$= \int_{\mathbb{T}^d} V(m_X^N, x) d(m_Y^N - m_X^N)(x).$$

By density of the measures of the form m_X^N and m_Y^N and by continuity of V, we deduce from (A.8) that, for any measure $m, m' \in \mathcal{P}(\mathbb{T}^d)$,

$$\left| U(m') - U(m) - \int_{\mathbb{T}^d} V(m, x) d(m' - m)(x) \right| \leqslant \mathbf{d}_1(m, m') w(\mathbf{d}_1(m, m')),$$

which shows that U is $\mathcal{C}^1$ in the sense of Definition 2.2.1 with $\frac{\delta U}{\delta m} = V$. $\square$

A.1.3 From Differentiability in m to Differentiability Along Random Variables

We now discuss the converse to Proposition A.1.1

Proposition A.1.2. *Assume that U satisfies the assumption of Definition 2.2.2. Then, U satisfies the differentiability property (A.5). Moreover, it holds that $D_m U(m, y) = \partial_\mu U(m)(y)$, for $m \in \mathcal{P}(\mathbb{T}^d)$ and $y \in \mathbb{T}^d$.*

Proof. We are given two random variables X and Y with values in the torus $\mathbb{T}^d$. By Definition 2.2.1,

$$U(\mathcal{L}(Y)) - U(\mathcal{L}(X))$$

$$= \int_0^1 \Big[\int_{\mathbb{T}^d} \frac{\delta U}{\delta m}\big(\lambda \mathcal{L}(Y) + (1-\lambda)\mathcal{L}(X), y\big) d\big(\mathcal{L}(Y) - \mathcal{L}(X)\big)(y)\Big] d\lambda$$

$$= \int_0^1 \mathbb{E}\Big[\frac{\delta U}{\delta m}\big(\lambda \mathcal{L}(Y) + (1-\lambda)\mathcal{L}(X), Y\big) - \frac{\delta U}{\delta m}\big(\lambda \mathcal{L}(Y) + (1-\lambda)\mathcal{L}(X), X\big)\Big] d\lambda$$

$$= \int_0^1 \int_0^1 \mathbb{E}\Big[D_y \frac{\delta U}{\delta m}\big(\lambda \mathcal{L}(Y) + (1-\lambda)\mathcal{L}(X)\big)\big(\lambda' \bar{Y} + (1-\lambda')\bar{X}\big)(\bar{Y} - \bar{X})\Big] d\lambda d\lambda',$$

where $\bar{X}$ and $\bar{Y}$ are $\mathbb{R}^d$-valued random variables that represent the $\mathbb{T}^d$-valued random variables X and Y, while $D_y[\delta U/\delta m](m, \cdot)$ is seen as a periodic function from $\mathbb{R}^d$ into $\mathbb{R}^{d \times d}$.

By uniform continuity of $D_m U = D_y[\delta U/\delta m]$ on the compact set $\mathcal{P}(\mathbb{T}^d) \times \mathbb{T}^d$, we deduce that

$$U\big(\mathcal{L}(Y)\big) - U\big(\mathcal{L}(X)\big) = \mathbb{E}\left[D_y \frac{\delta U}{\delta m}(\mathcal{L}(X))(\bar{X})(\bar{Y} - \bar{X})\right]$$
$$+ O\Big(\mathbb{E}[|\bar{X} - \bar{Y}|^2]^{1/2} w\big(\mathbb{E}[|\bar{X} - \bar{Y}|^2]^{1/2}\big)\Big), \tag{A.9}$$

for a function $w : \mathbb{R}_+ \to \mathbb{R}_+$ that tends to 0 in 0 (w being independent of X and Y) and where $|O(r)| \leqslant |r|$. Above, we used the fact that $\mathbf{d}_1(\mathcal{L}(X), \mathcal{L}(Y)) \leqslant \mathbb{E}[|\bar{X} - \bar{Y}|^2]^{1/2}$.

Let now $Z_\lambda = \tau_{\lambda(\bar{Y} - \bar{X})}(X)$, for $\lambda \in [0, 1]$, so that $Z_{\lambda + \varepsilon} = \tau_{\varepsilon(\bar{Y} - \bar{X})}(Z_\lambda)$, for $0 \leqslant \lambda \leqslant \lambda + \varepsilon \leqslant 1$. Then, $(\lambda + \varepsilon)\bar{Y} + [1 - (\lambda + \varepsilon)]\bar{X}$ and $\lambda\bar{Y} + (1 - \lambda)\bar{X}$ are representatives of $Z_{\lambda + \varepsilon}$ and Z_λ and the distance between both reads

$$\big|(\lambda + \varepsilon)\bar{Y} + [1 - (\lambda + \varepsilon)]\bar{X} - \lambda\bar{Y} - (1 - \lambda)\bar{X}\big| = \varepsilon|\bar{Y} - \bar{X}|.$$

Therefore, by (A.9),

$$\frac{d}{d\lambda} U(Z_\lambda) = \mathbb{E}\left[D_y \frac{\delta U}{\delta m}(\mathcal{L}(Z_\lambda))(Z_\lambda)(\bar{Y} - \bar{X})\right], \quad \lambda \in [0, 1].$$

Integrating with respect to $\lambda \in [0, 1]$, we get (A.5) with the identity $\partial_\mu U(m)(y) = D_m U(m, y)$. $\qquad \square$

A.2   TECHNICAL REMARKS ON DERIVATIVES

Here we collect several results related to the notion of derivative described in Definition 2.2.1.

The first one is a quantified version of Proposition 2.2.3.

Proposition A.2.1.   *Assume that $U : \mathbb{T}^d \times \mathcal{P}(\mathbb{T}^d) \to \mathbb{R}$ is $\mathcal{C}^1$, that, for some $n \in \mathbb{N}$, $U(\cdot, m)$ and $\frac{\delta U}{\delta m}(\cdot, m, \cdot)$ are in $\mathcal{C}^{n+\alpha}$ and in $\mathcal{C}^{n+\alpha} \times \mathcal{C}^2$ respectively, and that there exists a constant C_n such that, for any $m, m' \in \mathcal{P}(\mathbb{T}^d)$,*

$$\left\|\frac{\delta U}{\delta m}(\cdot, m, \cdot)\right\|_{(n+\alpha, 2)} \leqslant C_n, \tag{A.10}$$

and

$$\left\| U(\cdot, m') - U(\cdot, m) - \int_{\mathbb{T}^d} \frac{\delta U}{\delta m}(\cdot, m, y) d(m' - m)(y) \right\|_{n+\alpha} \tag{A.11}$$
$$\leqslant C_n \mathbf{d}_1^2(m, m').$$

Fix $m \in \mathcal{P}(\mathbb{T}^d)$ *and let* $\phi \in L^2(m, \mathbb{R}^d)$ *be a vector field. Then,*

$$\left\| U(\cdot, (id + \phi)\sharp m) - U(\cdot, m) - \int_{\mathbb{T}^d} D_m U(\cdot, m, y) \cdot \phi(y) \, dm(y) \right\|_{n+\alpha} \tag{A.12}$$
$$\leqslant C_n' \|\phi\|_{L^2(m)}^2,$$

where C_n' *depends only on* C_n.

Below, we give conditions that ensure that (A.11) holds true.

Proof.   Using (A.11) we obtain

$$\left\| U(\cdot, (id + \phi)\sharp m) - U(\cdot, m) \right.$$
$$\left. - \int_{\mathbb{T}^d} \frac{\delta U}{\delta m}(\cdot, m, y) d((id + \phi)\sharp m - m)(y) \right\|_{n+\alpha} \tag{A.13}$$
$$\leqslant C_n \mathbf{d}_1^2\big(m, (id + \phi)\sharp m\big) \leqslant C_n \|\phi\|_{L^2(m)}^2.$$

Using the regularity of $\frac{\delta U}{\delta m}$, we obtain, for an $\{1, \cdots, d\}$-valued tuple ℓ of length $|\ell| \leqslant n$ and for any $x \in \mathbb{T}^d$, (omitting the dependence with respect to m for simplicity):

$$\left| \int_{\mathbb{T}^d} D_x^\ell \frac{\delta U}{\delta m}(x, y) d\{(id + \phi)\sharp m\}(y) - \int_{\mathbb{T}^d} D_x^\ell \frac{\delta U}{\delta m}(x, y) dm(y) \right.$$
$$\left. - \int_{\mathbb{T}^d} D_x^\ell D_m U(x, y) \cdot \phi(y) \, dm(y) \right|$$
$$= \left| \int_{\mathbb{T}^d} \left(D_x^\ell \frac{\delta U}{\delta m}(x, y + \phi(y)) - D_x^\ell \frac{\delta U}{\delta m}(x, y) \right. \right.$$
$$\left. \left. - D_x^\ell D_m U(x, y) \cdot \phi(y) \right) dm(y) \right|$$
$$= \left| \int_0^1 \int_{\mathbb{T}^d} \left(D_x^\ell D_y \frac{\delta U}{\delta m}(x, y + s\phi(y)) - D_x^\ell D_m U(x, y) \right) \cdot \phi(y) \, dm(y) ds \right|$$
$$= \left| \int_0^1 \int_0^1 \int_{\mathbb{T}^d} s \big(D_x^\ell D_y D_m U(x, y + st\phi(y))\phi(y) \big) \cdot \phi(y) \, dm(y) \, dsdt \right|$$
$$\leqslant C_n \|\phi\|_{L^2(m)}^2,$$

where we used (A.10) in the last line.

Coming back to (A.13), this shows that

$$\left\|D^\ell U\big(\cdot, (id + \phi)\sharp m\big) - D^\ell U(\cdot, m) - \int_{\mathbb{T}^d} D_x^\ell D_m U(\cdot, y) \cdot \phi(y)\; dm(y)\right\|_\infty$$
$$\leqslant C_n\|\phi\|_{L^2(m)}^2,$$

which proves (A.12) but with $\alpha = 0$.

The proof of the Hölder estimate goes along the same line: if $x, x' \in \mathbb{T}^d$, then

$$\left| \int_{\mathbb{T}^d} D_x^\ell \frac{\delta U}{\delta m}(x, y)d\{(id + \phi)\sharp m\}(y) - \int_{\mathbb{T}^d} D_x^\ell \frac{\delta U}{\delta m}(x, y)dm(y) \right.$$
$$- \int_{\mathbb{T}^d} D_x^\ell D_m U(x, y) \cdot \phi(y)\; dm(y)$$
$$- \left(\int_{\mathbb{T}^d} D_x^\ell \frac{\delta U}{\delta m}(x', y)d\{(id + \phi)\sharp m\}(y) - \int_{\mathbb{T}^d} D_x^\ell \frac{\delta U}{\delta m}(x', y)dm(y) \right.$$
$$\left. \left. - \int_{\mathbb{T}^d} D_x^\ell D_m U(x', y) \cdot \phi(y)\; dm(y) \right) \right|$$
$$= \left| \int_{\mathbb{T}^d} \left(D_x^\ell \frac{\delta U}{\delta m}(x, y + \phi(y)) - D_x^\ell \frac{\delta U}{\delta m}(x, y) - D_x^\ell D_m U(x, y) \right) \cdot \phi(y)\; dm(y) \right.$$
$$\left. - \int_{\mathbb{T}^d} \left(D_x^\ell \frac{\delta U}{\delta m}(x', y + \phi(y)) - D_x^\ell \frac{\delta U}{\delta m}(x', y) - D_x^\ell D_m U(x', y) \cdot \phi(y) \right)\; dm(y) \right|$$
$$= \left| \int_0^1 \int_{\mathbb{T}^d} \left(D_x^\ell D_y \frac{\delta U}{\delta m}(x, y + s\phi(y)) - D_x^\ell D_m U(x, y) \cdot \phi(y) \right)\; dm(y)ds \right.$$
$$\left. - \int_0^1 \int_{\mathbb{T}^d} \left(D_x^\ell D_y \frac{\delta U}{\delta m}(x', y + s\phi(y)) - D_x^\ell D_m U(x', y) \right) \cdot \phi(y)\; dm(y)ds \right|.$$

Hence,

$$\left| \int_{\mathbb{T}^d} D_x^\ell \frac{\delta U}{\delta m}(x, y)d\{(id + \phi)\sharp m\}(y) - \int_{\mathbb{T}^d} D_x^\ell \frac{\delta U}{\delta m}(x, y)dm(y) \right.$$
$$- \int_{\mathbb{T}^d} D_x^\ell D_m U(x, y) \cdot \phi(y)\; dm(y)$$
$$- \left(\int_{\mathbb{T}^d} D_x^\ell \frac{\delta U}{\delta m}(x', y)d\{(id + \phi)\sharp m\}(y) - \int_{\mathbb{T}^d} D_x^\ell \frac{\delta U}{\delta m}(x', y)dm(y) \right.$$
$$\left. \left. - \int_{\mathbb{T}^d} D_x^\ell D_m U(x', y) \cdot \phi(y)\; dm(y) \right) \right|$$

$$= \left| \int_0^1 \int_0^1 \int_{\mathbb{T}^d} s \Big[\Big(D_x^\ell D_y D_m U \big(x, y + st\phi(y) \big) \right.$$

$$\left. - D_x^\ell D_y D_m U \big(x', y + st\phi(y) \big) \Big) \phi(y) \Big] \cdot \phi(y) \, dm(y) \, ds \, dt \right|$$

$$\leqslant C_n |x - x'|^\alpha \|\phi\|_{L^2(m)}^2.$$

This shows that

$$\left\| \int_{\mathbb{T}^d} \frac{\delta U}{\delta m}(\cdot, m, y) d\big[\{ (id + \phi) \sharp m \} - m \big](y) - \int_{\mathbb{T}^d} D_m U(\cdot, m, y) \cdot \phi(y) \, dm(y) \right\|_{n+\alpha}$$

$$\leqslant C_n \|\phi\|_{L^2(m)}^2.$$

Plugging this inequality into (A.13) shows the result.   □

We now give conditions under which (A.11) holds.

Proposition A.2.2.   *Assume that $U : \mathbb{T}^d \times \mathcal{P}(\mathbb{T}^d) \to \mathbb{R}$ is C^1 and that, for some $n \in \mathbb{N}^*$,*

$$\mathrm{Lip}_n \left(\frac{\delta U}{\delta m} \right) \leqslant C_n.$$

Then, for any $m, m' \in \mathcal{P}(\mathbb{T}^d)$, we have

$$\left\| U(\cdot, m') - U(\cdot, m) - \int_{\mathbb{T}^d} \frac{\delta U}{\delta m}(\cdot, m, y) d(m' - m)(y) \right\|_{n+\alpha} \leqslant C_n \mathbf{d}_1^2(m, m').$$

We refer to Section 2.3 for the definition of Lip_n, see, for instance, **(HF1(n))**.

Proof.   We only show the Holder regularity: the L^∞ estimates go along the same line and are simpler. For any $\ell \in \mathbb{N}^d$ with $|\ell| \leqslant n$ and any $x, x' \in \mathbb{T}^d$, we have

$$\left| D_x^\ell U(x, m') - D_x^\ell U(x, m) - \int_{\mathbb{T}^d} D_x^\ell \frac{\delta U}{\delta m}(x, m, y) d(m' - m)(y) \right.$$

$$\left. - \left(D_x^\ell U(x', m') - D_x^\ell U(x', m) - \int_{\mathbb{T}^d} D_x^\ell \frac{\delta U}{\delta m}(x', m, y) d(m' - m)(y) \right) \right|$$

$$\leqslant \int_0^1 \left| \int_{\mathbb{T}^d} \left(D_x^\ell \frac{\delta U}{\delta m} \big(x, (1-s)m + sm', y \big) - D_x^\ell \frac{\delta U}{\delta m}(x, m, y) \right. \right.$$

$$\left. \left. - \left[D_x^\ell \frac{\delta U}{\delta m} \big(x', (1-s)m + sm', y \big) - D_x^\ell \frac{\delta U}{\delta m}(x', m, y) \right] \right) d(m' - m)(y) \right| ds$$

$$\leqslant \sup_{s,y} \left| D_y D_x^\ell \frac{\delta U}{\delta m}(x, (1-s)m + sm', y) - D_y D_x^\ell \frac{\delta U}{\delta m}(x, m, y) \right.$$
$$\left. - \left[D_y D_x^\ell \frac{\delta U}{\delta m}(x', (1-s)m + sm', y) - D_y D_x^\ell \frac{\delta U}{\delta m}(x', m, y) \right] \right| \mathbf{d}_1(m, m')$$
$$\leqslant \operatorname{Lip}_n \left(\frac{\delta U}{\delta m} \right) |x - x'|^\alpha \mathbf{d}_1^2(m, m').$$

This proves our claim. $\qquad\square$

Proposition A.2.3. *Let* $U : \mathcal{P}(\mathbb{T}^d) \to \mathbb{R}$ *be* $\mathcal{C}^2$ *and satisfy, for any* $m, m' \in \mathcal{P}(\mathbb{T}^d)$,

$$\left| U(m') - U(m) - \int_{\mathbb{T}^d} \frac{\delta U}{\delta m}(m, y) d(m' - m)(y) \right.$$
$$\left. - \frac{1}{2} \int_{\mathbb{T}^d} \int_{\mathbb{T}^d} \frac{\delta^2 U}{\delta m^2}(m, y, y') d(m' - m)(y) d(m' - m)(y') \right| \qquad \text{(A.14)}$$
$$\leqslant \mathbf{d}_1^2(m, m') w\big(\mathbf{d}_1(m, m')\big),$$

where $w(r) \to 0$ *as* $r \to 0$, *together with*

$$\left\| \frac{\delta U}{\delta m}(m, \cdot) \right\|_3 + \left\| \frac{\delta^2 U}{\delta m^2}(m, \cdot, \cdot) \right\|_{(2,2)} \leqslant C_0.$$

Then, for any $m \in \mathcal{P}(\mathbb{T}^d)$ *and any vector field* $\phi \in L^3(m; \mathbb{R}^d)$, *we have*

$$\left| U\big((id + \phi)\sharp m\big) - U(m) - \int_{\mathbb{T}^d} D_m U(m, y) \cdot \phi(y) \, dm(y) \right.$$
$$- \frac{1}{2} \int_{\mathbb{T}^d} D_y D_m U(m, y) \phi(y) \cdot \phi(y) \, dm(y)$$
$$\left. - \frac{1}{2} \int_{\mathbb{T}^d} \int_{\mathbb{T}^d} D_{mm}^2 U(m, y, y') \phi(y) \cdot \phi(y') \, dm(y) dm(y') \right|$$
$$\leqslant \|\phi\|_{L_m^3}^2 \tilde{w}(\|\phi\|_{L_m^3}),$$

where the modulus $\tilde{w}$ *depends on* w *and on* C_0.

Proof. We argue as in Proposition A.2.1. By assumption, we have

$$\left| U\big((id + \phi)\sharp m\big) - U(m) - \int_{\mathbb{T}^d} \frac{\delta U}{\delta m}(m, y) d\big(\{(id + \phi)\sharp m\} - m\big)(y) \right.$$
$$\left. - \frac{1}{2} \int_{\mathbb{T}^d} \int_{\mathbb{T}^d} \frac{\delta^2 U}{\delta m^2}(m, y, z) d\big(\{(id + \phi)\sharp m\} - m\big)(y) d\big(\{(id + \phi)\sharp m\} - m\big)(z) \right|$$
$$\leqslant \mathbf{d}_1^2\big(m, (id + \phi)\sharp m\big) w\big(\mathbf{d}_1(m, (id + \phi)\sharp m)\big) \leqslant \|\phi\|_{L_m^3}^2 w(\|\phi\|_{L_m^3}).$$

Now

$$\int_{\mathbb{T}^d} \frac{\delta U}{\delta m}(m,y) d\Big[\{(id+\phi)\sharp m\} - m\Big](y)$$

$$= \int_{\mathbb{T}^d} \left(\frac{\delta U}{\delta m}(m,y+\phi(y)) - \frac{\delta U}{\delta m}(m,y) \right) dm(y)$$

$$= \int_{\mathbb{T}^d} \left(D_y \frac{\delta U}{\delta m}(m,y) \cdot \phi(y) + \frac{1}{2}\big(D_y^2 \frac{\delta U}{\delta m}(m,y)\phi(y)\big) \cdot \phi(y) + O(|\phi(y)|^3) \right) dm(y)$$

$$= \int_{\mathbb{T}^d} \left(D_m U(m,y) \cdot \phi(y) + \frac{1}{2}\big(D_y D_m U(m,y)\phi(y)\big) \cdot \phi(y) + O(|\phi(y)|^3) \right) dm(y),$$

where

$$\int_{\mathbb{T}^d} |O(|\phi(y)|^3)| \; dm(y) \leqslant \|D_y^2 D_m U\|_\infty \int_{\mathbb{T}^d} |\phi(y)|^3 dm(y) \leqslant C_0 \|\phi\|_{L_m^3}^3.$$

Moreover,

$$\int_{\mathbb{T}^d}\int_{\mathbb{T}^d} \frac{\delta^2 U}{\delta m^2}(m,y,z) d\Big[\{(id+\phi)\sharp m\} - m\Big](y) d\Big[\{(id+\phi)\sharp m\} - m\Big](z)$$

$$= \int_{\mathbb{T}^d}\int_{\mathbb{T}^d} \left(\frac{\delta^2 U}{\delta m^2}(m,y+\phi(y),z+\phi(z)) - \frac{\delta^2 U}{\delta m^2}(m,y+\phi(y),z) \right.$$

$$\left. - \frac{\delta^2 U}{\delta m^2}(m,y,z+\phi(z)) + \frac{\delta^2 U}{\delta m^2}(m,y,z) \right) dm(y)dm(z)$$

$$= \int_{\mathbb{T}^d}\int_{\mathbb{T}^d} \left(D_{y,z}^2 \frac{\delta^2 U}{\delta m^2}(m,y,z)\phi(y) \cdot \phi(z) \right.$$

$$\left. + O\big(|\phi(y)|^2|\phi(z)| + |\phi(y)||\phi(z)|^2\big) \right) dm(y)dm(z)$$

$$= \int_{\mathbb{T}^d}\int_{\mathbb{T}^d} \left(D_{mm}^2 U(m,y,z)\phi(y) \cdot \phi(z) \right.$$

$$\left. + O\big(|\phi(y)|^2|\phi(z)| + |\phi(y)||\phi(z)|^2\big) \right) dm(y)dm(z),$$

where

$$\int_{\mathbb{T}^d} \Big|O\big(|\phi(y)|^2|\phi(z)| + |\phi(y)||\phi(z)|^2\big)\Big| dm(y)dm(z)$$

$$\leqslant \sup_m \|D_{mm}^2 U(m,\cdot,\cdot)\|_{(\mathcal{C}^1)^2} \|\phi\|_{L_m^3}^3 \leqslant C_0 \|\phi\|_{L_m^3}^3.$$

Putting the above estimates together gives the result. $\qquad\square$

We complete the section by giving conditions under which inequality (A.14) holds:

Proposition A.2.4. *Assume that the mapping* $\mathcal{P}(\mathbb{T}^d) \ni m \mapsto \frac{\delta^2 U}{\delta m^2}(m, \cdot, \cdot)$ *is continuous from* $\mathcal{P}(\mathbb{T}^d)$ *into* $(\mathcal{C}^2(\mathbb{T}^d))^2$ *with a modulus* w. *Then* (A.14) *holds true.*

Proof. We have

$$
U(m') - U(m)
$$
$$
= \int_0^1 \int_{\mathbb{T}^d} \frac{\delta U}{\delta m}((1-s)m + sm', y)d(m'-m)(y)
$$
$$
= \int_{\mathbb{T}^d} \frac{\delta U}{\delta m}(m, y)d(m'-m)(y)
$$
$$
+ \int_0^1 \int_0^1 \int_{\mathbb{T}^d} s\frac{\delta^2 U}{\delta m^2}((1-s\tau)m + s\tau m', y, y')d(m'-m)(y)d(m'-m)(y').
$$

Hence

$$
\left| U(m') - U(m) - \int_{\mathbb{T}^d} \frac{\delta U}{\delta m}(m, y)d(m'-m)(y) \right.
$$
$$
\left. -\frac{1}{2} \int_{\mathbb{T}^d} \int_{\mathbb{T}^d} \frac{\delta^2 U}{\delta m^2}(m, y, y')d(m'-m)(y)d(m'-m)(y') \right|
$$
$$
\leqslant \mathbf{d}_1(m, m')^2
$$
$$
\times \int_0^1 \int_0^1 s \left\| D_{yy'}^2 \frac{\delta^2 U}{\delta m^2}((1-s\tau)m + s\tau m', \cdot, \cdot) - D_{yy'}^2 \frac{\delta^2 U}{\delta m^2}(m, \cdot, \cdot) \right\|_\infty d\tau \, ds
$$
$$
\leqslant \mathbf{d}_1(m, m')^2 w(\mathbf{d}_1(m, m')).
$$

A.3   VARIOUS FORMS OF ITÔ'S FORMULA

A.3.1   Itô–Wentzell Formula

We here give a short reminder of the so-called Itô–Wentzell formula, which we alluded to several times in the text. We refer the reader to Chapter 3 in Kunita's monograph [68] for a complete account, see in particular Theorem 3.3.1 therein. We emphasize that our presentation is slightly different from [68], as it is tailor-made to the framework considered throughout the book.

The statement is as follows. Let $(\Omega, \mathcal{A}, \mathbb{P})$ be a complete probability space equipped with two d-dimensional Brownian motions $(B_t)_{t\geqslant 0}$ and $(W_t)_{t\geqslant 0}$ and endowed with the completion $(\mathcal{F}_t)_{t\geq 0}$ of the filtration generated by $(B_t)_{t\geq 0}$ and $(W_t)_{t\geq 0}$. Let $(\Psi_t(\cdot))_{t\geqslant 0}$ be a continuous process with values in the space $\mathcal{C}^2(\mathbb{R}^d)$ of twice continuously differentiable functions from $\mathbb{R}^d$ to $\mathbb{R}$, equipped with the collection of seminorms $\{\|\ell\|_{\mathcal{C}^2(K)} := \sup_{x \in K}(|\ell(x)| + |D\ell(x)| + |D^2\ell(x)|), \ell \in$

$\mathcal{C}^2(\mathbb{R}^d)\}$ indexed by the compact subsets K of $\mathbb{R}^d$. Assume that, for any $x \in \mathbb{R}^d$, $(\Psi_t(x))_{t\geq 0}$ may be expanded under the following Itô form:

$$\Psi_t(x) = \Psi_0(x) + \int_0^t f_s(x)\, ds + \sum_{k=1}^d \int_0^t g_s^k(x) dW_s^k, \quad t \geq 0,$$

where $(f_t(\cdot))_{t\geq 0}$ and $((g_t^k(\cdot))_{t\geq 0})_{k=1,\cdots,d}$ are progressively measurable processes with values in $\mathcal{C}^2(\mathbb{R}^d)$ such that, for any compact subsets $K \subset \mathbb{R}^d$, with probability 1 under $\mathbb{R}^d$, for all $T > 0$,

$$\int_0^T \|f_s(\cdot)\|_{\mathcal{C}^1(K)}\, ds < \infty, \quad \sum_{k=1}^d \int_0^T \|g_s^k(\cdot)\|_{\mathcal{C}^2(K)}^2\, ds < \infty.$$

Consider in addition an m-dimensional Itô process $(X_t)_{t\geq 0}$ with the following expansion:

$$dX_t = b_t\, dt + \sum_{k=1}^d \sigma_t^k dB_t^k + \sum_{k=1}^d \zeta_t^k dW_t^k, \quad t \geq 0,$$

where $(b_t)_{t\geq 0}$, $((\sigma_t^k)_{t\geq 0})_{k=1,\cdots,d}$ and $((\zeta_t^k)_{t\geq 0})_{k=1,\cdots,d}$ are progressively measurable processes with values in $\mathbb{R}^m$ satisfying with probability 1:

$$\forall T > 0, \quad \int_0^T |b_s|ds < \infty, \quad \sum_{k=1}^d \int_0^T \left(|\sigma_s^k|^2 + |\zeta_s^k|^2\right) ds < \infty.$$

Then, $(\Psi_t(X_t))_{t\geq 0}$ is an Itô process. With probability 1, it expands, for all $t \geq 0$, as:

$$d\Psi_t(X_t) = \left[D\Psi_t(X_t) \cdot b_t + \frac{1}{2}\sum_{k=1}^d \operatorname{trace}\left(D^2\Psi_t(X_t)\left(\sigma_t^k \otimes \sigma_t^k + \zeta_t^k \otimes \zeta_t^k\right)\right) \right.$$

$$\left. + f_t(X_t) + \sum_{k=1}^d Dg_t^k(X_t) \cdot \zeta_t^k \right] dt$$

$$+ \sum_{k=1}^d D\Psi_t(X_t) \cdot \left(\sigma_t^k dB_t^k + \zeta_t^k dW_t^k\right) + \sum_{k=1}^d g_t^k(X_t) dW_t^k.$$

Above, $\sigma_t^k \otimes \sigma_t^k$ denotes the $m \times m$ matrix with entries $((\sigma_t^k)_i(\sigma_t^k)_j)_{i,j=1,\cdots,m}$, and similarly for $((\zeta_t^k)_i(\zeta_t^k)_j)_{i,j=1,\cdots,m}$. The proof follows from the combination of Exercise 3.1.5 and Theorem 3.3.1 in [68].

A.3.2   Chain Rule for Function of a Measure Argument

Let U be a function satisfying the same assumption as in Definition 2.4.4 and, for a given $t_0 \in [0, T]$, $(\tilde{m}_t)_{t \in [t_0, T]}$ be an adapted process with paths in $\mathcal{C}^0([t_0, T], \mathcal{P}(\mathbb{T}^d))$ such that, with probability 1, for any smooth test function $\varphi \in \mathcal{C}^n(\mathbb{T}^d)$,

$$d_t \left[\int_{\mathbb{T}^d} \varphi(x) d\tilde{m}_t(x) \right] = \left\{ \int_{\mathbb{T}^d} [\Delta\varphi(x) - \beta_t(x + \sqrt{2}(W_t - W_{t_0})) \right.$$
$$\left. \cdot D\varphi(x)] d\tilde{m}_t(x) \right\} dt, \qquad (A.15)$$

for $t \in [t_0, T]$, for some adapted process $(\beta_t)_{t \in [t_0, T]}$, with paths in the space $\mathcal{C}^0([t_0, T], [\mathcal{C}^0(\mathbb{T}^d)]^d)$, such that

$$\text{essup}_{\omega \in \Omega} \sup_{t \in [t_0, T]} \|\beta_t\|_0 < \infty,$$

so that, by Lebesgue's dominated convergence theorem,

$$\lim_{h \to 0} \mathbb{E}\left[\sup_{s, t \in [t_0, T], |t - s| \leqslant h} \|\beta_s - \beta_t\|_0 \right] = 0.$$

In other words, $(\tilde{m}_t)_{t \in [t_0, T]}$ stands for the flow of conditional marginal laws of $(\tilde{X}_t)_{t \in [t_0, T]}$ given $\mathcal{F}_T$, where $(\tilde{X}_t)_{t \in [t_0, T]}$ solves the stochastic differential equation:

$$d\tilde{X}_t = -\beta_t(\tilde{X}_t + \sqrt{2}(W_t - W_{t_0})) \, dt + \sqrt{2} dB_t, \quad t \in [t_0, T],$$

$\tilde{X}_{t_0}$ being distributed according to m_{t_0} conditional on $\mathcal{F}_T$. In particular, there exists a deterministic constant C such that, with probability 1, for all $t_0 \leqslant t \leqslant t + h \leqslant T$,

$$\mathbf{d}_1(\tilde{m}_{t+h}, \tilde{m}_t) \leqslant C\sqrt{h}.$$

Given some $t \in [t_0, T]$, we denote by $m_t = (\cdot \mapsto \cdot + \sqrt{2}(W_t - W_{t_0}))\sharp\tilde{m}_t$ the push-forward of $\tilde{m}_t$ by the application $\mathbb{T}^d \ni x \mapsto x + W_t - W_{t_0} \in \mathbb{T}^d$ (so that $m_{t_0} = \tilde{m}_{t_0}$). Equivalently, $(m_t)_{t \in [t_0, T]}$ is the flow of conditional marginal laws of $(X_t)_{t \in [t_0, T]}$ given $\mathcal{F}_T$, where $(X_t)_{t \in [t_0, T]}$ solves the stochastic differential equation:

$$dX_t = -\beta_t(X_t) \, dt + \sqrt{2} dB_t + \sqrt{2} dW_t, \quad t \in [t_0, T],$$

X_{t_0} being distributed according to m_{t_0} conditional on $\mathcal{F}_T$.

We then have the local Itô–Taylor expansion:

Lemma A.3.1. *Under the above assumption, we can find a family of real-valued random variables* $(\varepsilon_{s,t})_{s,t\in[t_0,T]:s\leqslant t}$ *such that*

$$\lim_{h\searrow 0}\sup_{s,t\in[t_0,T]:|s-t|\leqslant h}\mathbb{E}\big[|\varepsilon_{s,t}|\big]=0,$$

and, for any $t\in[t_0,T]$,

$$
\begin{aligned}
\frac{1}{h}\Big[&\mathbb{E}\big[U\big(t+h,x+\sqrt{2}(W_{t+h}-W_{t_0}),m_{t+h}\big)\\
&-U\big(t+h,x+\sqrt{2}(W_{t+h}-W_{t_0}),m_t\big)|\mathcal{F}_t\big]\Big]\\
=&\,\Delta_x U\big(t,x+\sqrt{2}(W_t-W_{t_0}),m_t\big)\\
&+2\int_{\mathbb{T}^d}\mathrm{div}_y\big[D_m U\big]\big(t,x+\sqrt{2}(W_t-W_{t_0}),m_t,y\big)dm_t(y)\\
&-\int_{\mathbb{T}^d}D_m U\big(t,x+\sqrt{2}(W_t-W_{t_0}),m_t,y\big)\cdot\beta_t(y)dm_t(y)\\
&+2\int_{\mathbb{T}^d}\mathrm{div}_x\big[D_m U\big]\big(t,x+\sqrt{2}(W_t-W_{t_0}),m_t,y\big)dm_t(y)\\
&+\int_{[\mathbb{T}^d]^2}\mathrm{Tr}\big[D_{mm}^2 U\big]\big(t,x+\sqrt{2}(W_t-W_{t_0}),m_t,y,y'\big)dm_t(y)dm_t(y')\\
&+\varepsilon_{t,t+h}.
\end{aligned}
$$

Proof. Without any loss of generality, we assume that $t_0=0$. Moreover, throughout the analysis, we shall use the following variant of (5.25): For two random processes $(\gamma_t)_{t\in[0,T]}$ and $(\gamma'_t)_{t\in[0,T]}$, with paths in $\mathcal{C}^0([0,T],\mathcal{C}^0(E))$ and $\mathcal{C}^0([0,T],F)$ respectively, where E is a compact metric space (the distance being denoted by d_E) and F is a metric space (the distance being denoted by d_F), satisfying

$$\mathrm{essup}_{\omega\in\Omega}\sup_{t\in[0,T]}\|\gamma_t\|_0<\infty,$$

it must hold that

$$\lim_{h\searrow 0}\sup_{s,t\in[0,T]:|s-t|\leqslant h}\mathbb{E}\big[|\eta_{s,t}|\big]=0,$$

$$\text{with } \eta_{s,t}=\sup_{r\in[s,t]}\sup_{x,y\in E:d_E(x,y)\leqslant\sup_{u\in[s,t]}d_F(\gamma'_u,\gamma'_s)}\big|\gamma_r(y)-\gamma_s(x)\big|. \tag{A.16}$$

The proof of (A.16) holds in two steps. The first one is to show, by a compactness argument, that, $\mathbb{P}$-almost surely, $\sup_{s,t\in[0,T]:|s-t|\leqslant h}|\eta_{s,t}|$ tends to 0; the claim then follows from the dominated convergence theorem.

Now, for given $t\in[0,T)$ and $h\in(0,T-t]$, we let $\delta_h W_t:=W_{t+h}-W_t$ and $\delta_h m_t:=m_{t+h}-m_t$. By Taylor–Lagrange's formula, we can find some random variable λ with values in $[0,1]^1$ such that

$$
\begin{aligned}
U\big(t+h&, x+\sqrt{2}W_{t+h}, m_{t+h}\big) - U\big(t+h, x+\sqrt{2}W_t, m_t\big) \\
&= \sqrt{2} D_x U\big(t+h, x+\sqrt{2}W_t, m_t\big) \cdot \delta_h W_t \\
&\quad + \int_{\mathbb{T}^d} \frac{\delta U}{\delta m}\big(t+h, x+\sqrt{2}W_t, m_t, y\big) d\big(\delta_h m_t\big)(y) \\
&\quad + D_x^2 U\big(t+h, x+\sqrt{2}W_t + \sqrt{2}\lambda\delta_h W_t, m_t + \lambda\delta_h m_t\big) \cdot (\delta_h W_t)^{\otimes 2} \\
&\quad + \sqrt{2}\int_{\mathbb{T}^d} D_x \frac{\delta U}{\delta m}\big(t+h, x+\sqrt{2}W_t + \sqrt{2}\lambda\delta_h W_t, m_t \\
&\quad\quad + \lambda\delta_h m_t, y\big) \cdot \delta_h W_t d\big(\delta_h m_t\big)(y) \\
&\quad + \frac{1}{2}\int_{[\mathbb{T}^d]^2} \frac{\delta^2 U}{\delta m^2}\big(t+h, x+\sqrt{2}W_t + \sqrt{2}\lambda\delta_h W_t, m_t \\
&\quad\quad + \lambda\delta_h m_t, y, y'\big) d\big(\delta_h m_t\big)(y) d\big(\delta_h m_t\big)(y') \\
&=: T_h^1 + T_h^2 + T_h^3 + T_h^4 + T_h^5,
\end{aligned}
\tag{A.17}
$$

where we used the dot "$\cdot$" to denote the inner product in Euclidean spaces. Part of the analysis relies on the following decomposition. Given a bounded and Borel measurable function $\varphi : \mathbb{T}^d \to \mathbb{R}$, it holds that

$$
\begin{aligned}
\int_{\mathbb{T}^d} &\varphi(y) d\big(\delta_h m_t\big)(y) \\
&= \int_{\mathbb{T}^d} \varphi(y) dm_{t+h}(y) - \int_{\mathbb{T}^d} \varphi(y) dm_t(y) \\
&= \int_{\mathbb{T}^d} \varphi\big(y+\sqrt{2}W_{t+h}\big) d\tilde{m}_{t+h}(y) - \int_{\mathbb{T}^d} \varphi\big(y+\sqrt{2}W_t\big) d\tilde{m}_t(y) \\
&= \int_{\mathbb{T}^d} \varphi\big(y+\sqrt{2}W_{t+h}\big) d\big(\tilde{m}_{t+h} - \tilde{m}_t\big)(y) \\
&\quad + \int_{\mathbb{T}^d} \Big[\varphi\big(y+\sqrt{2}W_{t+h}\big) - \varphi\big(y+\sqrt{2}W_t\big)\Big] d\tilde{m}_t(y) \\
&= \int_{\mathbb{T}^d} \varphi\big(y+\sqrt{2}W_{t+h}\big) d\big(\tilde{m}_{t+h} - \tilde{m}_t\big)(y) \\
&\quad + \int_{\mathbb{T}^d} \Big[\varphi\big(y+\sqrt{2}\delta_h W_t\big) - \varphi(y)\Big] dm_t(y).
\end{aligned}
\tag{A.18}
$$

---

[1]The fact that λ is a random variable may be justified as follows. Given a continuous mapping φ from $\mathbb{T}^d \times \mathcal{P}(\mathbb{T}^d)$ into $\mathbb{R}$ and two random variables (X, m) and (X', m') with values in $(\mathbb{R}^d, \mathcal{P}(\mathbb{T}^d))$ such that the mapping $[0,1] \ni c \mapsto \varphi(cX' + (1-c)X, cm' + (1-c)m)$ vanishes at least once, the quantity $\lambda = \inf\{c \in [0,1] : \varphi(cX' + (1-c)X, cm' + (1-c)m) = 0\}$ defines a random variable since $\{\lambda > c\} = \cup_{n\in\mathbb{N}\setminus\{0\}} \cap_{c'\in\mathbb{Q}\in[0,c]} \{\varphi(c'X' + (1-c')X, c'm' + (1-c')m)\varphi(X, m) > 1/n\}$.

Likewise, whenever φ is a bounded Borel measurable mapping from $[\mathbb{T}^d]^2$ into $\mathbb{R}$, it holds that

$$
\int_{[\mathbb{T}^d]^2} \varphi(y, y') d(\delta_h m_t)(y) d(\delta_h m_t)(y')
$$

$$
= \int_{[\mathbb{T}^d]^2} \varphi(y + \sqrt{2} W_{t+h}, y') d(\tilde{m}_{t+h} - \tilde{m}_t)(y) d(\delta_h m_t)(y')
$$

$$
+ \int_{[\mathbb{T}^d]^2} \left[\varphi(y + \sqrt{2} \delta_h W_t, y') - \varphi(y, y') \right] dm_t(y) d(\delta_h m_t)(y')
$$

$$
= \int_{[\mathbb{T}^d]^2} \varphi(y + \sqrt{2} W_{t+h}, y' + \sqrt{2} W_{t+h})
$$

$$
d(\tilde{m}_{t+h} - \tilde{m}_t)(y) d(\tilde{m}_{t+h} - \tilde{m}_t)(y')
$$

$$
+ \int_{[\mathbb{T}^d]^2} \left[\varphi(y + \sqrt{2} W_{t+h}, y' + \sqrt{2} \delta_h W_t) \right.
$$

$$
\left. - \varphi(y + \sqrt{2} W_{t+h}, y') \right] d(\tilde{m}_{t+h} - \tilde{m}_t)(y) dm_t(y') \tag{A.19}
$$

$$
+ \int_{[\mathbb{T}^d]^2} \left[\varphi(y + \sqrt{2} \delta_h W_t, y' + \sqrt{2} W_{t+h}) \right.
$$

$$
\left. - \varphi(y, y' + \sqrt{2} W_{t+h}) \right] dm_t(y) d(\tilde{m}_{t+h} - \tilde{m}_t)(y')
$$

$$
+ \int_{[\mathbb{T}^d]^2} \left[\varphi(y + \sqrt{2} \delta_h W_t, y' + \sqrt{2} \delta_h W_t) - \varphi(y + \sqrt{2} \delta_h W_t, y') \right.
$$

$$
\left. - \varphi(y, y' + \sqrt{2} \delta_h W_t) + \varphi(y, y') \right] dm_t(y) dm_t(y').
$$

We now proceed with the analysis of (A.17). We start with T_h^1. It is pretty clear that

$$
\mathbb{E}\left[T_h^1 | \mathcal{F}_t \right] = 0. \tag{A.20}
$$

Look at now the term T_h^2. Following (A.18), write it

$$
T_h^2 = \int_{\mathbb{T}^d} \frac{\delta U}{\delta m}(t + h, x + \sqrt{2} W_t, m_t, y + \sqrt{2} W_{t+h}) d(\tilde{m}_{t+h} - \tilde{m}_t)(y)
$$

$$
+ \int_{\mathbb{T}^d} \left[\frac{\delta U}{\delta m}(t + h, x + \sqrt{2} W_t, m_t, y + \sqrt{2} \delta_h W_t) \right.
$$

$$
\left. - \frac{\delta U}{\delta m}(t + h, x + \sqrt{2} W_t, m_t, y) \right] dm_t(y) \tag{A.21}
$$

$$
=: T_h^{2,1} + T_h^{2,2}.
$$

By the PDE satisfied by $(\tilde{m}_t)_{t\in[t_0,T]}$, we have

$$
\begin{aligned}
T_h^{2,1} = \int_t^{t+h} ds \int_{\mathbb{T}^d} \Delta_y \frac{\delta U}{\delta m}\big(t+h, x+\sqrt{2}W_t, m_t, y+\sqrt{2}W_{t+h}\big) d\tilde{m}_s(y) \\
- \int_t^{t+h} ds \int_{\mathbb{T}^d} D_y \frac{\delta U}{\delta m}\big(t+h, x+\sqrt{2}W_t, m_t, y+\sqrt{2}W_{t+h}\big) \qquad (\text{A.22}) \\
\cdot \beta_s\big(y + \sqrt{2}W_s\big) d\tilde{m}_s(y).
\end{aligned}
$$

Therefore, taking the conditional expectation, dividing by h and using the fact that m_t is the push-forward of $\tilde{m}_t$ by the mapping $\mathbb{T}^d \ni x \mapsto x + \sqrt{2}W_t$ (take note that the measures below are m_t and not $\tilde{m}_t$), we can write

$$
\begin{aligned}
\frac{1}{h}\mathbb{E}\big[T_h^{2,1}|\mathcal{F}_t\big] = \int_{\mathbb{T}^d} \Delta_y \frac{\delta U}{\delta m}\big(t, x+\sqrt{2}W_t, m_t, y\big) dm_t(y) \\
- \int_{\mathbb{T}^d} D_m U\big(t, x+\sqrt{2}W_t, m_t, y\big) \cdot \beta_t(y) dm_t(y) + \varepsilon_{t,t+h},
\end{aligned}
$$

where, as in the statement, $(\varepsilon_{s,t})_{0\leqslant s\leqslant t\leqslant T}$ is a generic notation for denoting a family of random variables that satisfies

$$
\lim_{h\searrow 0} \sup_{|t-s|\leqslant h} \mathbb{E}\big[|\varepsilon_{s,t}|\big] = 0. \qquad (\text{A.23})
$$

Here we used the trick explained in (A.16) to prove (A.23). The application of (A.16) is performed in two steps. The first one is to choose $E = [\mathbb{T}^d]^2$, $F = \mathbb{R}^d$, $\gamma_t(x,y) = \Delta_y \frac{\delta U}{\delta m}(t, x, m_t, y)$ and $\gamma_t' = W_t$ and then $\gamma_t(x,y) = D_y \frac{\delta U}{\delta m}(t, x, m_t, y) \cdot \beta_t(y)$ and $\gamma_t' = W_t$. Then, by a first application of (A.16), we can write

$$
\begin{aligned}
T_h^{2,1} = \int_t^{t+h} ds \int_{\mathbb{T}^d} \Delta_y \frac{\delta U}{\delta m}\big(s, x+\sqrt{2}W_s, m_s, y+\sqrt{2}W_s\big) d\tilde{m}_s(y) \\
- \int_t^{t+h} ds \int_{\mathbb{T}^d} D_y \frac{\delta U}{\delta m}\big(s, x+\sqrt{2}W_s, m_s, y+\sqrt{2}W_s\big) \\
\cdot \beta_s\big(y + \sqrt{2}W_s\big) d\tilde{m}_s(y) + h\varepsilon_{t,t+h}.
\end{aligned}
$$

Then, we can apply (A.16) once again with

$$
\begin{aligned}
\gamma_s(x) = \int_{\mathbb{T}^d} \Delta_y \frac{\delta U}{\delta m}\big(s, x+\sqrt{2}W_s, m_s, y+\sqrt{2}W_s\big) d\tilde{m}_s(y) \\
- \int_{\mathbb{T}^d} D_y \frac{\delta U}{\delta m}\big(s, x+\sqrt{2}W_s, m_s, y+\sqrt{2}W_s\big) \cdot \beta_s\big(y + \sqrt{2}W_s\big) d\tilde{m}_s(y),
\end{aligned}
$$

and γ' constant. Using Itô's formula to handle the second term in (A.21), we get in a similar way

$$\frac{1}{h}\mathbb{E}\big[T_h^2|\mathcal{F}_t\big] = 2\int_{\mathbb{T}^d} \Delta_y \frac{\delta U}{\delta m}\big(t, x + \sqrt{2}W_t, m_t, y\big)dm_t(y)$$

$$- \int_{\mathbb{T}^d} D_m U\big(t, x + \sqrt{2}W_t, m_t, y\big) \cdot \beta_t(y)dm_t(y) + \varepsilon_{t,t+h}.$$

(A.24)

Turn now to T_h^3 in (A.17). Using again (A.16), it is quite clear that

$$\frac{1}{h}\mathbb{E}\big[T_h^3|\mathcal{F}_t\big] = \Delta_x U(t, x + \sqrt{2}W_t, m_t) + \varepsilon_{t,t+h}.$$ (A.25)

We now handle T_h^4. Following (A.18), we write

$$T_h^4 = \sqrt{2}\int_{\mathbb{T}^d}\Big[D_x \frac{\delta U}{\delta m}\big(t + h, x + \sqrt{2}W_t + \sqrt{2}\lambda\delta_h W_t, m_t$$

$$+ \lambda\delta_h m_t, y + \sqrt{2}W_{t+h}\big) \cdot \delta_h W_t\Big]d\big(\tilde{m}_{t+h} - \tilde{m}_t\big)(y)$$

$$+ \sqrt{2}\int_{\mathbb{T}^d}\Big[D_x \frac{\delta U}{\delta m}\big(t + h, x + \sqrt{2}W_t + \sqrt{2}\lambda\delta_h W_t, m_t + \lambda\delta_h m_t, y + \sqrt{2}\delta_h W_t\big)$$

$$- D_x \frac{\delta U}{\delta m}\big(t + h, x + \sqrt{2}W_t + \sqrt{2}\lambda\delta_h W_t, m_t + \lambda\delta_h m_t, y\big)\Big]$$

$$\cdot \delta_h W_t dm_t(y) =: T_h^{4,1} + T_h^{4,2}.$$

Making use of the forward Fokker–Planck equation for $(\tilde{m}_t)_{t\in[t_0,T]}$ as in the proof of (A.24), we get that

$$\frac{1}{h}\mathbb{E}\big[T_h^{4,1}|\mathcal{F}_t\big] = \varepsilon_{t,t+h}.$$

Now, by Taylor–Lagrange's formula, we can find another $[0, 1]$-valued random variable λ' such that

$$T_h^{4,2} = 2\int_{\mathbb{T}^d}\Big[D_y D_x \frac{\delta U}{\delta m}\big(t + h, x + \sqrt{2}W_t + \sqrt{2}\lambda\delta_h W_t, m_t$$

$$+ \lambda\delta_h m_t, y + \sqrt{2}\lambda'\delta_h W_t\big) \cdot (\delta_h W_t)^{\otimes 2}\Big]dm_t(y).$$

And, then, invoking once again the trick (A.16), but with $E = \mathbb{T}^d \times \mathcal{P}(\mathbb{T}^d) \times \mathbb{T}^d$, $F = \mathbb{R}^d \times \mathcal{P}(\mathbb{T}^d)$, $\gamma_t(x, m, y) = D_y D_x \frac{\delta U}{\delta m}(t, x, m, y)$ and $\gamma_t' = (W_t, m_t)$, we get

$$
\begin{aligned}
\frac{1}{h}\mathbb{E}\big[T_h^4 | \mathcal{F}_t\big] &= \frac{1}{h}\mathbb{E}\big[T_h^{4,2} | \mathcal{F}_t\big] + \varepsilon_{t,t+h} \\
&= 2 \int_{\mathbb{T}^d} \operatorname{div}_y \Big[D_x \frac{\delta U}{\delta m}\Big](t, x + \sqrt{2}W_t, m_t, y) dm_t(y) + \varepsilon_{t,t+h} \quad \text{(A.26)} \\
&= 2 \int_{\mathbb{T}^d} \operatorname{div}_x \Big[D_y \frac{\delta U}{\delta m}\Big](t, x + \sqrt{2}W_t, m_t, y) dm_t(y) + \varepsilon_{t,t+h}.
\end{aligned}
$$

It finally remains to handle T_h^5. Thanks to (A.19), we write

$$
\begin{aligned}
T_h^5 &= \frac{1}{2} \int_{[\mathbb{T}^d]^2} \frac{\delta^2 U}{\delta m^2}\big(t + h, x + \sqrt{2}W_t + \sqrt{2}\lambda \delta_h W_t, m_t + \lambda \delta_h m_t, y \\
&\quad + \sqrt{2}W_{t+h}, \ y' + \sqrt{2}W_{t+h}\big) d(\tilde{m}_{t+h} - \tilde{m}_t)(y) d(\tilde{m}_{t+h} - \tilde{m}_t)(y') \\
&\quad + \frac{1}{2} \int_{[\mathbb{T}^d]^2} \Big[\frac{\delta^2 U}{\delta m^2}\big(t + h, x + \sqrt{2}W_t + \sqrt{2}\lambda \delta_h W_t, m_t \\
&\quad + \lambda \delta_h m_t, y + \sqrt{2}\delta_h W_t, \ y' + \sqrt{2}W_{t+h}\big) \\
&\quad - \frac{\delta^2 U}{\delta m^2}\big(t + h, x + \sqrt{2}W_t + \sqrt{2}\lambda \delta_h W_t, m_t + \lambda \delta_h m_t, y, y' \\
&\quad + \sqrt{2}W_{t+h}\big)\Big] dm_t(y) d(\tilde{m}_{t+h} - \tilde{m}_t)(y') \\
&\quad + \frac{1}{2} \int_{[\mathbb{T}^d]^2} \Big[\frac{\delta^2 U}{\delta m^2}\big(t + h, x + \sqrt{2}W_t + \sqrt{2}\lambda \delta_h W_t, m_t \quad \text{(A.27)} \\
&\quad + \lambda \delta_h m_t, y + \sqrt{2}W_{t+h}, y' + \sqrt{2}\delta_h W_t\big) \\
&\quad - \frac{\delta^2 U}{\delta m^2}\big(t + h, x + \sqrt{2}W_t + \sqrt{2}\lambda \delta_h W_t, m_t + \lambda \delta_h m_t, y \\
&\quad + \sqrt{2}W_{t+h}, y'\big)\Big] d(\tilde{m}_{t+h} - \tilde{m}_t)(y) dm_t(y') \\
&\quad + \frac{1}{2} \int_{[\mathbb{T}^d]^2} \Big[\frac{\delta^2 U}{\delta m^2}\big(t + h, x + \sqrt{2}W_t + \sqrt{2}\lambda \delta_h W_t, m_t \\
&\quad + \lambda \delta_h m_t, y, y' + \sqrt{2}\delta_h W_t, y' + \sqrt{2}\delta_h W_t\big) \\
&\quad - \frac{\delta^2 U}{\delta m^2}\big(t + h, x + \sqrt{2}W_t + \sqrt{2}\lambda \delta_h W_t, m_t + \lambda \delta_h m_t, y + \sqrt{2}\delta_h W_t, y'\big) \\
&\quad - \frac{\delta^2 U}{\delta m^2}\big(t + h, x + \sqrt{2}\lambda \delta_h W_t, m_t + \lambda \delta_h m_t, y, y' + \sqrt{2}\delta_h W_t\big)
\end{aligned}
$$

$$+ \frac{\delta^2 U}{\delta m^2}\Big(t + h, x + \sqrt{2}W_t + \sqrt{2}\lambda\delta_h W_t, m_t + \lambda\delta_h m_t, y, y'\Big)\bigg] dm_t(y)dm_t(y')$$

$$=: \frac{1}{2}\Big(T_h^{5,1} + T_h^{5,2} + T_h^{5,3} + T_h^{5,4}\Big).$$

Making use of the Fokker–Planck equation satisfied by $(\tilde{m}_t)_{t\in[t_0,T]}$ together with the regularity assumptions of $\delta^2 U/\delta m^2$ in Definition 2.4.4, it is readily seen that

$$\frac{1}{h}\mathbb{E}\big[T_h^{5,1} + T_h^{5,2} + T_h^{5,3}|\mathcal{F}_t\big] = \varepsilon_{t,t+h}. \tag{A.28}$$

Focus now on $T_h^{5,4}$. With obvious notation, write it under the form

$$T_h^{5,4} =: T_h^{5,4,1} - T_h^{5,4,2} - T_h^{5,4,3} + T_h^{5,4,4}. \tag{A.29}$$

Performing a second-order Taylor expansion, we get

$$T_h^{5,4,1} = \int_{[\mathbb{T}^d]^2} \frac{\delta^2 U}{\delta m^2}\Big(t + h, x + \sqrt{2}W_t$$

$$+ \sqrt{2}\lambda\delta_h W_t, m_t + \lambda\delta_h m_t, y, y'\Big)dm_t(y)dm_t(y')$$

$$+ \int_{[\mathbb{T}^d]^2} \sqrt{2}D_y \frac{\delta^2 U}{\delta m^2}\Big(t + h, x + \sqrt{2}W_t + \sqrt{2}\lambda\delta_h W_t, m_t$$

$$+ \lambda\delta_h m_t, y, y'\Big)\cdot \delta_h W_t dm_t(y)dm_t(y')$$

$$+ \int_{[\mathbb{T}^d]^2} \sqrt{2}D_{y'} \frac{\delta^2 U}{\delta m^2}\Big(t + h, x + \sqrt{2}W_t + \sqrt{2}\lambda\delta_h W_t, m_t$$

$$+ \lambda\delta_h m_t, y, y'\Big)\cdot \delta_h W_t dm_t(y)dm_t(y')$$

$$+ \int_{[\mathbb{T}^d]^2} D_y^2 \frac{\delta^2 U}{\delta m^2}\Big(t + h, x + \sqrt{2}W_t + \sqrt{2}\lambda\delta_h W_t, m_t$$

$$+ \lambda\delta_h m_t, y, y'\Big)\cdot \big(\delta_h W_t\big)^{\otimes 2} dm_t(y)dm_t(y')$$

$$+ \int_{[\mathbb{T}^d]^2} D_{y'}^2 \frac{\delta^2 U}{\delta m^2}\Big(t + h, x + \sqrt{2}W_t + \sqrt{2}\lambda\delta_h W_t, m_t$$

$$+ \lambda\delta_h m_t, y, y'\Big)\cdot \big(\delta_h W_t\big)^{\otimes 2} dm_t(y)dm_t(y')$$

$$+ \int_{[\mathbb{T}^d]^2} 2D_y D_{y'} \frac{\delta^2 U}{\delta m^2}\Big(t + h, x + \sqrt{2}W_t + \sqrt{2}\lambda\delta_h W_t, m_t$$

$$+ \lambda\delta_h m_t, y, y'\Big)\cdot \big(\delta_h W_t\big)^{\otimes 2} dm_t(y)dm_t(y') + h\varepsilon_{t,t+h}$$

$$=: T_h^{5,4,4} + I_h^1 + I_h^2 + J_h^1 + J_h^2 + J_h^{1,2} + h\varepsilon_{t,t+h}.$$

Similarly, we get

$$T_h^{5,4,2} = T_h^{5,4,4} + I_h^1 + J_h^1 + h\varepsilon_{t,t+h},$$
$$T_h^{5,4,3} = T_h^{5,4,4} + I_h^2 + J_h^2 + h\varepsilon_{t,t+h},$$

from which, together with (A.29), we deduce that

$$T_h^{5,4} = J_h^{1,2} + h\varepsilon_{t,t+h}, \tag{A.30}$$

and then, with (A.28),

$$\frac{1}{h}\mathbb{E}\big[T_h^5|\mathcal{F}_t\big] = \frac{1}{2h}\mathbb{E}\big[T_h^{5,4}|\mathcal{F}_t\big] + \varepsilon_{t,t+h}$$
$$= \int_{[\mathbb{T}^d]^2} \mathrm{Tr}\Big[D_y D_{y'}\frac{\delta^2 U}{\delta m^2}\big(t, x + \sqrt{2}W_t, m_t, y, y'\big)\Big] \tag{A.31}$$
$$dm_t(y)dm_t(y') + \varepsilon_{t,t+h}.$$

From (A.17), (A.20), (A.24), (A.25), (A.26), and (A.31), we deduce that

$$\frac{1}{h}\Big[\mathbb{E}\big[U\big(t+h, x+\sqrt{2}W_{t+h}, m_t\big) - U\big(t+h, x+\sqrt{2}W_t, m_t\big)|\mathcal{F}_t\big]\Big]$$
$$= \Delta_x U\big(t, x+\sqrt{2}W_t, m_t\big) + 2\int_{\mathbb{T}^d} \mathrm{div}_y\big[D_m U\big]\big(t, x+\sqrt{2}W_t, m_t, y\big)dm_t(y)$$
$$- \int_{\mathbb{T}^d} D_m U\big(t, x+\sqrt{2}W_t, m_t, y\big)\cdot\beta_t(y)dm_t(y)$$
$$+ 2\int_{\mathbb{T}^d} \mathrm{div}_x\big[D_m U\big]\big(t, x+\sqrt{2}W_t, m_t, y\big)dm_t(y)$$
$$+ \int_{[\mathbb{T}^d]^2} \mathrm{Tr}\Big[D_{mm}^2 U\big(t, x+\sqrt{2}W_t, m_t, y, y'\big)\Big]dm_t(y)dm_t(y') + \varepsilon_{t,t+h},$$

which completes the proof. □

We now deduce from the previous lemma an expansion for $(U(t, x, m_t))_{t\in[0,T]}$.

Theorem A.1. *Under the assumption stated at the beginning of Subsection A.3.2, the process $(U(t, x, m_t))_{t\in[0,T]}$ expands, for any $x \in \mathbb{T}^d$, as a semi-*

martingale. Namely, with probability 1 under $\mathbb{P}$, *for all* $t \in [0, T]$,

$$
d_t U(t, x, m_t) = \left\{ \partial_t U(t, x, m_t) + \int_{\mathbb{T}^d} \left[2\mathrm{div}_y \left[D_m U \right](t, x, m_t, y) \right. \right.
$$
$$
\left. - D_m U(t, x, m_t, y) \cdot \beta_t(y) \right] dm_t(y)
$$
$$
+ \int_{\mathbb{T}^d} \int_{\mathbb{T}^d} \mathrm{Tr}\left[D_{mm}^2 U \right](t, x, m_t, y, y') dm_t(y) dm_t(y') \right\} dt
$$
$$
+ \sqrt{2} \left(\int_{\mathbb{T}^d} D_m U(t, x, m_t, y) dm_t(y) \right) \cdot dW_t.
$$

It is worth noting that the variable x in $(U(t, x, m_t))_{t \in [0,T]}$ may be replaced by a stochastic process $(X_t)_{t \in [0,T]}$ with a semi-martingale expansion, in which case the global expansion of $(U(t, X_t, m_t))_{t \in [0,T]}$ may be easily derived by combining the above formula with Itô–Wentzell's formula.

Proof. The strategy is to apply Lemma A.3.1. To do so, observe that, the variable x being frozen, we are led back to the case when U is independent of x. So, with the same notation as in Lemma A.3.1, we get

$$
\mathbb{E}\left[U(t + h, x, m_{t+h}) - U(t + h, x, m_t) | \mathcal{F}_t \right]
$$
$$
= \left\{ 2 \int_{\mathbb{T}^d} \mathrm{div}_y \left[D_m U \right](t, x, m_t, y) dm_t(y) \right.
$$
$$
- \int_{\mathbb{T}^d} D_m U(t, x, m_t, y) \cdot \beta_t(y) dm_t(y) \tag{A.32}
$$
$$
\left. + \int_{[\mathbb{T}^d]^2} \mathrm{Tr}\left[D_{mm}^2 U \right](t, x, m_t, y, y') dm_t(y) dm_t(y') + \varepsilon_{t,t+h} \right\} h.
$$

Of course, this gives the absolutely continuous part only in the semi-martingale expansion of $(U(t, x, m_t))_{t \in [0,T]}$. To compute the martingale part, one must revisit the proof of Lemma A.3.1. Going back to (A.17), we know that, in our case, T_h^1, T_h^3, and T_h^4 are 0 (as everything works as if U was independent of x).
Now, denoting by $(\eta_{s,t})_{s,t \in [0,T]: s \leqslant t}$ a family of random variables satisfying

$$
\lim_{h \searrow 0} \frac{1}{h} \sup_{s,t \in [0,T]: |s-t| \leqslant h} \mathbb{E}\left[|\eta_{s,t}|^2 \right] = 0, \tag{A.33}
$$

we can write, by (A.21) and (A.22):

$$
T_h^2 = \sqrt{2} \left(\int_{\mathbb{T}^d} D_y \frac{\delta U}{\delta m}(t, x, m_t, y) dm_t(y) \right) \cdot \delta_h W_t + \eta_{t,t+h}
$$

Moreover, by (A.27) and (A.30):

$$T_h^5 = \eta_{t,t+h},$$

proving that

$$U(t+h, x, m_{t+h}) - \mathbb{E}\big[U(t+h, x, m_{t+h})|\mathcal{F}_t\big]$$
$$= \sqrt{2}\bigg(\int_{\mathbb{T}^d} D_y \frac{\delta U}{\delta m}(t, x, m_t, y) dm_t(y)\bigg) \cdot \delta_h W_t + \eta_{t,t+h},$$

for some family $(\eta_{s,t})_{s,t\in[0,T]:s\leqslant t}$ that must satisfy (A.33). With such a decomposition, it holds that $\mathbb{E}[\eta_{t,t+h}|\mathcal{F}_t] = 0$. Therefore, for any $t \in [0,T]$ and any partition $0 = r_0 < r_1 < r_2 < \cdots < r_N = t$, we have

$$\sum_{i=0}^{N-1}\bigg(U(r_{i+1}, x, m_{r_{i+1}}) - \mathbb{E}\big[U(r_{i+1}, x, m_{r_{i+1}})|\mathcal{F}_{r_i}\big]\bigg)$$
$$= \sum_{i=0}^{N-1}\bigg[\sqrt{2}\bigg(\int_{\mathbb{T}^d} D_y \frac{\delta U}{\delta m}(r_i, x, m_{r_i}, y) dm_{r_i}(y)\bigg) \cdot (W_{r_{i+1}} - W_{r_i}) + \eta_{r_i, r_{i+1}}\bigg],$$

with the property that

$$\mathbb{E}\big[\eta_{r_i, r_{i+1}}|\mathcal{F}_{r_i}\big] = 0, \quad \mathbb{E}\big[|\eta_{r_i, r_{i+1}}|^2\big] \leqslant \pi_{r_i, r_{i+1}}|r_{i+1} - r_i|,$$

where $\lim_{h\searrow 0} \sup_{(s,t)\in[0,T]^2:|s-t|\leqslant h} \pi_{s,t} = 0$. By a standard computation of conditional expectation, we have that

$$\lim_{\delta\to 0} \mathbb{E}\bigg[\bigg|\sum_{i=0}^{N-1} \eta_{r_i, r_{i+1}}\bigg|^2\bigg] = 0,$$

where δ stands for the mesh of the partition $r_0, r_1, \ldots, r_N$. As a consequence, the following limit holds true in L^2:

$$\lim_{\delta\searrow 0} \sum_{i=0}^{N-1}\bigg(U(r_{i+1}, x, m_{r_{i+1}}) - \mathbb{E}\big[U(r_{i+1}, x, m_{r_{i+1}})|\mathcal{F}_{r_i}\big]\bigg)$$
$$= \sqrt{2}\int_0^t D_m U(s, x, m_s, y) \cdot dW_s.$$

Together with (A.32), we deduce that

$$
\lim_{\delta \searrow 0} \sum_{i=0}^{N-1} \Big(U(r_{i+1}, x, m_{r_{i+1}}) - U(r_{i+1}, x, m_{r_i}) \Big)
$$

$$
= \int_0^t \left\{ \int_{\mathbb{T}^d} \left[2\mathrm{div}_y \left[D_m U \right](s, x, m_s, y) - D_m U(s, x, m_s, y) \cdot \beta_s(y) \right] dm_s(y) \right.
$$

$$
\left. + \int_{\mathbb{T}^d} \int_{\mathbb{T}^d} \mathrm{Tr} \left[D_{mm}^2 U \right](s, x, m_s, y, y') dm_s(y) dm_s(y') \right\} ds
$$

$$
+ \sqrt{2} \int_0^t D_m U(s, x, m_s, y) \cdot dW_s.
$$

Finally, with probability 1, for all $t \in [0, T]$,

$$
U(t, x, m_t) - U(0, x, m_0)
$$

$$
= \int_0^t \left\{ \partial_t U(s, x, m_s) \right.
$$

$$
+ \int_{\mathbb{T}^d} \left[2\mathrm{div}_y \left[D_m U \right](s, x, m_s, y) - D_m U(s, x, m_s, y) \cdot \beta_s(y) \right] dm_t(y)
$$

$$
\left. + \int_{\mathbb{T}^d} \int_{\mathbb{T}^d} \mathrm{Tr} \left[D_{mm}^2 U \right](t, x, m_t, y, y') dm_t(y) dm_t(y') \right\} dt
$$

$$
+ \sqrt{2} \left(\int_{\mathbb{T}^d} D_m U(t, x, m_t, y) dm_t(y) \right) \cdot dW_t,
$$

which completes the proof. $\square$

References

[1] Achdou, Y., and Capuzzo-Dolcetta, I. (2010). Mean field games: numerical methods. *SIAM J. Numer. Anal.*, 48, 1136–1162.

[2] Achdou, Y., Camilli, F., and Capuzzo-Dolcetta, I. (2013). Mean field games: convergence of a finite difference method. *SIAM J. Numer. Anal.*, 51, 2585–2612.

[3] Achdou, Y., and Porretta, A. (2016). Convergence of a finite difference scheme to weak solutions of the system of partial differential equations arising in mean field games. *SIAM J. Numer. Anal.*, 54, 161–186.

[4] Achdou, Y., Buera, F. J., Lasry, J. M., Lions, P. L., and Moll, B. (2014). Partial differential equation models in macroeconomics. *Philos. Trans. R. Soc. A: Math. Phys. Engin. Sci.*, 372 (2028), 20130397.

[5] Achdou, Y., Han, J., Lasry, J. M., Lions, P. L., and Moll, B. (2014). Heterogeneous agent models in continuous time. https://economics.yale.edu/sites/default/files/moll_131105_0.pdf/

[6] Aiyagari, S. R. (1994). Uninsured idiosyncratic risk and aggregate saving. *Q. J. Econ.*, 109 (3), 659–684.

[7] Ambrosio, L., Gigli, N., and Savaré, G. (2008). *Gradient Flows in Metric Spaces and in the Space of Probability Measures.* Second edition. Lectures in Mathematics ETH Zürich. Basel: Birkhäuser Verlag.

[8] Ambrosio, L., and Feng, J. (2014). On a class of first order Hamilton–Jacobi equations in metric spaces. *J. Differ. Eq.*, 256(7), 2194–2245.

[9] Ajtai, M., Komlos, J., and Tusnády, G. (1984). On optimal matchings. *Combinatorica*, 4 (4), 259–264.

[10] Aumann R. (1964). Markets with a continuum of traders. *Econometrica*, 32(1/2), 39–50.

[11] Başar, T., and Olsder, G. (1998). *Dynamic Noncooperative Game Theory.* *SIAM*.

[12] Bardi, M. (2012). Explicit solutions of some linear-quadratic mean field games. *Networks Heterog. Media*, 7(2), 243–261.

[13] Bardi, M., and Feleqi, E. (2016). Nonlinear elliptic systems and mean field games. *NoDEA Nonlinear Differ. Eq. Appl.*, 23, p. 44.

[14] Bensoussan, A., and Frehse, J. (2002). Smooth solutions of systems of quasilinear parabolic equations. *ESAIM: Control Optim. Calc. Variat.*, 8, 169–193.

[15] Bensoussan, A., and Frehse, J. (2012). Control and Nash games with mean field effect. *Chin. Ann. Math. B*, 34 (2), 161–192.

[16] Bensoussan, A., Frehse J., and Yam, S.C.P. (2013). *Mean field games and mean field type control theory.* Springer Briefs in Mathematics. New York: Springer Science+Business Media.

[17] Bensoussan, A., Frehse J., and Yam, S.C.P. (2015). The master equation in mean field theory. *J. Math. Pures Appl.*, 103, 1441–1474.

[18] Bensoussan, A., Frehse J., and Yam, S.C.P. (2017). On the interpretation of the master equation. *Stoc. Proc. App.*, 127, 2093–2137.

[19] Bewley, T. (1986). Stationary monetary equilibrium with a continuum of independently fluctuating consumers. In *Contributions to Mathematical Economics in Honor of Gerard Debreu*, ed. Werner Hildenbrand and Andreu Mas-Collel. Amsterdam: North-Holland.

[20] Bogachev, V. I., Krylov, N. V., and Röckner, M. (2009). Elliptic and parabolic equations for measures. *Russ. Math. Surv.* 64 (6), 973.

[21] Bressan, A., and Shen W. (2004). Small BV solutions of hyperbolic non-cooperative differential games. *SIAM J. Control Optim.* 43, 104–215.

[22] Bressan, A., and Shen W. (2004). Semi-cooperative strategies for differential games. Int. *J. Game Theory*, 32, 561–593.

[23] Buckdahn, R., Cardaliaguet, P., and Rainer, C. (2004). Nash equilibrium payoffs for nonzero-sum stochastic differential games. *SIAM J. Control Optim.*, 43 (2), 624–642.

[24] Buckdahn, R., Li, J., Peng, S., and Rainer, C. (2017). Mean-field stochastic differential equations and associated PDEs. *Ann. Probab.*, 45, 824–878.

[25] Carmona, R., and Delarue, F. (2013). Probabilist analysis of Mean-Field Games. *SIAM J. Control Optim.*, 51(4), 2705–2734.

[26] Carmona, R., and Delarue, F. (2015). Forward–backward stochastic differential equations and controlled McKean Vlasov dynamics. *Ann. Probab.*, 43, 2647–2700.

[27] Carmona R., and Delarue F. (2014). The master equation for large population equilibriums. In *Stochastic Analysis and Applications*. D. Crisan, B. Hambly, T. Zariphopoulou eds., New York: Springer Science+Business Media, 77–128.

[28] Carmona, R., Delarue, F., and Lachapelle, A. (2013). Control of McKean–Vlasov dynamics versus mean field games. *Math. Finan. Econ.*, 7 (2), 131–166.

[29] Carmona, R., Delarue, F., and Lacker, D. (2016). Probabilistic analysis of mean field games with a common noise. *Ann. Probab.*, 44, 3740–3803.

[30] Case, J. (1969). Toward a theory of many player differential games. *SIAM J. Control*, 7(2), 179–197.

[31] Chassagneux, J. F., Crisan, D., and Delarue, F. (2014). Classical solutions to the master equation for large population equilibria. arXiv preprint arXiv:1411.3009.

[32] Crandall, M.G., and Lions, P.-L. (1983). Viscosity solutions of Hamilton–Jacobi equations. *Trans. Am. Math. Soc.*, 277 (1), 1–42.

[33] Crandall M.G., Ishii H., and Lions P.-L. (1992). User's guide to viscosity solutions of second order partial differential equations. *Bull. Amer. Soc.*, 27, 1–67.

[34] Dereich, S., Scheutzow, M., and Schottstedt, R. (2013). Constructive quantization: approximation by empirical measures. In *Annales de l'IHP, Probabiliés et Statistiques*, 49(4), 1183–1203.

[35] Evans, L.C., and Souganidis, P.E. (1984). Differential games and representation formulas for solutions of Hamilton–Jacobi Equations. *Indiana Univ. Math. J.*, 282, 487–502.

[36] Feng, J., and Katsoulakis, M. (2009). A comparison principle for Hamilton–Jacobi equations related to controlled gradient flows in infinite dimensions. *Arch. Ration. Mech. Anal.*, 192(2), 275–310.

[37] Ferreira R., and Gomes, D. (2018). Existence of weak solutions to stationary mean-field games through variational inequalities. To appear in *SIAM J. Math. Anal.*, 50(6), 5969–6006.

[38] Fischer, M. (2017). On the connection between symmetric N-player games and mean field games. *Ann. Appl. Probab.*, 27(2), 757–810.

[39] Fleming, W. H. (1961). The convergence problem for differential games. *J. Math. Anal. Appl.*, 3, 102–116.

[40] Fleming, W.H., and Souganidis, P.E. (1989). On the existence of value functions of two-player, zero-sum stochastic differential games. *Indiana Univ. Math. J.* 38(2), 293–314.

[41] Fleming, W. H., and Soner, H. M. (2006). *Controlled Markov Processes and Viscosity Solutions*, Second edition. New York: Springer Science+Business Media.

[42] Fournier, N., and Guillin, A. (2015). On the rate of convergence in Wasserstein distance of the empirical measure. *Probab. Theory Relat. Fields*, 162(3), 707–738.

[43] Friedman, A. (1972). Stochastic differential games. *J. Differ. Eq.*, 11(1), 79–108.

[44] Gangbo, W., and Swiech, A. (2015). Metric viscosity solutions of Hamilton–Jacobi equations depending on local slopes. *Calc. Variat. Partial Differ. Eq.*, 54, 1183–1218.

[45] Gangbo, W., and Swiech, A. (2015). Existence of a solution to an equation arising from the theory of mean field games. *J. Differ. Eq.*, 259, 6573–6643.

[46] Gomes, D. A., and Mitake, H. (2015). Existence for stationary mean-field games with congestion and quadratic Hamiltonians. *NoDEA Nonlinear Differ. Eq. Appl.*, 22(6), 1897–1910.

[47] Gomes, D. A., Patrizi, S., and Voskanyan, V. (2014). On the existence of classical solutions for stationary extended mean field games. *Nonlinear Anal.*, 99, 49–79.

[48] Gomes, D.A., Pimentel, E., and Voskanyan, V. (2016). *Regularity Theory for Mean-Field Game Systems*. SpringerBriefs in Mathematics. New York: Springer Science+Business Media.

[49] Gomes, D., and Saude J. (2014). Mean field games models-a brief survey. *Dyn. Games Appl.*, 4(2), 110–154.

[50] Green, E. J. (1984). Continuum and finite-player noncooperative models of competition. *Econometrica*, 52(4), 975–993.

[51] Guéant, O., Lasry, J.-M, and Lions, P.-L. (2011). Mean field games and applications. *Paris-Princeton Lectures on Mathematical Finance 2010*. Tankov, Peter; Lions, Pierre-Louis; Laurent, Jean-Paul; Lasry, Jean-Michel; Jeanblanc, Monique; Hobson, David; Guéant, Olivier; Crépey, Stéphane; Cousin, Areski. Berlin: Springer, pp. 205–266.

[52] Huang, M. (2010). Large-population LQG games involving a major player: The Nash certainty equivalence principle. *SIAM J. Control Optim.*, 48, 3318–3353.

[53] Huang, M., Malhamé, R.P., and Caines, P.E. (2006). Large population stochastic dynamic games: closed-loop McKean–Vlasov systems and the Nash certainty equivalence principle. *Commun. Infor. Syst.*, 6, 221–252.

[54] Huang, M., Caines, P.E., and Malhamé, R.P. (2007). Large-population cost-coupled LQG problems with nonuniform agents: individual-mass behavior and decentralized ϵ-Nash equilibria. *IEEE Trans. Autom. Control*, 52(9), 1560–1571.

[55] Huang, M., Caines, P.E., and Malhamé, R.P. (2007). The Nash Certainty Equivalence Principle and McKean–Vlasov Systems: an Invariance Principle and Entry Adaptation. 46th IEEE Conference on Decision and Control, 121–123.

[56] Huang, M., Caines, P.E., and Malhamé, R.P. (2007). An invariance principle in large population stochastic dynamic games. *J. Syst. Sci. Complex.*, 20(2), 162–172.

[57] Huang, M., Caines, P. E., and Malhamé, R. P. (2010). The NCE (mean field) principle with locality dependent cost interactions. *IEEE Trans. Autom. Control*, 55(12), 2799–2805.

[58] Huggett, M. (1993). The risk-free rate in heterogeneous-agent incomplete-insurance economies. *J. Econ. Dyn. Control*, 17(5-6), 953–969.

[59] Isaacs R. (1965). *Differential Games*. New York: John Wiley & Sons.

[60] Jordan, R., Kinderlehrer, D., and Otto, F. (1998). The variational formulation of the Fokker–Planck equation. *SIAM J. Math. Anal.*, 29 (1), 1–17.

[61] Kolokoltsov, V.N. (2010). *Nonlinear Markov Processes and Kinetic Equations*. Cambridge: Cambridge University Press.

[62] Kolokoltsov, V. N., Li, J., and Yang, W. (2011). Mean field games and nonlinear Markov processes. Preprint arXiv:1112.3744.

[63] Kolokoltsov, V. N., Troeva, M., and Yang, W. (2014). On the rate of convergence for the mean-field approximation of controlled diffusions with large number of players. *Dyn. Games Appl.*, 4, 208–230.

[64] Kononenko, A. F. (1976). Equilibrium positional strategies in nonantagonistic differential games. *Dokl. Akad. Nauk SSSR*, 231(2), 285–288.

[65] Kruscll, P., and Smith, Jr, A. A. (1998). Income and wealth heterogeneity in the macroeconomy. *J. Polit. Econ.*, 106(5), 867–896.

[66] Krylov, N. V. (1980). *Controlled Diffusion Processes*. New York: Springer-Verlag.

[67] Krylov, N. (2011). On the Itô–Wentzell formula for distribution-valued processes and related topics. *Probab. Theory Relat. Fields*, 150, 295–319.

[68] Kunita, H. (1990). *Stochastic Flows and Stochastic Differential Equations.* Cambridge: Cambridge University Press.

[69] Lacker, D. (2016). A general characterization of the mean field limit for stochastic differential games. *Probab. Theory Relat. Fields*, 165, 581–648.

[70] Ladyženskaja O.A., Solonnikov, V.A, and Ural'ceva, N.N. (1967). *Linear and Quasilinear Equations of Parabolic Type.* Translations of Mathematical Monographs, Vol. 23. Providence, RI: American Mathematical Society.

[71] Lasry, J.-M., and Lions, P.-L. (2006). Jeux á champ moyen. I. Le cas stationnaire. *C. R. Math. Acad. Sci. Paris*, 343(9), 619–625.

[72] Lasry, J.-M., and Lions, P.-L. (2006). Jeux á champ moyen. II. Horizon fini et contrôle optimal. *C. R. Math. Acad. Sci. Paris*, 343(10), 679–684.

[73] Lasry, J.-M., and Lions, P.-L. (2007). Large investor trading impacts on volatility. *Ann. Inst. H. Poincaré Anal. Non Linéaire*, 24(2), 311–323.

[74] Lasry, J.-M., and Lions, P.-L. (2007). Mean field games. *Jpn. J. Math.*, 2(1), 229–260.

[75] Lieberman, G. M. (1996). *Second Order Parabolic Differential Equations.* Singapore: World Scientific.

[76] Lions, P.-L. Cours au Collège de France. www.college-de-france.fr.

[77] McKean, H.P. (1967). Propagation of chaos for a class of non linear parabolic equations. In Lecture Series in Differential Equations, Vol. 7. New York: Springer-Verlag. 41–57.

[78] Mas-Colell A. (1984). On a theorem of Schmeidler. *J. Math. Econ.*, 3, 201–206.

[79] Méléard, S. (1996). Asymptotic behaviour of some interacting particle systems; McKean–Vlasov and Boltzmann models. Probabilistic models for nonlinear partial differential equations (Montecatini Terme, 1995), 42–95, Lecture Notes in Math., 1627, Berlin: Springer-Verlag.

[80] Mischler, S., and Mouhot, C. (2013). Kac's Program in Kinetic Theory. *Inventiones mathematicae*, 193, 1–147.

[81] Mischler, S., Mouhot, C., and Wennberg, B. (2015). A new approach to quantitative propagation of chaos for drift, diffusion and jump processes. *Probab. Theory Relat. Fields*, 161, 1–59.

[82] Nash, J. (1951). Non-cooperative games. *Ann. Math.*, 286–295.

[83] Otto, F. (2001). The geometry of dissipative evolution equations: the porous medium equation. *Commun. Partial Differ. Eq.*, 26(1-2), 101–174.

[84] Pardoux, E., and Răşcanu, A. (2014). *Stochastic Differential Equations, Backward SDEs, Partial Differential Equations*. New York: Springer Science+Business Media.

[85] Peng, S. (1992). Stochastic Hamilton Jacobi Bellman equations. *SIAM J. Control Optim.*, 30, 284–304.

[86] Peng, S., and Wu, Z. (1999). Fully coupled forward–backward stochastic differential equations and applications to optimal control. *SIAM J. Control Optim.*, 37, 825–843.

[87] Pontryagin, N.S. (1968). Linear differential games I and II. *Soviet Math. Doklady*, 8(3-4), 769–771 & 910,912.

[88] Pham, H. (2009). *Continuous-Time Stochastic Control and Optimization with Financial Applications*. Berlin: Springer-Verlag.

[89] Rachev, S.T., and Schendorf, L. R. (1998). *Mass Transportation problems*. Vol. I: Theory; Vol. II: Applications. New York: Springer-Verlag.

[90] Rashid, S. (1983). Equilibrium points of non-atomic games: asymptotic results. *Econ. Lett.*, 12(1), 7–10.

[91] Schmeidler, D. (1973). Equilibrium points of nonatomic games. *J. Stat. Phys.*, 7, 295–300.

[92] Sznitman, A.-S. (1989). *Topics in Propagation of Chaos*. Cours de l'Ecole d'été de Saint-Flour. Lecture Notes in Mathematics, Vol. 1464. New York: Springer-Verlag.

[93] Sorin, S. (1992). Repeated games with complete information. *Handbook of Game Theory with Economic Applications*, 1, 71–107.

[94] Starr, A. W., and Ho, Y.-C. (1969). Nonzero-sum differential. *J. Optim. Theory Appl.*, 3(3), 184–206.

[95] Villani, C. (2008). *Optimal Transport, Old and New*. Springer Verlag, 2008.

[96] Von Neumann, J., and Morgenstern, O. (2007). *Theory of Games and Economic Behavior*. Princeton, NJ: Princeton University Press.

[97] Zhang, J. (2017). *Backward Stochastic Differential Equations*. New York: Springer Science+Business Media.

Index